Pattern Classification

Springer
London
Berlin
Heidelberg
New York
Barcelona
Hong Kong
Milan
Paris
Singapore
Tokyo

Shigeo Abe

Pattern Classification

Neuro-fuzzy Methods and Their Comparison

With 124 Figures

 Springer

Shigeo Abe, DrEng
Graduate School of Science and Technology, Kobe University, Rokkodai,
Nada, Kobe 657-8501, Japan

ISBN 1-85233-352-9 Springer-Verlag London Berlin Heidelberg

British Library Cataloguing in Publication Data
Abe, Shigeo
    Pattern classification : Neuro-fuzzy methods and their
    comparison
    1. Pattern recognition systems 2. Neural networks (Computer
    science) 3. Fuzzy systems
    I. Title
    006.4
ISBN 1852333529

Library of Congress Cataloging-in-Publication Data
Abe, Shigeo.
    Pattern classification : neuro-fuzzy methods and their comparison / Shigeo Abe.
      p. cm.
    ISBN 1-85233-352-9
    1. Neual networks (Computer science) 2. Fuzzy systems. 3. Pattern recognition
systems. I. title.
    QA76.87. A323 2001
    006.3'2—dc21
                                                                00-044672

Typesetting: Camera ready by author
Printed and bound at the Athenæum Press Ltd., Gateshead, Tyne & Wear
69/3830-543210 Printed on acid-free paper SPIN 10774083

Preface

Multilayer neural networks and fuzzy systems are widely applied to pattern classification and function approximation. Although the two approaches are proved to be mathematically equivalent in that they are convertible [1], they have their advantages and disadvantages [2].

Multilayer neural networks can learn from data: just gather data and train the network using the back-propagation algorithm; then we can get the classifier or the function approximator that we want. However, if the trained network does not perform as we have expected, we need to retrain the network changing the network size, the number of hidden neurons, or adding new training data. This is because, through training, the algorithm for pattern classification or function approximation is acquired in the network weights that connect between layers of neurons and we cannot tell the meaning of each weight separately. In addition, since the training is slow, repetition of training is prohibitive especially for a large size network with a large number of training data.

On the other hand, fuzzy rules need to be acquired by interviewing experts. In addition, for a complicated system expert knowledge that is obtained by intuition and experience is difficult to express in a rule format. Thus rule acquisition requires much time. But once the fuzzy rules are obtained, analysis of the fuzzy system is relatively easy.

In developing fuzzy systems we need to divide the input space in advance. Thus as the number of input variables increases, the number of rules to be defined increases exponentially. This is called the curse of dimensionality. In addition, since the input space is divided into subregions that are parallel to input axes, performance of fuzzy systems is usually inferior to that of multilayer neural networks.

According to these, multilayer neural networks have been applied to the problems where domain knowledge is scarce but numerical data can be acquired easily, while fuzzy systems have been applied to the problems where domain knowledge is abundant but numerical data are difficult to obtain.

To compensate their deficiencies and enhance their applicability to real world problems, extensive research has been conducted to fuse these technologies [3, 4, 5, 6]. For instance, fuzzy rules with initial parameter values are imbedded into a multilayer neural network and then the parameter values are determined by training the network by the back-propagation algorithm [7]. Or multilayer neural networks are enhanced to treat linguistic information as well as numerical data [8, 9, 10, 11]. One approach transforms numerical data into fuzzy data [8, 9, 10] and the other approach allows mixture of numerical data and fuzzy data [11].

In this book, however, we take different approaches. Our goals are to realize, without fusing both technologies, multilayer neural networks with high speed training capability and trainable fuzzy systems with performance comparable to that of multilayer neural networks.

To realize high-speed training of multilayer neural network classifiers, we train the network based on the synthesis principle [2]. We represent each class by the center of the training data belonging to a class and determine the hyperplanes that separate each class in the input space. Then according to whether the center is on the positive or negative side of the hyperplane, we determine the target values of the hidden neurons. Training of the network is now expressed by two sets of inequalities and we solve the sets successively.

To realize trainable fuzzy systems free from the curse of dimensionality, we define variable-size fuzzy regions according to the existence of data in the input space. To realize performance comparable to that of multilayer neural networks we do not limit fuzzy regions to hyperboxes that are parallel to input axes. There is a tradeoff between interpretability and performance of fuzzy systems. We put an emphasis on performance that is important in applying fuzzy systems to real world problems.

In addition, to speed up training, we do not resort to the steepest descent method on which the back-propagation algorithm is based. In training fuzzy classifiers, we define the initial fuzzy rules and then we tune their membership functions. Namely, when the slope of a membership function is increased or decreased, the correctly classified data become misclassified or misclassified data become correctly classified. Increasing or decreasing the slope, we count the net increase of the recognition rate of the training data at the points where the recognition rate changes, and set the slope to the value that maximizes the recognition rate. We iterate tuning until the recognition rate is not improved. Similarly, we tune the locations of the membership functions. Since this method directly maximizes the recognition rate, tuning is very efficient and leads to high classification performance.

Our emphasis is on the applicability of fuzzy classifiers to real world problems. Therefore, using several benchmark data sets including real world data sets we make detailed comparisons on training time and classification performance among nearest neighbor classifiers, multilayer neural networks, support vector machines for pattern classification, and fuzzy classifiers with different architectures.

This book consists of two parts: the first part from Chapters 1 to 12 discusses pattern classification and the second part from Chapters 13 to 16 discusses function approximation. Chapter 1 explains the pattern classification task and clarifies the problems to be considered in developing pattern classification systems. Chapter 2 discusses the neural network classifier and explain the conventional back-propagation algorithm and the training algorithm based on the synthesis principle of the neural network classifier; namely training is done by solving two sets of linear inequalities successively.

Chapter 3 discusses the support vector machines for pattern classification which are based on statistical learning theory [12]. The input space is mapped into the dot product space called feature space that is generated by the polynomials of input variables. In the feature space, the classifier is constructed so that the training data of one class are separated by the hyperplane with the maximum margin from those of the remaining classes. Training results in solving a quadratic optimization problem.

Conventional fuzzy classifiers divide the ranges of the input variables into several sub-intervals that correspond to fuzzy sets, and define a one-dimensional membership function for each interval. In this book, instead of using one-dimensional membership functions, we use multi-dimensional membership functions. In Chapter 4, first we define one-dimensional membership functions, i.e., triangular, trapezoidal, and bell-shaped membership functions. Then we extend them to multi-dimensional membership functions, i.e., (truncated) rectangular pyramidal membership functions, (truncated) polyhedral pyramidal membership functions, and bell-shaped membership functions and clarify their relations.

Their definitions are based on the shapes of the membership functions. If we name the membership functions according to the shape of the fuzzy regions, they are, respectively, hyperbox, polyhedral, and ellipsoidal membership functions. (In this book we simply say polyhedral regions or ellipsoidal regions instead of saying hyper-polyhedral regions or hyper-ellipsoidal regions.)

Since in this book, we only define membership functions for the if-part of the fuzzy rule, we call the fuzzy rules with (truncated) rectangular pyramidal membership functions the hyperbox fuzzy rules, the fuzzy rules with (truncated) polyhedral pyramidal membership functions the polyhedral fuzzy rules, and the fuzzy rule with bell-shaped membership functions ellipsoidal fuzzy rules. As a special case we call the fuzzy rules with rectangular pyramidal membership functions the fuzzy rules with pyramidal membership functions. We also call fuzzy classifiers with hyperbox fuzzy rules fuzzy classifiers with hyperbox regions and so on.

In generating fuzzy rules, overlaps between classes need to be resolved. This can be done before and after rule generation, and during rule generation. We call the former static rule generation and the latter dynamic rule generation. In Chapter 5, we discuss static generation of fuzzy rules with pyramidal membership functions, polyhedral fuzzy rules, and ellipsoidal fuzzy rules and their training architecture. In Chapter 6, we discuss clustering techniques that are used to divide class data into clusters and in Chapter 7 we discuss how to tune membership functions of the fuzzy rules that are generated by static rule generation.

In the previous chapters we assume that the training data are 100% correct. But in a real world situation, outliers may be included in the training data and if included they degrade performance of classifiers considerably. In Chapter 8 we discuss robust pattern classification using ellipsoidal fuzzy rules when outliers are included.

In Chapter 9 we discuss dynamic generation of two types of hyperbox fuzzy rules. One method resolves overlaps between hyperboxes of different classes by expanding or contracting hyperboxes. The other method resolves overlaps by defining overlapping regions as an inhibition hyperbox. In Chapter 10 we compare classification performance and the computation time of the fuzzy classifiers discussed in this book, the multilayer neural network classifier, the support vector machine, and the nearest neighbor classifier for several benchmark data sets.

Performance of pattern classification depends heavily on the input features used. Chapter 11 discusses the methods for extracting and selecting important features from the given input features. Feature extraction extracts important features by linear or nonlinear transformation of the original input features. The two-layer neural network that is the first to hidden layers of the auto-associative three-layer neural network is used as a feature extractor. In this chapter, we discuss fast training of two-layer neural networks by solving inequalities as discussed in Chapter 2 and demonstrate the effectiveness for some benchmark data sets.

Feature selection selects important features from the original features. In latter part of this chapter we discuss the selection criteria based on the degree of overlap between class regions approximated by hyperboxes or ellipsoids and show the usefulness using several benchmark data sets.

In developing a pattern classification system, usually we prepare two data sets: one for training and the other for testing to check generalization ability. Since the training data needs to be unbiased representation of the events, the characteristics of the two sets need to be similar. To realize this, random division of the data is used. In Chapter 12, the similarity of two data sets is measured by the similarity of centers and covariance matrices of the two. Then the division algorithm is proposed and the effectiveness of the method for two benchmark data sets is shown.

From Chapters 13 to 16 we discuss function approximation using fuzzy rules extracted from data. In Chapter 13 we survey function approximation methods using neural networks, conventional fuzzy systems, and fuzzy systems with learning capability.

Similar to pattern classification, we define the membership functions for only the if-parts of the fuzzy rules. Thus we can classify fuzzy rules for function approximation into hyperbox fuzzy rules with (truncated) rectangular pyramidal membership functions and ellipsoidal fuzzy rules with bell-shaped membership functions. For function approximation, we do not use polyhedral fuzzy rules. In Chapter 14 we discuss function approximators with hyperbox fuzzy rules and two types of fuzzy function approximators with ellipsoidal fuzzy rules: FACG in which the Takagi-Sugeno type model and the center-of-gravity defuzzification are used, and FALC in which output is synthesized by the linear combination of the degrees of membership.

In Chapter 15 we discuss fuzzy rule generation by preclustering and post-clustering for FACG and FALC and tuning of membership functions. Then we compare performance of fuzzy function approximators and multilayer neural networks. In Chapter 16, we discuss robust fuzzy function approximation when outliers are included.

Acknowledgments

This book is based mostly on my research efforts on neural networks and fuzzy systems over the past 10 years. I express my sincere thanks to my collaborators while I was in Hitachi Research Laboratory, Hitachi, Ltd., the faculty members of Kobe University, and undergraduate and graduate students of Kobe University who were interested in my work, created new algorithms, and developed their computer programs.

I am grateful to Mr. Y. Kobayashi of Hitachi Research Laboratory, Hitachi, Ltd. for providing the hiragana data, Professor N. Matsuda of Kawasaki Medical School for providing the blood cell data, to Mr. P. M. Murphy and Mr. D. W. Aha of the University of California at Irvine for organizing the data bases including the thyroid data (ftp://ftp.ics.uci.edu/pub/machine-learning-databases/) and to Dr. B. Schölkopf et al. for providing the support vector machine software (http://svm.cs.rhbnc.ac.uk/).

Kobe, September 2000   *Shigeo Abe*

Contents

Part III. Appendices

Nomenclature

Instead of saying hyper-ellipsoids, hyper-polyhedrons, etc., we simply say ellipsoids, polyhedrons, etc. except for hyperboxes and hyperplanes. Throughout the book, we use lowercase bold letters to denote vectors and uppercase italic letters to denote matrices. In the following, we list the symbols used throughout the book and those specific to chapters.

A^{-1} : the inverse of matrix A
A^t : the transpose of matrix A
m : the number of input variables
n : the number of classes
$\|\mathbf{x}\|$: the Euclidean norm of vector $\mathbf{x}$
X_i : the set for class i training data
$|X_i|$: the number of data in the set X_i
$\mathbf{x}$: the m-dimensional input vector

Chapter 2   Multilayer Neural Network Classifiers

α : the learning rate, p. 34
$\mathbf{b}$: the margin vector, p. 40
β : the momentum coefficient, p. 35
δ_d : the margin parameter, p. 36
$\varepsilon(2)$: the tolerance of convergence for the hidden neuron outputs, p. 39
$\varepsilon(3)$: the tolerance of convergence for the output neuron outputs, p. 39

Chapter 3   Support Vector Machines

C : the upper bound, p. 53
γ : the parameter for slope control, p. 55

Chapter 5   Static Fuzzy Rule Generation

$\alpha_{ij}(>0)$: the tuning parameter for cluster ij, p. 83
A_{ij} : the fuzzy region for cluster ij, p. 81
$\mathbf{c}_{ij}$: the center vector of cluster ij, p. 90
η : the precision parameter, p. 92
ε : the minimum edge length, p. 83
ij : the jth cluster for class i, p. 81
$m_{ij}(\mathbf{x})$: the membership function of cluster ij for input $\mathbf{x}$, p. 84

N_c : the minimum number of misclassifications for postclustering, p. 106

Q_{ij} : the $m \times m$ covariance matrix of cluster ij, p. 91

X_{ij} : the subset of the training data that are included in cluster ij, p. 82

Chapter 6   Clustering

N_{max} : the upper bound of the number of data belonging to each cluster, p. 114

N_{min} : the lower bound of the number of data belonging to each cluster, p. 114

θ : the maximum average edge length of a hyperbox for fuzzy min-max clustering algorithm, p. 115

Chapters 7   Tuning Membership Functions

$\beta_{ij}(l)$: the maximum value of $V_{ij}(\mathbf{x})$ which is smaller than $U_{ij}(l)$, p. 124

δ : the parameter to control the margin of α_{ij} setting, p. 125

δ^c : the parameter to control the margin of c_{ij} setting, p. 138

Dec(l) : the number of misclassified data that are correctly classified when the value of α_{ij} is in $(L_{ij}(l), L_{ij}(l-1)]$, p. 128

$\gamma_{ij}(l)$: the minimum value of $K_{ij}(\mathbf{x})$ which is larger than $L_{ij}(l)$, p. 124

Inc(l) : the number of misclassified data that are correctly classified when α_{ij} is in $[U_{ij}(l-1), U_{ij}(l))$, p. 128

$K_{ij}(\mathbf{x})$: the upper bound of α_{ij} that makes misclassified $\mathbf{x}$ become correctly classified, p. 128

$L_{ij}(\mathbf{x})$: the lower bound of α_{ij} to keep $\mathbf{x}$ correctly classified, p. 122

$L_{ij}(l)$: the lower bound of α_{ij} that allows $l-1$ correctly classified data to be misclassified, p. 122

l_M : the maximum allowable number of misclassifications plus 1 for controlling α_{ij}, p. 124

l_{M_c} : the maximum allowable number of misclassifications plus 1 for controlling c_{ij}, p. 133

$U_{ij}(\mathbf{x})$: the upper bound of α_{ij} to keep $\mathbf{x}$ correctly classified, p. 122

$U_{ij}(l)$: the upper bound of α_{ij} that allows $l-1$ correctly classified data to be misclassified, p. 122

$V_{ij}(\mathbf{x})$: the lower bound of α_{ij} that makes misclassified $\mathbf{x}$ correctly classified, p. 128

X : the training data that are correctly classified using the set of fuzzy rules $\{R_{ij}\}$, p. 126

Y : the training data misclassified using the set of fuzzy rules $\{R_{ij}\}$, p. 126

Chapter 8   Robust Pattern Classification

β_{ij} : the interclass tuning parameter between classes i and j, p. 165

u_k : the deviation ratio for datum $\mathbf{x}_k$, p. 162

ε_i : the threshold of deviation ratios for class i, p. 164

Chapter 9   Dynamic Fuzzy Rule Generation

α : the expansion parameter for the fuzzy min-max classifier with inhibition, p. 186

γ : the sensitivity parameter of the membership functions, p. 178

θ : the maximum average edge length of a hyperbox for the fuzzy min-max classifier, p. 179

Chapter 11   Optimizing Features

β : the minimum change from the original exception ratio, p. 224

ε : the minimum edge length, p. 220

δ : the minimum change of exception ratio, p. 225

Chapters 13 to 16

A_i : the ith fuzzy region

$\mathbf{c}_i$: the center vector for the ith fuzzy rule

M : the number of training data

N : the number of fuzzy rules

o_i : the output value for the ith fuzzy rule

Q_i : the covariance matrix for the ith fuzzy rule

R : the maximum average one-dimensional distance for clustering

R_i : the ith fuzzy rule

S_i : the ith subset of the training inputs

$|S_i|$: the number of elements in the set S_i

y : the one-dimensional output variable

$\hat{y}$: the estimate of y

Part I

Pattern Classification

1. Introduction

1.1 Development of a Classification System

In pattern classification, an object is classified into one of the categories called classes using the features that well separate the classes. Consider recognizing a block of numerals written on an arbitrary part of an envelope by processing the scanned image. First, the location on which the numerals are written must be identified. Then the orientation of the numerals is adjusted, the size is normalized, and each numeral is extracted. Fig. 1.1 shows an example of an extracted numeral. To classify the extracted image into one of the numerals, i.e., 0, 1, ..., 9, we first need to transform the image into features. These features may be the average gray levels of subregions divided by a grid as shown in Fig. 1.1, or the two-dimensional central moments calculated from the gray levels [13], or the features extracted by preprocessing, such as the number of holes and the curvature at some points [14]. Then using these features the extracted image is classified into one of the numerals.

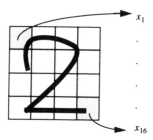

Fig. 1.1. Example features for numeral classification

The discussions of the first part of this book are focused on optimizing features and classifiers to realize high classification performance for a given classification problem. As for classifiers our attention is directed to neural networks and fuzzy systems with learning capability. Since there is no systematic way of optimizing features and classifiers, we need trials and errors in developing a classification system as follows:

- **Feature Determination**   Initially, we do not know what set of features is the best for the given classification problem. Thus, first we determine some appropriate set of features. Since it is relatively easy to delete redundant features but difficult to add necessary features, we set a sufficient number of features.
- **Data Gathering**   We gather samples of the features for each class and normalize the samples so that the range of the samples for each feature is [0, 1]. This is to make each feature that has different physical meaning have an equal weight. Normalization is a necessary step for neural networks since they do not have scale invariance. Some fuzzy systems have scale invariance and linear invariance as discussed in Section 11.1.
- **Feature Optimization**   As discussed in Chapter 11 features can be optimized either by feature selection or feature extraction. In feature extraction, original features are reduced into a small number of features by linear or nonlinear transformation. In feature selection we analyze the gathered samples or extracted features whether they can be classified correctly: we analyze the overlaps of class regions and delete the redundant features or if the input features are not sufficient to separate classes, we analyze the classes that are difficult to separate. If the features are not sufficient to separate classes, we add some features to guarantee class separability.
- **Division of Data**   As discussed in Chapter 12, we divide the samples into the training data set and test data set. Since the training data set needs to be representation of the events that will occur, the training data set should not be biased. Thus, the data set is divided into the training data set and the test data set so that their characteristics become similar.
- **Classifier Evaluation**   We select a classifier from among, for example, neural network classifiers (Chapter 2) and fuzzy classifiers (Chapters 5 and 9), train the classifier using the training data set and evaluate the classifier using the test data set. If the performance is not sufficient, we change parameters of the classifier, add more samples for training and testing, or change classifiers.
- **Field Test**   We implement the classifier into the field and evaluate performance. If the performance is not satisfactory, we tune the classifier using the field data.

1.2 Optimum Features

For a given classification problem we need to determine an initial set of features. There is no strategic way of determining the initial set since there are numerous set possibilities. Thus in this book we assume that we can set some appropriate initial set.

Using the initial set of features, we determine the optimum set of features by feature selection or feature extraction. Here the optimum set of features is the minimum number of features that realizes at least comparable class separability with the initial set. Reduction in the number of features is necessary to construct a compact classification system and to realize high classification performance. Feature selection selects optimum features from the initial set. On the other hand, feature extraction extracts features by some appropriate transformation of the initial set.

Now consider feature selection. Assume that we use a set of m features $x_1, x_2, \ldots, x_m$ for a two-class classification problem. Also, assume that the data belonging to each class exist in a corresponding ellipsoid. Fig. 1.2 shows this with two features, i.e., $m = 2$. In the x_1–x_2 space, Classes 1 and 2 are separated. Thus, if the class boundary is chosen as shown in the figure by some appropriate classifier, data belonging to one of the two classes can be correctly classified. But if we use only one feature x_1 or x_2, there is a region where two classes overlap. Thus, the data belonging to the overlapping region are misclassified. If the two classes are not separable by any one feature x_i ($i \in \{1, \ldots, m\}$), the set of features $\{x_1, x_2\}$ is optimum in the sense that they are the minimum number of features that separate data belonging to Classes 1 and 2 correctly. In this case, the addition of features does not contribute to improving separability.

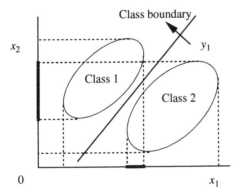

Fig. 1.2. Example of separable classes

When some class boundaries overlap in the initial feature space, the overlap does exist in its subspace. The idea of feature selection, in such a situation, is to eliminate redundant features that do not contribute in maintaining the initial class separability. Consider the case shown in Fig. 1.3. In this case, the regions for Classes 1 and 2 overlap. If we delete x_1 or x_2, the overlap still remains but the interval of overlap is larger when x_1 is deleted. Thus if we can define the degree of overlap or degree of class inseparability, we can delete features that do not increase the degree of overlap. In Chapter 11, as a measure for the degree of overlap, the exception ratio is defined and used to determine the optimum features.

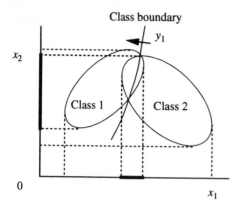

Fig. 1.3. Examples of inseparable classes

If the redundant features work to deteriorate classification performance, deleting these features may improve classification performance. But in general, as will be clear from the above discussion, the optimum set of features by feature selection does not necessary leads to improving the classification performance. Rather, the set contributes in realizing a compact classification system.

In the discussion of feature selection, we assume that data of each class exist in the corresponding ellipsoid. But in actual situations, we get only the sample data and even if the sample data of each class are within the corresponding ellipsoid, there is no guarantee that these ellipsoids are boundaries of the two classes. Thus, the results depend on how well samples are collected to represent the real data distribution.

Feature extraction linearly or nonlinearly transforms original features into reduced ones. Consider linear transformation. Assume that in the previous examples shown in Figs. 1.2 and 1.3, only features x_1 and x_2 are available. Further, assume that the class boundary is linear. Then the best feature is orthogonal to the boundary. Namely, y_1 in the figures. Since multilayer neural networks are known to learn separating hyperplanes [2], in Section 11.2.3 extraction of features based on separating hyperplanes is discussed.

Since class separability does not change by linear transformation of features, improvement in classification performance may not be significant, although some classifiers may exploit class separability of the transformed features more than that of the initial features. Nonlinear transformation of features, on the other hand, changes class separability, and thus improvement of classification performance is possible. In addition, robust features can be obtained. For example, in recognizing numerals, classification by the gray levels of the grid given by Fig. 1.1 is vulnerable to translation, scaling, and rotation. The two-dimensional moments calculated from the gray levels are translation and scale invariant. Thus using these features a robust classification system with smaller number of features is obtained [15, 16].

Support vector machines, which are based on theoretical learning theory [12], defy the concept of the optimum set of features. In the support vector machine, a multi-class problem is converted into multiple two-class problems and for each two-class problem a high-dimensional feature space is generated by the polynomials of the original features, that is, by nonlinear transformation. Instead of selecting the optimum set of features from the polynomials, the optimal separating hyperplane for classification is constructed in the high-dimensional feature space. The optimal separating hyperplane is determined by solving the quadratic optimization problem with the number of variables equal to that of the training data. But since each term in the polynomials need not be calculated the amount of calculations does not increase considerably by the introduction of the high-dimensional feature space.

1.3 Classifiers

Among various neural networks, multilayer neural networks are known to have relatively good classification performance for a wide range of applications [2, 17]. On the other hand, performance of conventional fuzzy classifiers is usually inferior to that of multilayer neural networks, although fuzzy classifiers are more easily analyzed than multilayer neural networks. To improve classification performance comparable to that of multilayer neural networks, fuzzy classifiers with learning capability have been developed. In this section, we overview neural network classifiers, conventional fuzzy classifiers, and fuzzy classifiers with learning capability.

1.3.1 Neural Network Classifiers

Fig. 1.4 (a) shows a typical neuron model used in neural networks. The output function or activation function of neurons is modeled by a nonlinear function $f(u)$ with saturation where u is the input to the neuron. Figs. 1.4 (b) to (e) show examples of activation functions: (b) the threshold function which is used for the Hopfield neural network [2] for combinatorial optimization or associative memories, (c) the sigmoid function which is used for multilayer neural networks, (d) the Gaussian function which is used for radial basis function neural networks, and (e) the piecewise linear function which is used for the Hopfield neural network [2].

The sigmoid function is given by

$$z = \frac{1}{1 + \exp\left(-\dfrac{u}{T}\right)}, \tag{1.1}$$

where u and z are the input and output of the neuron, respectively and T is used to control the slope of the function and is usually set to 1. The output range is $[0, 1]$.

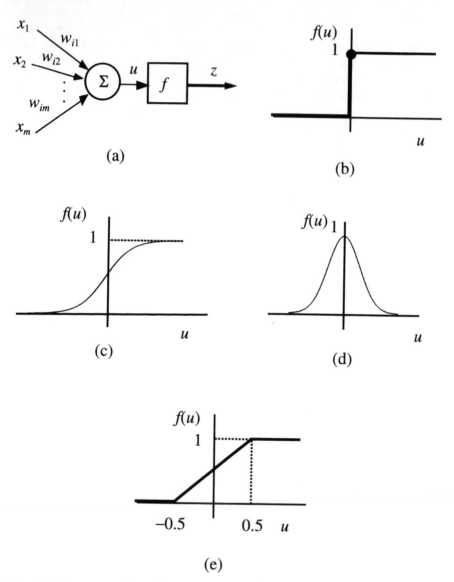

(a)

(b)

(c)

(d)

(e)

Fig. 1.4. Neuron model.   (a) Structure. (b) Threshold function.   (c) Sigmoid function.   (d) Gaussian function. (e) Piecewise linear function

The sigmoid function is a global function in which a wide domain of the input space affects the output, while the Gaussian function is a local function. Synapses connect from inputs to a neuron and between neurons, and for each synapse a weight is assigned. A positive weight works to activate the connected neuron, and a negative weight works to deactivate the connected neuron. When a signal is transmitted via a synapse, it is multiplied by the associated weight. Thus if several synapses are connected to a neuron, the signals are multiplied by the associated weights while they are transmitted through the synapses and they are summed and inputted to the neuron.

By changing activation functions and network connections, various neural network models can be defined. To put neural networks into work, the weights need to be determined and the determination is called learning or training. There are two types of learning methods: supervised learning in which the target class for an input is assigned; and unsupervised learning in which the target class for an input is not assigned.

Among neural network classifiers, multilayer neural network classifiers are successfully applied to a number of applications. Multilayer neural networks consist of several layers, i.e., the input layer, the hidden layers, and the output layer. In this book, we count the input layer as the number of layers. Thus, the neural network shown in Fig. 1.5 is a three-layer neural network with one hidden layer. In [18] three-layer neural networks are shown to classify any training data set if no identical data are included in different classes. The back-propagation algorithm which is based on the steepest descent method is widely used for training. But since training is slow, various speedup methods have been developed. In Chapter 2, speedup of training by solving inequalities is discussed.

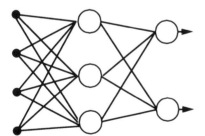

Fig. 1.5. Example of a three-layer neural network

Radial basis function neural network classifiers have three layers. Instead of the sigmoid function, the Gaussian function shown in Fig. 1.4 (d) is used for the activation function of the hidden neurons, and the activation function of the output neurons is linear. Since radial basis function neural networks have local activation functions, in general training of the networks is faster than that of multilayer neural networks but the generalization ability which is defined as the classification performance for the data not used for training is lower. In Chapter 7, we show how the fuzzy classifier that does not have the linear output layer of the radial basis function neural network classifier can improve the generalization ability.

1.3.2 Conventional Fuzzy Classifiers

In conventional fuzzy classifiers, the ranges of input variables are divided into subregions in advance and for each grid region, a fuzzy rule is defined. Consider a two-class classification problem with two inputs. The range of each input variable is divided into several subintervals. Suppose inputs x_1 and x_2 are divided into two subintervals as shown in Fig. 1.6. Then we obtain four subregions. We call the subregions fuzzy regions. Then a fuzzy set with a membership function is defined for each subinterval. Let them be S (Small) and L (Large) and the associated membership functions be given by Fig. 1.7. The membership function defines the degree that the input belongs to the fuzzy set. The range of the degree of membership is $[0, 1]$ and when the degree of membership is 1, the input completely belongs to the fuzzy set and when it is 0, the input does not belong to the fuzzy set.

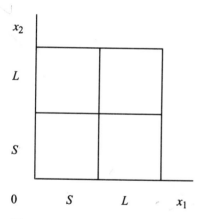

Fig. 1.6. Division of the input space

The fuzzy rules are defined, for instance, as follows:

If x_1 is small and x_2 is large then **x** belongs to Class 1,

If x_1 is large and x_2 is small then **x** belongs to Class 2,

where the if-part is connected by the AND operator and each fuzzy rule is connected by the OR operator. These fuzzy rules are defined according to the experts' knowledge.

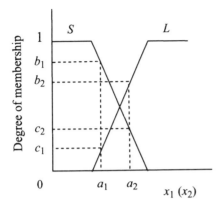

Fig. 1.7. Definition of membership functions

Classification using fuzzy rules is done in the following way. Let the input be $x_1 = a_1$ and $x_2 = a_2$. From Fig. 1.7, the degree of membership of a_1 for the fuzzy set S is b_1 and that of a_2 for L is b_2. Considering the AND operation in the if-part is minimum, the degree of membership of the input (a_1, a_2) for the first fuzzy rule is $\min(b_1, b_2) = b_2$. Similarly, the degree of membership for the input (a_1, a_2) for the second fuzzy rule is $\min(c_1, c_2) = c_1$. Considering the OR operation among the fuzzy rules is maximum, $\max(b_2, c_1) = b_2$. Thus the degree of membership of the Class 1 fuzzy rule is larger than that of Class 2. Therefore the input (a_1, a_2) is classified into Class 1.

One of the advantages of the conventional fuzzy classifiers over neural networks is that the experts' knowledge can be expressed by if-then rules and is readily understandable. But the disadvantages are that usually it is difficult to acquire knowledge from experts and, if acquired, performance of the resulting fuzzy classifier is far from satisfaction, usually inferior to that of the neural network classifiers. To ease the difficulty of rule extraction and to improve the performance, much work has been done to utilize learning capability of neural networks [2]. A generalized neural network in which fuzzy rule prototypes are imbedded is trained using numerical data. After training, fuzzy rules are extracted and these fuzzy rules are tuned to improve a recognition rate [7, 19]. Another method of rule generation is to use genetic algorithms [20, 21, 22].

The major reasons that performance of conventional fuzzy classifiers is inferior to that of neural network classifiers are that the division of the input space needs to be done beforehand and that the divided regions are rectangles which are parallel to the input axes. Usually we do not know to what extent we need to divide the input space, and the size of the division should be determined not for the range of the input variable but for each class that is approximated. In addition, if some of the input variables are correlated, it is inadequate to approximate the region by a rectangle which is parallel to the input axes. To solve the problems, several fuzzy classifiers with learning capability have been developed. These fuzzy classifiers extract fuzzy rules with variable-size fuzzy regions from data and fuzzy regions are not necessarily rectangles parallel to the input axes. In the following, fuzzy classifiers with learning capability are overviewed.

1.3.3 Fuzzy Classifiers with Learning Capability

In fuzzy classifiers, a set of fuzzy rules is defined for each class. In classification, the degree of membership of an unknown input for each fuzzy rule is calculated and the input is classified into the class associated with the fuzzy rule with the highest degree of membership. In this chapter, we overview various fuzzy classifiers, their fuzzy rule generation methods, and fuzzy rule tuning methods to improve classification performance.

Definition of Fuzzy Classifiers. Now assume that we have a set of m-dimensional input data and that each input datum is labeled with a class number $i \in \{1, \dots, n\}$. Then to develop the fuzzy classifier that classifies these data into correct classes, first we divide the set into training and test data sets, where the training data set is used for developing a fuzzy classifier and the test data set is used for testing the performance of the classifier.

In developing a fuzzy classifier we need to define fuzzy rules for each class. Here we assume that several fuzzy rules are defined for class i as follows:

$$R_{ij}: \quad \text{If } \mathbf{x} \text{ is } A_{ij} \quad \text{then } \mathbf{x} \text{ belongs to class } i, \tag{1.2}$$

where $\mathbf{x}$ is the m-dimensional input vector, A_{ij} is the jth fuzzy region for class i defined in the input space.

According to the shapes of the fuzzy regions A_{ij}, fuzzy classifiers can be classified into (see Fig. 1.8):

- fuzzy classifiers with hyperbox regions;
- fuzzy classifiers with polyhedral regions; and
- fuzzy classifiers with ellipsoidal regions.

In fuzzy classifiers with hyperbox regions, class regions are approximated by hyperboxes with each surface parallel to one of the input variables. Associated with the fuzzy region, a membership function discussed in Chapter 4 is defined. We define the membership function of $\mathbf{x}$ for A_{ij} as $m_{ij}(\mathbf{x})$. For example, the membership function for the one-dimensional hyperbox (i.e., interval) is defined as shown in Fig. 1.9. When input x is in the interval $[a, b]$, i.e., in the hyperbox, the degree of membership is 1: input x completely belongs to this hyperbox. If input x moves away from the hyperbox, the degree of membership decreases and reaches 0, meaning that x does not belong to this hyperbox. For the given input vector, the degrees of membership for all the fuzzy rules are calculated and the input vector is classified into the class with the maximum degree of membership.

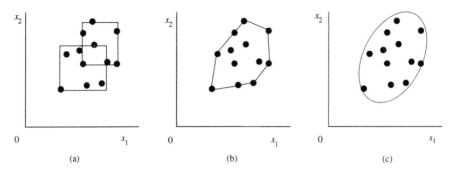

Fig. 1.8. Fuzzy regions. (**a**) Hyperbox. (**b**) Polyhedron. (**c**) Ellipsoid

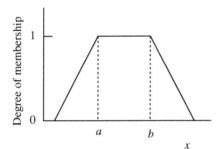

Fig. 1.9. Membership function for the one-dimensional hyperbox

Each type of fuzzy classifier has variations according to the variations of membership functions, rule generation methods, and tuning methods.

Fuzzy Rule Generation. For a given classification problem, there is no systematic way of selecting the optimum classifier. Therefore we assume that we have selected one classifier. Then the next problem is how many fuzzy rules must be generated to realize sufficient recognition rates for both the training data and test data sets. This depends on how data for each class are gathered and how data between different classes overlap in the input space. If class i is approximated by one fuzzy region A_{i1} and for any data $\mathbf{x}$ belonging to class i, the degree of membership for A_{i1}, $m_{i1}(\mathbf{x})$, is the largest, the class i data are correctly classified. Thus one fuzzy rule is sufficient for class i. But if this does not hold, we need to define more than one fuzzy rule to resolve overlaps between classes and then to improve recognition rates. In general there are two ways to solve this problem. One is to detect overlaps and, if they exist, to generate additional fuzzy rules or modify existing fuzzy rules to resolve the overlaps. We call this method the dynamic clustering method. The other is to generate fuzzy rules without considering overlaps and then to tune membership functions for overlap resolution. In the latter method, class data are divided into clusters in advance (preclustering) or after rule generation (postclustering).

There are several ways to resolve overlaps between classes by dynamic clustering. The general flow of rule generation is as follows:

1. Generate a fuzzy rule using all or part of the training data included in a class. If there are no data remained to generate rules, go to Step 3.
2. Check whether the fuzzy region defined in Step 1 overlaps with other fuzzy regions defined previously. If there is no overlap, go to Step 1. Otherwise resolve the overlap by modifying fuzzy rules and go to Step 1.
3. Tune membership functions so that the recognition rate of the training data is improved.

Some algorithms, e.g. [23], may need additional data to tune membership functions.

In the preclustering method, first, the training data belonging to a class are divided into several clusters and then for each cluster a fuzzy rule is generated. While in the postclustering method, first, one fuzzy rule is generated for each class and then if the recognition rate is not sufficient, fuzzy rules are generated to resolve overlaps between classes.

The general flow of rule generation by preclustering is as follows:

1. Cluster training data for each class into several clusters using some clustering technique.
2. Define a fuzzy rule for each cluster using the training data included in the cluster.
3. Tune membership functions to improve the recognition rate of the training data.

In preclustering, since the clustering does not directly connected to the classifier performance, usually we need to repeat the above procedure to obtain a best recognition rate for the classifier.

The general flow of rule generation by postclustering is as follows:

1. Define one fuzzy rule for each class using all the data belonging to the class.
2. Tune membership functions to improve the recognition rate of the training data.
3. If the recognition rate of the training data is sufficient, stop rule generation. If not, go to Step 4.
4. If the number of the data belonging to a class that are misclassified into another class exceeds a specified number, define the cluster that includes these data, and define a fuzzy rule for the cluster. If no new rule is defined, stop generating fuzzy rules.
5. Tune the newly defined fuzzy rules so that the recognition rate of the training data is improved. Go to Step 4.

In postclustering, since we can generate fuzzy rules for the misclassified training data, clustering can be done if necessary. If fuzzy classifiers can implement preclustering, it is also possible to implement postclustering. Which of the clustering techniques should be selected depends on the classifier architecture and the classification problem to be solved.

Fuzzy Rule Tuning. After rule generation by preclustering or postclustering, we can improve the recognition rate by tuning fuzzy rules, i.e., locations and slopes of the membership functions using the training data. By some dynamic clustering methods, the recognition rate of the training data is 100%. Therefore, to tune the membership function, additional training data are necessary.

In general, there are two ways to tune membership functions: a direct method and an indirect method. The direct method tunes membership functions so that the recognition rate of the training data is improved and the indirect method tune membership functions so that the objective function related to the improvement of the recognition rate is optimized.

First we explain the concept of the direct method. Consider a two-class problem with a one-dimensional triangular membership function with each class as shown in Fig. 1.10. Datum 1 belongs to Class 1 but is misclassified into Class 2, because the degree of membership of Datum 1 for Class 2 is larger than that for Class 1. This misclassification can be resolved if the slope of the Class 1 membership function is decreased or that of the Class 2 is increased as shown in the dotted lines.

By changing slopes, the correctly classified data may be misclassified. But we can calculate the ranges of slopes that make correctly classified data remain correctly classified. In Fig. 1.10, if Datum 1 belongs to Class 2, it is correctly classified. It remains to be correctly classified so long as the degree of membership of Datum 1 for Class 2 is larger than that for Class 1. Thus if we change the slopes of the membership functions so that the misclassified data are correctly classified in the ranges that do not make correctly classified data be misclassified, the recognition rate of the training data is improved. We tune the slope of each membership function successively. This method does not allow the correctly classified data to be misclassified. But we can allow correctly classified data to be misclassified if the total recognition rate is improved. The tuning method thus far discussed directly improves the recognition rate.

Similar to the method for tuning slopes, we can tune the locations of the membership functions, if the one-dimensional membership functions are defined on the orthogonal axes. Tuning of locations is done in one axis at a time as shown in Fig. 1.11. In Fig. 1.11 (a) fuzzy regions are rectangles, and in Fig. 1.11 (b) they are ellipsoids. In Fig. 1.11 (b) we tune the membership function in the directions of the principal axes.

Fig. 1.12 shows the same situation as that shown in Fig. 1.10. Instead of tuning slopes, by tuning locations we can improve the recognition rate of the training data. Suppose Datum 1 belonging to Class 1 is misclassified into Class 2. This misclassification is resolved if we move the membership function for Class 2 or Class 1 to the right as shown in the dotted lines.

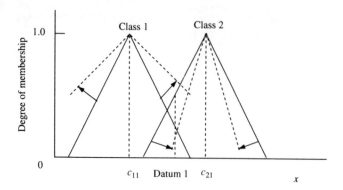

Fig. 1.10. Tuning slopes of the membership functions

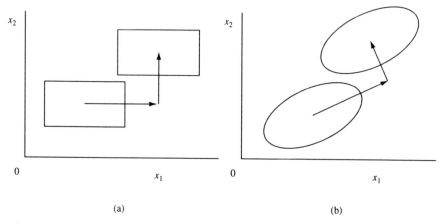

(a)   (b)

Fig. 1.11. Tuning locations of the membership function. **(a)** One-dimensional membership functions are defined on the input axes. **(b)** One-dimensional membership functions are defined on the orthogonal axes not parallel to input axes

The difference between tuning slopes and locations is that in the latter the range of the location to make misclassified datum be correctly classified is a finite interval. In the figure, if the center of the Class 1 membership function passes over Datum 1, the degree of membership begins to decrease. Thus the range of the location is a finite interval. Similar to tuning of slopes, we need to calculate the ranges of locations that make correctly classified data remain correctly classified. Then we can improve the recognition rate of the training data by tuning locations of the membership functions one axis at a time allowing formerly correctly classified data to be misclassified.

In the indirect method, we define an objective function such as:

$$E = \sum_{\mathbf{x} \in X_i} (m_{ij}(\mathbf{x}) - 1)^2 + \sum_{\mathbf{x} \notin X_i} m_{ij}^2(\mathbf{x}), \tag{1.3}$$

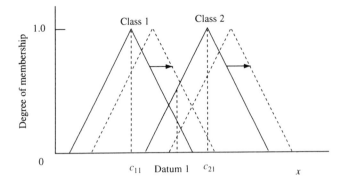

Fig. 1.12. Tuning locations of the membership functions

where X_i denotes the set of the training data belonging to class i. By applying the steepest descent method to (1.3) with respect to the locations and slopes, we can tune the locations and slopes. Although this method can be applied to the continuous membership functions, the tuning process is slow compared to the direct method.

1.4 Evaluation

We evaluate classification performance of a classifier by the recognition rate R given by

$$R = \frac{100M_c}{M} \ (\%), \tag{1.4}$$

where M_c is the number of correctly classified data and M is the number of classified data. The recognition rate of the training data shows how well the classifier fit to the training data. However, this does not guarantee that the classifier performs well to the unknown data. Thus in developing a classifier, usually we prepare two data sets: the training data set for training the classifier; and the test data set for evaluating classification performance for the untested data. We call the classification performance for the untested data generalization ability.

Excessive fitting to the training data may worsen the recognition rate of the test data, and thus it may lead to low generalization ability for the unknown data. In general, as training proceeds the recognition rate of the training data increases. At the early stage of training, this is also the case for the test data, but as training proceeds, the recognition rate of the test data begins to decrease. This phenomenon is called overfitting. The simplest way to avoid overfitting to the training data is to stop training when the recognition rate of the test data begins to decrease (see Fig. 1.13). Since this strategy exploits the test data while training, the recognition rate of the test data may be biased. Thus if we use this strategy, it is advised to divide the gathered data into the three data sets: the training data set for training the classifier; the validation data set to stop training; and the test data set for estimating the generalization ability.

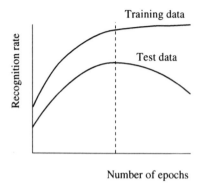

Fig. 1.13. Termination of training to prevent overfitting

Since the classifier performance depends on the training, (validation), and test data sets, special care must be taken to gather the data that generate these data sets. They need to represent the data that will occur. For pattern classification, data that are near the boundary of classes need to be gathered. The division of the gathered data into training and test data also influences the performance of the classifier. Since the training data set needs to reflect the data that will occur, the division must be done so that each data set has similar statistical characteristics. For this purpose, random division of data is used. In Chapter 12, we discuss one approach to divide the gathered data into the two sets with similar characteristics.

The leave-one-out method trains the classifier with $M - 1$ training data, where M is the number of the gathered data; classifies the one datum excluded from the training data; repeats this M times changing the datum excluded from the training data; and calculates the recognition rate of the M data excluded from the training data. Since the effect of the data division to the recognition rate is minimized, this method is used to compare generalization ability of different classifiers. But since the leave-one-out method is time consuming, this method can be use for selecting the suitable classifier for a given classification problem when the number of gathered data, M, is small. But it does not say anything about which of the M trained classifiers to choose.

1.5 Data Sets Used in the Book

The data sets used to evaluate the classifier performance in the book are the iris data [24, 25], the numeral data for license plate recognition [14], the thyroid data [26], and the blood cell data [27]. Table 1.1 shows the numbers of inputs, classes, training data, and test data of the benchmark data sets.

Table 1.1. Benchmark data specification

Data	Inputs	Classes	Training data	Test data
Iris	4	3	75	75
Numeral	12	10	810	820
Thyroid	21	3	3772	3428
Blood cell	13	12	3097	3100
Hiragana-50	50	39	4610	4610
Hiragana-105	105	38	8375	8356
Hiragana-13	13	38	8375	8356

The Fisher iris data are widely used for evaluating classification performance of classifiers. They consist of 150 data with four features and three classes; 50 data per class. We used the first 25 data of each class as the training data and the remaining 25 data of each class as the test data.

The numeral data were collected to identify Japanese license plates of running cars. An example of the license plates is shown in Fig. 1.14. They include numerals, hiragana and kanji characters. The original image taken from a TV camera was preprocessed and each numeral was transformed into 12 features such as the number of holes and the curvature of a numeral at some point.

Fig. 1.14. Example of a Japanese license plate

The thyroid data include 15 digital features and more than 92% of the data belong to one class. Thus the recognition rate smaller than 92% is useless.

The blood cell classification involves classifying optically screened white blood cells into 12 classes using 13 features. (Fig. 1.15 shows example images of white blood cells.) This is a very difficult problem; class boundaries for some classes are ambiguous because the classes are defined according to the growth stages of white blood cells.

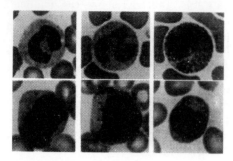

Fig. 1.15. Example images of white blood cells

Hiragana-50 and hiragana-105 data are gathered from Japanese license plates (An example of hiragana data is shown in the bottom left corner of the plate shown in Fig. 1.14). The original gray-scale images of hiragana characters were transformed into 5×10-pixel and 7×15-pixel images, respectively with the gray-scale range being from 0 to 255. Then by performing gray-scale shift, position shift, and random noise addition to the images, the training and test data were generated. Then for the hiragana-105 data to reduce the number of input variables, i.e., $7 \times 15 = 105$, the hiragana-13 data were generated by calculating the 13 central moments for the 7×15-pixel images [15, 16].

2. Multilayer Neural Network Classifiers

Any continuous function can be approximated with any accuracy by a three-layer neural network with an arbitrary number of hidden neurons. Likewise, any multi-dimensional data, whether they be linearly separable or not, can be classified correctly by a three-layer neural network. Thus three-layer neural networks are universal in function approximation and pattern classification. But since their training by the back-propagation algorithm is slow, numerous training algorithms have been developed.

In this chapter our discussions are focussed on fast training of multilayer neural network classifiers based on synthesis principles. First we define multilayer neural networks, and then we discuss the principles that synthesize multilayer neural network classifiers for a given data set. Finally, we discuss the two training methods: one is the well-known back-propagation algorithm and the other is based on the synthesis principle and trains the network by successively solving the sets of inequalities that specify the weights between layers. We compare their training time and the generalization ability for several benchmark data.

2.1 Three-layer Neural Networks

Fig. 2.1 shows a three-layer neural network with one hidden layer. In the figure, the input is fed from the left and each layer is called an input layer, a hidden layer, and an output layer. We may have more than one hidden layer. The input layer consists of input neurons and a bias neuron whose input is constant (usually 1). The hidden layer consists of hidden neurons and a bias neuron and the output layer consists of output neurons. The number of output neurons is the number of classes for pattern classification and the number of outputs to be synthesized for function approximation. The input neurons and the hidden neurons, and the hidden neurons and the output neurons are fully connected by weights. The input and output of the ith neuron of the kth layer are denoted by $x_i(k)$ and $z_i(k)$, respectively, and the weight between the ith neuron of the kth layer and the jth neuron of the $(k + 1)$st layer is denoted by $w_{ji}(k)$.

Inputs to the input layer are outputted by the input neuron without change, and the output of the bias neuron is 1. Namely,

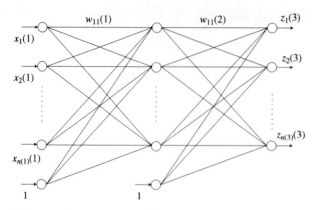

Fig. 2.1. Structure of a three-layer neural network

$$z_j(1) = \begin{cases} x_j(1) & \text{for } j = 1,\dots,n(1), \\ 1 & \text{for } j = n(1)+1, \end{cases} \tag{2.1}$$

where $n(1)$ is the number of input neurons.

Outputs of the input (hidden) neurons and the bias neuron are multiplied by the weights and their sums are inputted to the hidden (output) neurons as follows:

$$x_i(k+1) = \mathbf{w}_i^t(k)\,\mathbf{z}(k) \quad \text{for } i = 1,\dots,n(k+1), \tag{2.2}$$

where $k = 1, 2$, $n(k+1)$ is the number of the $(k+1)$st layer neurons, and

$$\begin{aligned}
\mathbf{x}(k) &= [x_1(k),\dots,x_{n(k)}(k)]^t, \\
\mathbf{z}(k) &= [z_1(k),\dots,z_{n(k)}(k), z_{n(k)+1}(k)]^t, \\
\mathbf{w}_i(k) &= [w_{i1}(k),\dots,w_{i,n(k)}(k), w_{i,n(k)+1}(k)]^t \\
&\quad \text{for } i = 1,\dots,n(k+1).
\end{aligned} \tag{2.3}$$

The output function of the hidden (output) neurons is given by the sigmoid function shown in Fig. 2.2. Namely, their outputs are given by

$$z_i(k+1) = \frac{1}{1 + \exp\left(-\dfrac{x_i(k+1)}{T}\right)} \quad \text{for } i = 1,\dots,n(k+1), \tag{2.4}$$

where T is a positive parameter to control the slope of the sigmoid function and usually $T = 1$.

Let for the training inputs $\mathbf{x}_1,\dots,\mathbf{x}_M$, where M is the number of training data, the desired outputs be $\mathbf{s}_1,\dots,\mathbf{s}_M$, respectively. Then the training of the network is to determine all the weights $\mathbf{w}_i(k)$ so that for the training input $\mathbf{x}_i$, the output becomes $\mathbf{s}_i$. Thus we want to determine the weights $\mathbf{w}_i(k)$ so that the sum of squared errors between the target values and the network outputs is minimized:

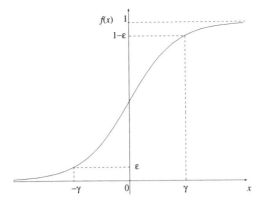

Fig. 2.2. Sigmoid function

$$E = \frac{1}{2} \sum_{l=1}^{M} (\mathbf{z}_l(3) - \mathbf{s}_l)^t (\mathbf{z}_l(3) - \mathbf{s}_l), \tag{2.5}$$

where $\mathbf{z}_l(3)$ is the network output for $\mathbf{x}_l$.

Or for the input-output pairs $(\mathbf{x}_l, \mathbf{s}_l)$, we determine the weights $\mathbf{w}_i(k)$ so that

$$|z_{lj}(3) - s_{lj}| \leq \varepsilon(3) \quad \text{for} \quad j = 1, \ldots, n(3) \tag{2.6}$$

are satisfied, where $\varepsilon(3) \, (> 0)$ is the tolerance of convergence for the output neuron outputs.

When the network is used for pattern classification, the ith output neuron corresponds to class i and for the training input $\mathbf{x}_l$ belonging to class i, the target values of the output neurons j $(j = 1, \ldots, n(3))$ are assigned as follows:

$$s_{lj} = \begin{cases} 1 & \text{for } j = i, \\ 0 & \text{for } j \neq i. \end{cases} \tag{2.7}$$

2.2 Synthesis Principles

Assuming the three-layer neural network with an arbitrary number of hidden neurons, Funahashi [28], Hornik et al. [29] and others proved separately that the network can approximate a real-valued continuous function with any precision. Thus the three-layer neural network is a universal approximator. But their proof is an existence proof and does not clarify how the network is synthesized for the given training data.

We can show that four-layer neural network classifiers can correctly classify any labeled data set that does not include the same data with different labels [2]. Thus they are universal classifiers. It is shown that there is a continuous class region that is not approximated by three-layer neural network classifiers [30, pp. 125–126], [31]. But under the assumption that there are no identical data in different classes, we can show that three-layer neural network classifiers are universal classifiers [18].

Unlike function approximation, universality can be proved through synthesizing a multilayer neural network classifier for a given data set [2, 18, 32, 33, 34]. In the following we discuss the synthesis principles based on [2, pp. 62–68] and [18].

Let us assume that training is converged under the convergence criterion given by (2.6). Then for the training datum belonging to class i, the outputs of the output neurons k $(k = 1, \ldots, n(3))$ satisfy

$$
\begin{aligned}
1 \geq z_k(3) \geq 1 - \varepsilon(3) \quad &\text{for} \quad k = i, \\
\varepsilon(3) \geq z_k(3) \geq 0 \quad &\text{for} \quad k \neq i,
\end{aligned}
\tag{2.8}
$$

where $\varepsilon(3)$ (> 0) is the tolerance of convergence for the output neuron outputs. By considering the saturation of the sigmoid function, transformation of (2.8) into the input side of the output neurons k $(k = 1, \ldots, n(3))$ gives

$$
\begin{aligned}
\infty \geq x_k(3) \geq \gamma(3) \quad &\text{for} \quad k = i, \\
-\gamma(3) \geq x_k(3) \geq -\infty \quad &\text{for} \quad k \neq i,
\end{aligned}
\tag{2.9}
$$

where

$$
\gamma(3) = -T \ln \left(\frac{1}{1 - \varepsilon(3)} - 1 \right) = T \ln \left(\frac{1}{\varepsilon(3)} - 1 \right).
\tag{2.10}
$$

Namely the finite intervals in the output range correspond to the infinite intervals in the input range. Then from (2.2) and (2.9), the weights between the ith output neuron and hidden neurons correspond to the coefficients of the hyperplane that separates class i data and data belonging to the remaining classes (see Fig. 2.3). In this way we can interpret the weights of the network as the coefficients of the hyperplane. By assuming $x_j(k + 1) = 0$ in (2.2),

$$
\mathbf{w}_j^t(k)\, \mathbf{z}(k) = 0
\tag{2.11}
$$

represents a hyperplane in the $n(k)$-dimensional space. The change of the weight $w_{j,n(k)+1}(k)$ associated with the bias neuron causes a parallel displacement of the hyperplane. From (2.4), the value of $z_j(k + 1)$ corresponding to $x_j(k + 1)$ satisfying (2.11), which is on the hyperplane, is $1/2$. We say the $n(k)$-dimensional point $(z_1(k), \ldots, z_{n(k)}(k))^t$ is on the positive side of the hyperplane if

$$
x_j(k + 1) > 0 \quad \text{or} \quad z_j(k + 1) > 1/2
\tag{2.12}
$$

and on the negative side if

$$
x_j(k + 1) < 0 \quad \text{or} \quad z_j(k + 1) < 1/2.
\tag{2.13}
$$

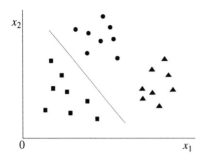

Fig. 2.3. Linearly separable data

From the above discussion, it is clear that two-layer neural network classifiers can classify correctly only linearly separable data as shown in Fig. 2.3.

A class is said to be singly separated by hyperplanes if all the training data in the class are on the same side of the hyperplanes and no training data in other classes exist in the separated region. If training data in a class are divided into subsets such that each subset of the data is singly separated by hyperplanes, the class is said to be plurally separated. We call these hyperplanes separating hyperplanes.

The following theorem holds for the synthesis of singly separated classes.

Theorem 2.2.1. *Consider classification of the $n(1)$-dimensional data into $n(3)$ classes. If $n(2)$ hyperplanes in the $n(1)$-dimensional space exist such that all the training data in any class can be singly separated by a subset of the $n(2)$ hyperplanes, the classifier can be synthesized by the three-layer neural network with $n(1)$ input, $n(2)$ hidden and $n(3)$ output neurons.*

Proof. Let the $n(2)$ hyperplanes be

$$x_j(2) = \mathbf{w}_j^t(1)\,\mathbf{z}(1) = 0 \quad \text{for} \quad j = 1, \ldots, n(2), \tag{2.14}$$

where $\mathbf{w}_j(1)$ and $\mathbf{z}(1)$ are the $(n(1)+1)$st vectors, $\mathbf{z}(1) = (z_1(1), \ldots, z_{n(1)}(1), 1)^t$, and $\mathbf{w}_j(1) = (w_{j1}(1), \ldots, w_{j,n(1)+1}(1))^t$.

Next, let $x_i(1) = z_i(1)\,(i = 1, \ldots, n(1))$, $x_i(1)$ be the input to the ith input neuron and $x_j(2)\,(j = 1, \ldots, n(2))$ be the jth hidden neuron input. Let the jth hidden neuron output be given by substituting $k = 1$ into (2.4), namely

$$z_j(2) = \begin{cases} \dfrac{1}{1 + \exp\left(-\dfrac{x_j(2)}{T}\right)} & \text{for } j = 1, \ldots, n(2), \\ 1 & \text{for } j = n(2) + 1. \end{cases} \tag{2.15}$$

According to the assumption, $z_j(2)$ corresponding to the training datum is either

$$z_j(2) > 1/2 \quad \text{or} \quad z_j(2) < 1/2. \tag{2.16}$$

Thus by multiplying the weight vector $\mathbf{w}_j(1)$ by a positive constant, we can set the value of $z_j(2)$ for the training datum as either

$$z_j(2) = 1 \quad \text{or} \quad z_j(2) = 0. \tag{2.17}$$

Now, to determine the weights between output and hidden neurons, let the input to the ith output neuron be given by

$$x_i(3) = \mathbf{w}_i^t(2)\,\mathbf{z}(2) \quad \text{for} \quad i = 1, \ldots, n(3). \tag{2.18}$$

The weight vector $\mathbf{w}_i(2)$ should be determined so that $z_i(3)$ is larger than or equal to $1 - \varepsilon(3)$ for the class i data and otherwise, smaller than or equal to $\varepsilon(3)$. From (2.9) and (2.18) this can be achieved when

$$x_i(3) = \mathbf{w}_i^t(2)\,\mathbf{z}(2) \geq \gamma(3) \quad \text{for class } i \text{ data}, \tag{2.19}$$
$$x_i(3) = \mathbf{w}_i^t(2)\,\mathbf{z}(2) \leq -\gamma(3) \quad \text{otherwise}. \tag{2.20}$$

If class i is separated by fewer than $n(2)$ hyperplanes, we can cancel the effect of the hyperplanes not contributing to the separation by setting the corresponding weights to zero. Thus we assume that class i is singly separated by $n(2)$ hyperplanes. According to this assumption the outputs $z_1(2), \ldots, z_{n(2)}(2)$ for the class i data are uniquely determined. Therefore, for class i data assume

$$\begin{aligned} (z_1(2), \ldots, z_s(2)) &= (1, \ldots, 1), \\ (z_{s+1}(2), \ldots, z_{n(2)}(2)) &= (0, \ldots, 0). \end{aligned} \tag{2.21}$$

To ensure that $x_i(3)$ is positive for the class i data, $w_{ij}(2)$ for $j = 1, \ldots, s$ need to be positive, and also to ensure that $x_i(3)$ is negative for the input other than class i, $w_{ij}(2)$ for $j = s + 1, \ldots, n(2)$ need to be negative. Since components of $\mathbf{z}(2)$ are 1 or 0, we set weights with the same magnitude as follows:

$$w_{ij}(2) = \begin{cases} 2\gamma(3) & \text{for } j = 1, \ldots, s, \\ -2\gamma(3) & \text{for } j = s + 1, \ldots, n(2). \end{cases} \tag{2.22}$$

Thus for the class i data (2.19) becomes

$$w_{i,n(2)+1}(2) \geq \gamma(3) - 2s\gamma(3), \tag{2.23}$$

and for input other than class i, (2.20) becomes

$$-\gamma(3) - 2\gamma(3) \sum_{j=1}^{s} z_j(2) + 2\gamma(3) \sum_{j=s+1}^{n(2)} z_j(2) \geq w_{i,n(2)+1}(2). \tag{2.24}$$

Since class i is singly separated, $z_j(2) = 0$ holds for some $j, j = 1, \ldots, s$, or $z_j(2) = 1$ for some j, $j = s + 1, \ldots, n(2)$. Therefore, if we set, for example, $w_{i,n(2)+1}(2) = \gamma(3)\,(1 - 2s)$, (2.23) and (2.24) hold. (Q.E.D.)

Theorem 2.2.2. *If class i is plurally separated, we can synthesize a four-layer neural network where the third layer classifies each of the separated regions and the fourth layer sums up the outputs of the third layer.*

Proof. Suppose class i is separated into n regions and suppose they are separately synthesized as discussed in Theorem 2.2.1. Let the corresponding outputs be $z_1(3), \ldots, z_n(3)$. Then their output range is either $1 - \varepsilon(3)$ to 1 or 0 to $\varepsilon(3)$. By using a positive number to multiply the weights connecting the second layer neurons and the first to nth neurons of the third layer, we can set their outputs as 1 or 0 as before. For an input corresponding to class i only one of the n neurons fires, and for an input corresponding to a class other than i, all the n neurons are inactive. Thus the following inequalities hold:

$$w_{j1}(3) + w_{j,n(3)+1}(3) \geq \gamma(3),$$
$$w_{j2}(3) + w_{j,n(3)+1}(3) \geq \gamma(3),$$
$$\cdots\cdots\cdots \tag{2.25}$$
$$w_{jn}(3) + w_{j,n(3)+1}(3) \geq \gamma(3),$$
$$w_{j,n(3)+1}(3) \geq -\gamma(3).$$

One solution of (2.25) is

$$w_{j1}(3) = w_{j2}(3) = \cdots = w_{j,n(3)}(3) = 2\gamma(3),$$
$$w_{j,n(3)+1}(3) = -\gamma(3). \tag{2.26}$$

(Q.E.D.)

As seen from the proof process, the solution that correctly classifies the given training data satisfies a set of inequalities given by (2.19) and (2.20). Thus if a small perturbation is added to the weights that satisfy (2.19) and (2.20), the modified weights can be a solution if they satisfy (2.19) and (2.20). Therefore if a solution exists, an infinite number of solutions can exist. Thus which solution is obtained is determined solely by the training method and initial weights.

Theorem 2.2.2 is a sufficient condition. Still, it is possible to synthesize a three-layer neural network for a plurally separated class by setting the coefficients of the separating hyperplanes to the weights between the input and the hidden neurons, as long as the class is separable from the remaining classes.

For example, consider the two-class problem shown in Fig. 2.4 (a). By lines P_1 and P_2, Class 1 is singly separated but Class 2 is plurally separated. When the positive sides of P_1 and P_2 are selected as indicated by the arrows in the figure, and the absolute values of the coefficients of the lines are set to be sufficiently large, the outputs of the two hidden neurons become as shown in Fig. 2.4 (b). Thus Classes 1 and 2 are separated by a single line P_1'. Therefore, although Class 2 is plurally separated, the neural network classifier can be synthesized in three layers.

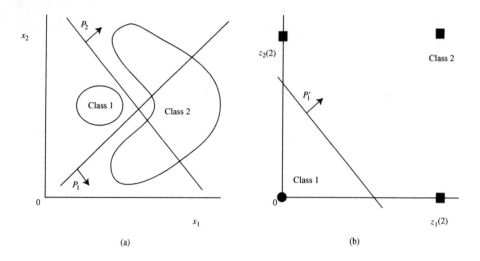

Fig. 2.4. A two-class problem which can be synthesized by a three-layer neural network. (a) Class regions in the input space and the two separating lines. (b) Hidden neuron outputs and the separating line

In [18], three-layer neural networks are proved, in a constructive way, to be universal classifiers. According to [18] we first need to determine the separating hyperplanes as follows:

Generation Algorithm.

1. Set a hyperplane in the input space so that training data belonging to only one class exists on the positive side of the hyperplane.
2. Delete the training data that are on the positive side of the hyperplane.
3. Iterate Steps 1 and 2 until the training data belonging to only one class exist.

The positive side does not have a special meaning. Namely, we may take the negative side in Steps 1 and 2. In Step 1, all the training data belonging to one class are not necessarily separated by a hyperplane. In addition, in the above procedure, instead of deleting the data on the positive side of the hyperplane in Step 2, we may limit the data in Step 1 to those that are in the intersection of the negative side regions of the previously selected hyperplanes.

We explain the procedure using the example shown in Fig. 2.4. In the figure, first we set the line P_1, which separates a part of Class 2 training data on the positive side of P_1 from the remaining training data, and delete the training data that are on the positive side of P_1. Then we set P_2 that separates Class 1 training data and the remaining Class 2 training data. Then we delete the Class 2 training data that are on the positive side of P_2. Since the remaining data belong to Class 1, we stop generating hyperplanes.

Now we need to consider whether the generation algorithm can generate a set of separating hyperplanes for any training data set. Assume that the same data do not exist in different classes. Then by repeating selection of external points in the training data set, we can always generate the set of separating hyperplanes [18]. In Fig. 2.5, Datum 1, which is an external point, is separated by line P_1, and it is deleted. Then, Datum 2 is separated by line P_2, and it is deleted. In this way, by selecting external points, we can always generate a set of separating hyperplanes.

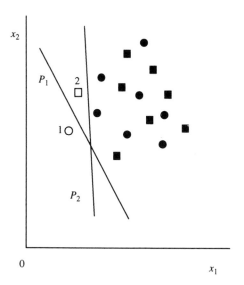

Fig. 2.5. Generation of separating hyperplanes

For the set of hyperplanes generated by the generation algorithm, the following theorem holds.

Theorem 2.2.3. *Setting the coefficients of the hyperplanes generated by the generation algorithm to the weights between the input to hidden neurons, the outputs of the hidden neurons for the training data are linearly separable.*

Proof. Let the number of hyperplanes be $n(2)$ and the values of weights between the input and hidden neurons be multiplied by a constant so that the values of hidden neuron outputs are 1 or 0. Then we prove that we can determine the weights between the hidden to output neuron for Class 1 so that for the Class 1 training data, the output of the output neuron is larger than or equal to $1 - \varepsilon(3)$ (the input of the output neuron is larger than or equal to $\gamma(3)$), and otherwise smaller than or equal to ε (the input of the output neuron is smaller than or equal to $-\gamma(3)$). In the following, the output neuron means the output neuron for Class 1.

If the Class 1 training data are on the positive side of the first hyperplane, for these data the output neuron input needs to be larger than or equal to $\gamma(3)$:

$$\sum_{j=1}^{n(2)} w_j z_j(2) + w_{n(2)+1} \geq \gamma(3), \qquad (2.27)$$

where w_j $(j = 1, \ldots, n(2) + 1)$ are weights between the hidden neurons and output neuron and $z_j(2)$ $(j = 1, \ldots, n(2))$ are hidden neuron outputs. For these training data, $z_1(2) = 1$ and $z_j(2) = 1$ or 0 $(j = 2, \ldots, n(2))$. Thus if

$$w_1 - \sum_{j=2}^{n(2)} |w_j| + w_{n(2)+1} \geq \gamma(3) \qquad (2.28)$$

is satisfied, (2.27) is also satisfied. Similarly, if the training data other than the Class 1 training data are on the positive side of the first hyperplane, for these data the output neuron input needs to be less than or equal to $-\gamma(3)$:

$$\sum_{j=1}^{n(2)} w_j z_j(2) + w_{n(2)+1} \leq -\gamma(3). \qquad (2.29)$$

Since $z_1(2) = 1$ and $z_j(2) = 1$ or 0 for $j = 2, \ldots, n(2)$, (2.29) is also satisfied when

$$w_1 + \sum_{j=1}^{n(2)} |w_j| + w_{n(2)+1} \leq -\gamma(3). \qquad (2.30)$$

Similarly, if the Class 1 training data are on the positive side of the ith hyperplane, these data must satisfy

$$\sum_{j=i}^{n(2)} w_j z_j(2) + w_{n(2)+1} \geq \gamma(3). \qquad (2.31)$$

Notice that $z_j(2) = 0$ for $j = 1, \ldots, i - 1$, $z_i(2) = 1$, and $z_j(2) = 1$ or 0 for $j = i + 1, \ldots, n(2)$. Thus if

$$w_i - \sum_{j=i+1}^{n(2)} |w_j| + w_{n(2)+1} \geq \gamma(3) \qquad (2.32)$$

is satisfied, (2.31) is also satisfied.

However, if the training data other than the Class 1 data are on the positive side of the hyperplane, the following equation must be satisfied for these data:

$$\sum_{j=i}^{n(2)} w_j z_j(2) + w_{n(2)+1} \leq -\gamma(3), \qquad (2.33)$$

where $z_j(2) = 0$ for $j = 1, \ldots, i-1$, $z_i(2) = 1$, and $z_j(2) = 1$ or 0 for $j = i+1, \ldots, n(2)$. Thus if

$$w_i + \sum_{j=i+1}^{n(2)} |w_j| + w_{n(2)+1} \leq -\gamma(3) \tag{2.34}$$

is satisfied, (2.33) is also satisfied.

If the Class 1 training data are on the positive side of the $n(2)$th hyperplane, these data must satisfy

$$w_{n(2)} + w_{n(2)+1} \geq \gamma(3). \tag{2.35}$$

Then since no Class 1 training data exist on the negative side of the hyperplane,

$$w_{n(2)+1} \leq -\gamma(3). \tag{2.36}$$

On the contrary, if the training data other than Class 1 data exist on the positive side of the $n(2)$th hyperplane,

$$w_{n(2)} + w_{n(2)+1} \leq -\gamma(3) \tag{2.37}$$

must be satisfied for these data. If the Class 1 training data exist on the negative side of the hyperplane,

$$w_{n(2)+1} \geq \gamma(3) \tag{2.38}$$

must be satisfied. But if the Class 1 training data do not exist on the negative side of the hyperplane (this happens when the number of classes is more than two), (2.36) must be satisfied.

Now we can solve (2.28), (2.30), (2.32), and (2.34)–(2.38) for $w_j(2)$ ($j = 1, \ldots, n(2) + 1$) by backward substitution. For (2.36) and (2.38), we set $w_{n(2)+1} = \gamma(3)$ and $-\gamma(3)$, respectively. Then Substituting $w_{n(2)+1}$ to (2.35) or (2.37), we can determine $w_{n(2)}$. In this way we can determine w_j ($j = 1, \ldots, n(2) + 1$). (Q.E.D.)

Now consider an example shown in Fig. 2.6 (a). Using the lines P_1 and P_2, the hidden neuron outputs become as shown in Fig. 2.6 (b). In this case, four layers are necessary to generate a neural network classifier.

Now according to the generation algorithm if we set lines P_1, P_2, and P_3 as shown in Fig. 2.7 (a), the hidden neuron outputs become as shown in Fig. 2.7 (b). Thus, the neural network classifier can be synthesized in three layers.

2.3 Training Methods

In training a multilayer neural network classifier for a given training data set, first we need to determine the structure of the network. Since the number of inputs and the number of classes are given, we need to determine the number of hidden layers and the number(s) of hidden neurons. According to Section

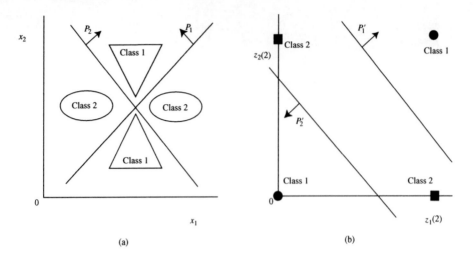

Fig. 2.6. A two-class problem which is synthesized by a four-layer neural network. (a) Class regions in the input space and the two separating lines. (b) Hidden neuron outputs and the separating lines

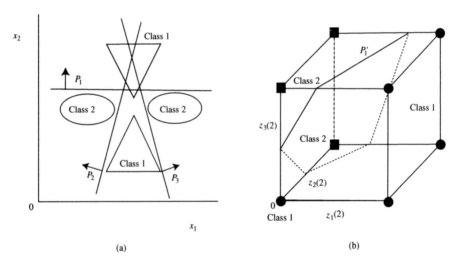

Fig. 2.7. A two-class problem which can be synthesized by a three-layer neural network. (a) Class regions in the input space and the three separating lines. (b) Hidden neuron outputs and the separating plane

2.2, one hidden layer is sufficient. Thus initially it may be advisable to set one hidden layer, and if the network does not converge we may try two hidden layers.

A number of methods have been developed to determine the number of hidden neurons [2]. One type of method is to add or delete hidden neurons during training and the other is to determine the optimal number by analyzing, for the training inputs, the hidden neuron outputs of the initially trained network. Or we may determine the number of hidden neurons heuristically using a rule of thumb, e.g. half the number of input variables, and do trials and errors until sufficient performance is obtained.

Pareceptrons, which are the origin of multilayer neural networks, have threshold functions shown in Fig. 1.4 (b) on page 8 as the output function. Since threshold functions are not differentiable, training is restricted to the weights between two layers. But as seen from the discussions in Section 2.2, two-layer neural networks are not universal; they can separate only linearly separable classes.

Training of multilayer perceptrons was possible by replacing the threshold functions with the differentiable sigmoid functions. In the back-propagation algorithm [35], the output errors of multilayer neural networks are calculated for an input-output pair, the weights are updated so that the errors are reduced, and the above procedure is repeated for all the input-output pairs until the network converges.

For many real world problems the back-propagation algorithm has been shown to be useful, but since the back-propagation training is slow, many high-speed training algorithms have been studied [2]. They are classified into acceleration within the framework of the back-propagation algorithm such as tuning the learning rate or the momentum coefficient, acceleration using the optimization methods such as the second-order optimization method [36], and acceleration based on the synthesis principle [37].

In [37], first a multilayer neural network classifier is trained by the back-propagation algorithm. Then the separating hyperplanes are extracted from the weights between input to hidden neurons, and using these weights the target values of the hidden neuron outputs are determined. Finally, the weights between the inputs to hidden neurons and the weights between the hidden to output neurons are tuned separately. Since the method uses the back-propagation algorithm, the training is still slow. In [38], using the Ho-Kashyap algorithm, two-layer neural networks are trained. In [39, 40], three-layer neural network classifiers are trained by successively solving two sets of inequalities. In the following we discuss the back-propagation algorithm and the training algorithm by solving two sets of inequalities.

2.4 Training by the Back-propagation Algorithm

In the back-propagation algorithm, for the training input vector $\mathbf{x}_l$ ($l = 1, \ldots, M$), the weights $w_{jk}(p-1)$ ($p = 2, 3$) are corrected in the direction that the following squared error is minimized:

$$E_l = \frac{1}{2}(\mathbf{z}_l(3) - \mathbf{s}_l)^t(\mathbf{z}_l(3) - \mathbf{s}_l). \tag{2.39}$$

Since the weights are corrected in the directions that the error (2.39) is minimized we call this method the steepest descent method.

The directions of the weights $w_{jk}(1)$ between the first and second layers and those of the weights $w_{jk}(2)$ between the second and third layers that minimize (2.39) are obtained by taking the partial derivatives of (2.39) with respect to these weights:

$$w_{jk}^{(N',l')}(p-1) = w_{jk}^{(N,L)}(p-1) - \alpha \frac{\partial E_l}{\partial w_{jk}^{(N,l)}(p-1)}$$

for $\quad p = 2, 3, \; k = 1, \ldots, n(p-1)+1, \; j = 1, \ldots, n(p),$ \hfill (2.40)

where

$$l' = \begin{cases} l+1 & \text{for } l < M, \\ 1 & \text{for } l = M, \end{cases} \qquad N' = \begin{cases} N & \text{for } l < M, \\ N+1 & \text{for } l + M, \end{cases} \tag{2.41}$$

and α is the learning rate and usually we set $0 < \alpha \leq 1$ so that oscillations due to large corrections of weights will not occur, and the minus sign for the last term in (2.40) is to get the direction that minimizes (2.39). Equation (2.41) means that after applying (2.40) for the M training data, N is increased by 1. Namely N denotes how many times the weights are updated using M training data, and we call N the number of epochs. The partial derivatives are evaluated using the weights determined by the training datum one step before; this means that we can modify each weight independently.

The procedure for the back-propagation algorithm using (2.40) and (2.41) is as follows.

1. Set random values to $w_{jk}^{(N,l)}(p-1)$ for $N = 1$ and $l = 1$.
2. Calculate the input and output values of each layer for the lth input datum using (2.2) to (2.4). We call this forward propagation.
3. For the output $z_{lj}(3)$ obtained by forward propagation of the lth input datum, if (2.6) is satisfied, we do not change weights. Otherwise, we determine $w_{jk}^{(N',l')}(p-1)$ by evaluating the second term on the right hand side of (2.40) using the input and output values of each layer determined by forward propagation. Since the error between the desired output and the network output is propagated from the output layer to the input layer, this process is called (error) back-propagation.
4. We calculate $w_{jk}^{(N,l)}(p-1)$ alternately repeating forward propagation (Step 2) and back-propagation (Step 3) for all M training data. When

(2.6) is satisfied for all M training data and no modification of weights is made, we terminate the calculation. Since there is no guarantee that the training terminates by this criterion, we further specify the maximum number of epochs. If the number of epochs exceeds the maximum number we terminate the training.

In (2.40) the weights are updated every time the one training datum is processed. We may delay the update of weights until all M data are processed by adding the weight changes to other variables. The former is called an online mode and the latter, a batch mode. The error function of the batch mode corresponds to (2.5). In general, it is said that the online mode can follow the fine changes of the outputs and thus is faster in convergence.

The weights are updated by (2.40) irrespective of the previous changes; the current changes may cancel the previous changes. To avoid this, the weight changes one step before are memorized and the weight changes are done incorporating the previous changes:

$$w_{jk}^{(N',l')}(p-1) = w_{jk}^{(N,L)}(p-1) - \alpha \frac{\partial E_l}{\partial w_{jk}^{(N,l)}(p-1)} + \beta \Delta w_{jk}^{(N,l)}(p-1)$$

$$\text{for} \quad p = 2, 3, \ k = 1, \ldots, n(p-1)+1, \ j = 1, \ldots, n(p), \quad (2.42)$$

where $\Delta w_{jk}^{(N,l)}(p-1)$ is a weight change for the $(l-1)$st training datum (the Mth datum for $l = 1$) and β is a momentum coefficient. But there is no guarantee that training is accelerated by the momentum term.

2.5 Training by Solving Inequalities

In training by the back-propagation algorithm, the target values of the output neuron outputs are set to 1 or 0 but there are no target values for the hidden neuron outputs. Thus the training is not efficient. To overcome this problem several approaches are developed to train the network layer by layer [39, 40, 41, 42, 43, 44, 45]. In [42], each layer is trained layer by layer using, as the objective function for each layer, the objective function for the discriminant analysis that maximizes the between-class scatter while keeping the within class scatter constant. In [43], the weights between the output and hidden layers and the outputs of the previous layer are determined by minimizing the sum of squared errors. Then the calculated outputs are used as the desired outputs for the hidden neurons. This method can be used both for pattern classification and function approximation.

If we limit applications to pattern classification, the situation becomes simpler. Since as discussed in Section 2.2 if the training data belonging to a class are separated by hyperplanes from other training data into a single region, a three-layer neural network classifier can be synthesized. Thus if we can determine the hyperplanes that separate each class, we can set target values of the hidden neuron outputs. Then we can determine the weights between input and hidden layers and the weights between hidden and output layers separately [39, 40].

In the following, assuming that each class is singly separated, we determine target values of hidden neuron outputs, derive the sets of inequalities that the three-layer neural network satisfies, and solve the sets by the Ho-Kashyap algorithm.

2.5.1 Setting of Target Values

We set the target values of hidden neuron outputs as follows. First, we calculate the center vector of each class using the training data included in the class, and we determine the separating hyperplane between two classes by the hyperplane that includes the middle point of the center vectors of the two classes and that is orthogonal to the line segment connecting these center vectors. Here, we define a positive margin parameter δ_d and if the distance from the center vector other than the two center vectors to the hyperplane is larger than δ_d, we assume that the associated class is also separated by the hyperplane, and iterate the above procedure until all the classes are separated by separating hyperplanes. In the following we show the algorithm of setting target values.

1. Calculate the center vector of class i $(i = 1, \ldots, n(3))$, $\mathbf{c}_i$, by

$$\mathbf{c}_i = \frac{1}{|X_i|} \sum_{\mathbf{x} \in X_i} \mathbf{x}, \tag{2.43}$$

   where X_i is the set of training data for class i and $|X_i|$ is the number of training data belonging to class i. Set $i = 1$, and $j = 2$.

2. For two classes i, j $(i < j)$, calculate the vector connecting $\mathbf{c}_j$ and $\mathbf{c}_i$, $\mathbf{n}_{ij}$, by

$$\mathbf{n}_{ij} = \frac{1}{2} (\mathbf{c}_j - \mathbf{c}_i), \tag{2.44}$$

   and determine the separating hyperplane P_{ij} that is orthogonal to $\mathbf{n}_{ij}$ and that includes the middle point of the two center vectors by

$$\left(\mathbf{x} - \frac{1}{2} (\mathbf{c}_j + \mathbf{c}_i) \right)^t \mathbf{n}_{ij} = 0, \tag{2.45}$$

   where the region in which the class i exists is on the negative side of the hyperplane and the region in which the class j exists is on the positive side.

3. If the distance between the center vector $\mathbf{c}_l$ $(l \neq i, j, l = 1, \ldots, n(3))$ for class l and the kth separating hyperplane P_{ij}, d, is less than or equal to δ_d:

$$d = \frac{\left|(\mathbf{c}_l - \frac{1}{2}(\mathbf{c}_j + \mathbf{c}_i))^t \mathbf{n}_{ij}\right|}{|\mathbf{n}_{ij}|} \le \delta_d, \tag{2.46}$$

we assume that class l is not separable by the hyperplane P_{ij} and set the target value of class l for the kth hyperplane P_{ij}, i.e., that of the kth hidden neuron output, $s_k(2)$, as dc where dc means either 1 or 0. When the distance is larger than δ_d, $s_k(2)$ is set to 1 when class l is on the positive side of the hyperplane and 0 otherwise.

4. If the number of center vectors in the region separated by the hyperplanes is equal to or less than 1, go to Step 5. If there are more than one center vector, select two center vectors, let the associated classes be i and j $(i < j)$, and go to Step 2.

5. Consider that dc takes on both 0 and 1 and check if different classes have the same target vector. If there is the same target vector, generate the separating hyperplane P_{ij} for the classes associated with the target vector in the similar way discussed in Steps 2 and 3. If there are no such classes, terminate the algorithm.

6. Calculate the distances between the generated hyperplane and the center vectors and generate the target values.

7. If the target vectors for all the classes are different considering dc takes on both 1 and 0, terminate the algorithm. If not, go to Step 5.

Figs. 2.8 and 2.9 show examples of applying the above algorithm to a classification problem with two inputs and five classes. Fig. 2.8 shows the separating hyperplanes generated after Step 4 is executed and Fig. 2.9 shows the separating hyperplanes generated when the algorithm is terminated. Table 2.1 lists the target values of the hidden neuron outputs. In the table, the results corresponding to Fig. 2.8 are the columns from $s_1(2)$ to $s_3(2)$. Considering dc takes on both 1 and 0, classes 2 and 4 are not separated. Thus in Fig. 2.9, P_{24} is added.

The target value of the output for the ith output neuron is set to 1 for the training data belonging to class i and 0 for the remaining training data. Namely,

$$s_j(3) = \begin{cases} 1 & \text{for } j = i, \\ 0 & \text{for } j \ne i. \end{cases} \tag{2.47}$$

Table 2.2 shows the target values for the example with two inputs and five classes.

2.5.2 Formulation of Training by Solving Inequalities

Since we have determined the target values of the hidden neuron outputs for the classes, we can describe the conditions that the weights between the input neurons and hidden neurons must satisfy.

For the weights connecting from the input neurons to the ith hidden neuron, the following inequalities must hold for all the training data:

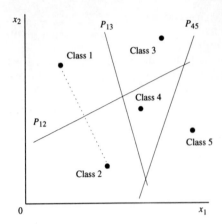

Fig. 2.8. Generation of separating hyperplanes [40, p. 278]

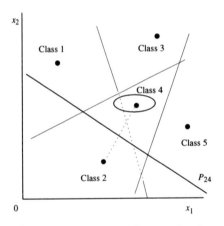

Fig. 2.9. Generation of separating hyperplanes considering class separability [40, p.278]

Table 2.1. Target values of hidden neurons [40, p.279]

Class	$s_1(2)$	$s_2(2)$	$s_3(2)$	$s_4(2)$
1	0	0	0	1
2	1	0	0	0
3	0	1	0	1
4	1	dc	0	1
5	1	1	1	1

Table 2.2. Target values of output neurons [40, p. 279]

Class	$s_1(3)$	$s_2(3)$	$s_3(3)$	$s_4(3)$	$s_5(3)$
1	1	0	0	0	0
2	0	1	0	0	0
3	0	0	1	0	0
4	0	0	0	1	0
5	0	0	0	0	1

$$|z_i(2) - s_i(2)| \leq \varepsilon(2) \qquad \text{for} \quad s_i(2) = 0, \ 1, \tag{2.48}$$

where $\varepsilon(2)$ denotes the tolerance of convergence for the hidden neuron outputs. When the value of $s_i(2)$ is dc, the above inequalities need not hold.

Using (2.4) (see Fig. 2.2), we can convert (2.48) that specify the hidden neuron outputs into the specifications of hidden neuron inputs:

$$\begin{aligned} \mathbf{z}^t(1)\,\mathbf{w}_i(1) &\leq -\gamma(2) \quad \text{for} \quad s_i(2) = 0, \\ \mathbf{z}^t(1)\,\mathbf{w}_i(1) &\geq \ \gamma(2) \quad \text{for} \quad s_i(2) = 1, \end{aligned} \tag{2.49}$$

where $\gamma(2)$ is given by

$$\gamma(2) = \ln\left(\frac{1}{\varepsilon(2)} - 1\right). \tag{2.50}$$

Similarly, for the output neuron outputs, the following inequalities must hold for all the training data:

$$|z_i(3) - s_i(3)| \leq \varepsilon(3) \qquad \text{for} \quad s_i(3) = 0, \ 1, \tag{2.51}$$

where $\varepsilon(3)$ is the tolerance of convergence for the output neuron outputs.

Using (2.4), the specifications of output neuron outputs given by (2.51) can be converted into the specifications of output neuron inputs as follows:

$$\begin{aligned} \mathbf{z}^t(2)\,\mathbf{w}_i(2) &\leq -\gamma(3), \qquad \text{for} \quad s_i(3) = 0, \\ \mathbf{z}^t(2)\,\mathbf{w}_i(2) &\geq \gamma(3), \qquad \text{for} \quad s_i(3) = 1, \end{aligned} \tag{2.52}$$

where $\gamma(3)$ is given by

$$\gamma(3) = \ln\left(\frac{1}{\varepsilon(3)} - 1\right). \tag{2.53}$$

2.5.3 Determination of Weights by Solving Inequalities

First we determine the target values of hidden neuron outputs and set the proper values to $\varepsilon(2)$ and $\varepsilon(3)$. Then we can determine the weights between the input and hidden neurons and the weights between the hidden and output neurons, solving (2.49) for $\mathbf{w}_i(1)$ first, and then (2.52) for $\mathbf{w}_i(2)$.

The inequalities can be solved by the quadratic programming techniques as discussed in Chapter 3. Here, we solved the inequalities applying the Ho-Kashyap algorithm [38, 46]. In this algorithm to solve $Y\mathbf{a} > 0$ for $\mathbf{a}$, where Y is an arbitrary matrix, a positive margin vector $\mathbf{b}$ is introduced to convert the original inequality to the equation $Y\mathbf{a} = \mathbf{b}$. Then $\mathbf{a}$ and $\mathbf{b}$ are corrected iteratively so that the squared error $(\mathbf{b} - Y\mathbf{a})^t (\mathbf{b} - Y\mathbf{a})$ is minimized. By correcting $\mathbf{b}$ so that the elements of $\mathbf{b}$ are non-decreasing, the obtained $\mathbf{a}$ is guaranteed to satisfy $Y\mathbf{a} > \mathbf{0}$. In the following we show the algorithm in detail.

1. Calculate the pseudo-inverse of Y, Y^+, by the singular value decomposition (see Appendix B.2) [47].
2. Set the initial vector to $\mathbf{b}$ so that $\mathbf{b}_0 > 0$.
3. Set the initial vector of $\mathbf{a}$ by $\mathbf{a}_0 = Y^+\mathbf{b}_0$.
4. By $\mathbf{e}_k = Y\mathbf{a}_k - \mathbf{b}_k$, calculate the iteration error vector $\mathbf{e}$, where k is the iteration number. If all the elements of $\mathbf{e}_k$ are zero or less than zero, terminate the algorithm. Otherwise go to the next step.
5. Calculate $\mathbf{b}_{k+1} = \mathbf{b}_k + \rho(\mathbf{e}_k + |\mathbf{e}_k|)$ where ρ is a constant which satisfies $0 < \rho < 1$ and $|\mathbf{e}_k| = (|e_{k1}|, \ldots, |e_{km}|)^t$.
6. Modify $\mathbf{a}$ by $\mathbf{a}_{k+1} = \mathbf{a}_k + \rho Y^+|\mathbf{e}_k|$ and go to Step 4.

By finite iterations of the algorithm, the errors of $Y\mathbf{a}$ and $\mathbf{b}$ converge and it is applicable to non-linearly separable problems.

To obtain the weights $\mathbf{w}_i(k)$ $(k = 1,2)$ that satisfy (2.49) and (2.52), we do the following.

For the lth $(l = 1, \ldots, M)$ training datum, we generate Y_{li} $(i = 1, \ldots, n(k)+1)$ by

$$Y_{li} = \begin{cases} z_i(k) & \text{for } s_i(k+1) = 1, \\ -z_i(k) & \text{for } s_i(k+1) = 0 \end{cases} \tag{2.54}$$

and the initial vector of $\mathbf{b}$ by

$$b_l = \gamma(k+1). \tag{2.55}$$

If the target value of the hidden neuron output is dc, we do not apply the above procedure. Using the solution $\mathbf{a}$ obtained by applying Y and $\mathbf{b}$ to the Ho-Kashyap algorithm, we obtain

$$w_{ij}(k) = a_j \qquad \text{for } j = 1, \ldots, n(k)+1. \tag{2.56}$$

By applying the above procedure to all the neurons in the hidden and output layers, we can determine the weights. To accelerate solving (2.52), we delete the training data that do not satisfy (2.49). Notice that when we obtain the weights between the hidden and output layers, $\mathbf{w}_i(2)$, we calculate $z_i(2)$ using $\mathbf{w}_i(1)$ and then generate Y_{li}. By this the exact values of the targets that are determined roughly as discussed in Section 2.5.1 are determined.

2.6 Performance Evaluation

We compared the training time and classification performance of training by the online-mode back-propagation (BP) algorithm and training by solving inequalities (SI) for the data sets listed in Table 1.1 on page 19, using a Sun UltraSPARC-IIi (335MHz) workstation.

For BP we set, by trial and error, the number of hidden neurons so that the high recognition rate of the test data was obtained. For SI, the number of hidden neurons was determined according to the value of the margin parameter δ_d. We set $\delta_d = 0.15$ so that the numbers of hidden neurons for BP and SI were similar. Table 2.3 lists the number of hidden neurons for BP and SI and the maximum number of epochs.

Table 2.3. The number of hidden neurons and the maximum number of epochs for training

Data	BP	SI	Epochs
Iris	3	2	1000
Numeral	6	11	1000
Blood cell	18	12	15000
Thyroid	3	2	4000
Hiragana-50	25	35	10000
Hiragana-105	25	—	10000
Hiragana -13	25	—	10000

For BP, we trained the network for the hiragana data three times changing the initial weights. Except for the hiragana data, we trained the network ten times. The tolerance of convergence $\varepsilon(3)$ was set to be 0.01, the learning rate and the momentum coefficient were set to be 1 and 0, respectively, and the training was finished when (2.6) was satisfied or the number of training epochs exceeded the prescribed maximum number of epochs. But for all the trials, (2.6) was not satisfied and training was continued until the maximum number of epochs was exceeded.

For SI, the tolerance of convergence for the output neuron outputs, $\varepsilon(3)$, was set to be 0.01 which was the same with that of BP. We evaluated the performance changing the tolerance of convergence for the hidden neuron outputs $\varepsilon(2)$.

2.6.1 Iris Data

Table 2.4 lists the recognition rates for the test (training) data, the numbers of epochs, and the average training time by BP. Each value in the rows of

Max., Min., and Avg. shows the value when the recognition rate of the test data was the maximum and minimum, and when the average was taken for all the trials, respectively.

Table 2.4. Performance for the iris data using BP

	Rates (%)	Epochs	Time (s)
Max.	98.67 (100)	1000	—
Min.	96.00 (96.00)	1000	—
Avg.	97.33 (97.87)	1000	0.85

Table 2.5 lists the results when the neural network was trained by SI. For the training data, the recognition rates were 98.67% (one misclassification) or 100% for different values of $\varepsilon(2)$. The maximum recognition rate of 97.33% (two misclassifications) was obtained for the test data when $\varepsilon(2) = 0.25$, 0.10, and 0.05. The maximum recognition rates of the both methods were comparable and on average an 11-times speedup of training was obtained by SI.

Table 2.5. Performance for the iris data using SI

$\varepsilon(2)$	Rates (%)	Time (s)
0.30	96.00 (98.67)	0.05
0.25	97.33 (98.67)	0.05
0.20	94.67 (98.67)	0.13
0.15	96.00 (100)	0.07
0.10	97.33 (100)	0.06
0.05	97.33 (100)	0.09

2.6.2 Numeral Data

Tables 2.6 lists the results when the network was trained by BP. The maximum recognition rate (3 misclassifications) and the average recognition rate (4.7 misclassifications) of the test data were comparable.

Tables 2.7 lists the results for SI. When the value of $\varepsilon(2)$ was small, the recognition rate of the test data (5 misclassifications) was comparable with

Table 2.6. Performance for the numeral data using BP

	Rates (%)	Epochs	Time (s)
Max.	99.63 (100)	1000	—
Min.	99.02 (99.88)	1000	—
Avg.	99.43 (99.96)	1000	37

the maximum recognition rate (3 misclassifications) of the test data by BP. The training was speeded up by 6 times on average.

Table 2.7. Performance for the numeral data using SI

$\varepsilon(2)$	Rates (%)	Time (s)
0.30	99.26 (100)	7
0.25	99.26 (100)	6
0.20	99.39 (99.88)	6
0.15	99.39 (100)	5
0.10	99.39 (99.88)	5
0.05	99.39 (99.88)	6

2.6.3 Thyroid Data

Table 2.8 lists the results by BP and Table 2.9 lists the results by SI. When $\varepsilon(2)$ was decreased in SI, the recognition rates for the training and test data were increased. The recognition rates of the test data when $\varepsilon(2) = 0.15$ and 0.05 were comparable with the maximum recognition rate by BP. The training was speeded up 8 times on average.

2.6.4 Blood Cell Data

Table 2.10 lists the results by BP. Blood cell classification was a very difficult classification problem and after 15000 epochs of training, the average recognition rate of the training data was still 94.28% and there was a performance gap between the maximum and average recognition rates.

Table 2.11 lists the results by SI. By SI, when $\varepsilon(2)$ was large, the recognition rates for the training and test data were high and for $\varepsilon(2) = 0.25$, the recognition rate of the test data was higher than the maximum recognition rate by BP and the training was speeded up by 29 times on average.

Table 2.8. Performance for the thyroid data using BP

	Rates (%)	Epochs	Time (s)
Max.	97.93 (99.15)	4000	—
Min.	97.05 (98.46)	4000	—
Avg.	97.53 (98.80)	4000	642

Table 2.9. Performance for the thyroid data using SI

$\varepsilon(2)$	Rates (%)	Time (s)
0.30	96.39 (97.27)	43
0.25	96.73 (97.77)	56
0.20	97.17 (98.17)	70
0.15	97.40 (98.32)	85
0.10	97.37 (<u>98.49</u>)	101
0.05	<u>97.43</u> (98.44)	124

Table 2.10. Performance for the blood cell data using BP

	Rates (%)	Epochs	Time (s)
Max.	91.42 (95.61)	15000	—
Min.	86.03 (91.12)	15000	—
Avg.	89.75 (94.28)	15000	4030

Table 2.11. Performance for the blood cell data using SI

$\varepsilon(2)$	Rates (%)	Time (s)
0.30	91.39 (<u>93.83</u>)	128
0.25	<u>91.65</u> (93.80)	135
0.20	91.06 (93.54)	139
0.15	91.22 (92.57)	148
0.10	90.38 (91.96)	146
0.05	89.54 (91.24)	145

2.6.5 Hiragana Data

Table 2.12 lists the results for the hiragana-50 data by BP and Table 2.13 lists the results by SI. The maximum and average recognition rates by BP were slightly better than those by SI for $\varepsilon(2) = 0.3$ both for the training and test data. For large $\varepsilon(2)$ the recognition rates for the training and test data were improved. The training was speeded up by 8 times.

Table 2.12. Performance for the hiragana-50 data using BP

	Rates (%)	Epochs	Time (s)
Max.	95.77 (98.92)	10000	—
Min.	94.58 (97.16)	10000	—
Avg.	95.34 (97.75)	10000	24865

Table 2.13. Performance for the hiragana-50 data using SI

$\varepsilon(2)$	Rates (%)	Time (s)
0.30	<u>95.12</u> (98.22)	2800
0.25	94.81 (<u>98.37</u>)	2697
0.20	94.07 (98.35)	2767
0.15	93.93 (<u>98.37</u>)	2940
0.10	93.62 (98.00)	3103
0.05	93.30 (97.80)	3944

Table 2.14 lists the results for the hiragana-105 data by BP. Training took almost 32 hours. The recognition rates for the test data were about 1% lower than those for the training data.

Table 2.15 lists the results for the hiragana-13 by BP. The hiragana-13 data were obtained by calculating the 13 central moments of the hiragana105 data. Thus the training time was reduced to one seventh with the comparable classification performance.

2.6.6 Discussions

Classification performance of the neural network trained by SI was comparable with that by BP and training was much faster.

Table 2.14. Performance for the hiragana-105 data using BP

	Rates (%)	Epochs	Time (s)
Max.	98.47 (99.52)	10000	—
Min.	98.37 (99.43)	10000	—
Avg.	98.41 (99.49)	10000	113970

Table 2.15. Performance for the hiragana-13 data using BP

	Rates (%)	Epochs	Time (s)
Max.	98.56 (99.64)	10000	—
Min.	98.46 (99.63)	10000	—
Avg.	98.51 (99.63)	10000	15380

When neural networks are trained by SI, as the value of $\varepsilon(2)$ becomes smaller, each component of the margin vector for the Ho-Kashyap algorithm becomes larger; thus the region that the inequalities satisfy becomes smaller. This means that it becomes more difficult to solve the inequalities. This was exemplified by the simulations; as the value of $\varepsilon(2)$ became smaller, the training time usually became longer.

For the iris and numeral data, which are easily classified, the value of $\varepsilon(2)$ did not have much influence on the recognition rates of the training data, and a smaller value of $\varepsilon(2)$ gave better recognition rates for the test data. For the blood cell and hiragana data, which were difficult to converge, the larger value of $\varepsilon(2)$ gave better recognition rates for the training and test data. This was a reasonable behavior. But for the thyroid data, which were difficult to converge, had the opposite tendency; the smaller value of $\varepsilon(2)$ gave better recognition rates for the training and test data.

In solving inequalities, we assumed that each class is separated into a single region. By clustering the training data for each class according to the generation algorithm, we can train the network when each class consists of several regions.

3. Support Vector Machines

Support vector machines are based on the theoretical learning theory developed by Vapnik [12], [17, pp. 92–129], [48], who defies the conventional belief that the optimal classification system can be developed using the optimally reduced features. In support vector machines, an n-class problem is converted into n two-class problems in which one class is separated from the remaining classes. For each two-class problem, the original input space is mapped into the high dimensional dot product space called feature space and in the feature space, the optimal hyperplane that maximizes the generalization ability from the stand point of the VC dimension is determined.

In this chapter we briefly summarize the architecture of the support vector machine and evaluate its classification performance as well as training time using the benchmark data.

3.1 Support Vector Machines for Pattern Classification

In training a classifier, usually we try to maximize classification performance for the training data. But if the classifier is too fit to the training data, the classification ability for unknown data, i.e., the generalization ability becomes worse. This phenomenon is called overfitting. Namely, there is a trade-off between the generalization ability and convergence of training. Various methods have been proposed to prevent overfitting [2, pp. 86–91].

Vapnik's support vector machines have been proposed to solve the overfitting problem from the standpoint of statistical learning theory [12], [17, pp. 92–129].

In training the support vector machine, an n-class problem is converted into n two-class problems. For each two-class problem, the m-dimensional input space $\mathbf{x}$ is mapped into the l-dimensional ($l \geq m$) feature space $\mathbf{z}$. Then in the feature space $\mathbf{z}$ the quadratic optimization problem is solved to separate two classes by the optimal separating hyperplane.

In the following, first we explain the training method of the support vector machine based on [12] and then show the performance evaluation results of the support vector machine for the benchmark data listed in Table 1.1 on page 19.

3.1.1 Conversion to Two-class Problems

For an n-class problem, we usually construct a classifier which classifies a datum into one of n classes. In support vector machines, to enhance linear separability, an n-class problem is converted into n two-class problems. Namely, in the ith two-class classifier the original n-class training data are relabeled as class i or not belonging to class i and are used for training. If only one two-class classifier classifies a datum into a definite class, the datum is classified into that class. Otherwise, the datum is unclassified.

In the following we consider constructing the support vector machine for a two-class problem.

3.1.2 The Optimal Hyperplane

Let m-dimensional inputs $\mathbf{x}_i$ $(i = 1, \ldots, M)$ belong to Class 1 or 2 and the associated labels be $y_i = 1$ for Class 1 and -1 for Class 2. If these data are linearly separable, we can determine the decision function:

$$D(\mathbf{x}) = \mathbf{w}^t \mathbf{x} + b, \tag{3.1}$$

where $\mathbf{w}$ is an m-dimensional vector, b is a scalar, and for $i = 1, \ldots, M$

$$\mathbf{w}^t \mathbf{x}_i + b \begin{cases} \geq +1 & \text{for} \quad y_i = +1, \\ \leq -1 & \text{for} \quad y_i = -1. \end{cases} \tag{3.2}$$

Equation (3.2) is equivalent to

$$y_i \left(\mathbf{w}^t \mathbf{x}_i + b\right) \geq 1 \quad \text{for} \quad i = 1, \ldots, M. \tag{3.3}$$

The hyperplane:

$$D(\mathbf{x}) = \mathbf{w}^t \mathbf{x} + b = c \quad \text{for} \quad -1 < c < 1 \tag{3.4}$$

forms a separating hyperplane that separate $\mathbf{x}_i$ $(i = 1, \ldots, M)$. When $c = 0$, the separating hyperplane is in the middle of the two separating hyperplanes with $c = 1$ and -1. The distance between the separating hyperplane and the training datum nearest to the hyperplane is called the margin. Assuming that the hyperplanes $D(\mathbf{x}) = 1$ and -1 includes at least one training datum, the hyperplane $D(\mathbf{x}) = 0$ has the maximum margin for $-1 < c < 1$. The region $\{\mathbf{x} \mid -1 \leq D(\mathbf{x}) \leq 1\}$ is the generalization region for the decision function.

Fig. 3.1 shows two decision functions that satisfy (3.3). Thus there are an infinite number of decision functions that satisfy (3.3), and thus separating hyperplanes. The generalization ability depends on the location of the separating hyperplane.

The hyperplane with the largest margin is called the optimal separating hyperplane (see Fig. 3.1). According to Vapnik's theory, the classification error is bounded with the probability of at least $1 - \eta$ by

$$R(\mathbf{w}, b) \leq R_{\text{emp}}(\mathbf{w}, b) + \phi, \tag{3.5}$$

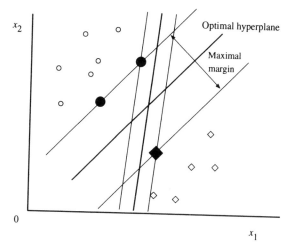

Fig. 3.1. Optimal separating hyperplane in a two-dimensional space

where $R_{emp}(\mathbf{w}, b)$ is the empirical risk (classification error) for the M training data and ϕ is the confidence interval (classification error) for the unknown data:

$$\phi = \frac{\varepsilon}{2}\left(1 + \sqrt{1 + \frac{4R_{emp}(\mathbf{w}, b)}{\varepsilon}}\right), \tag{3.6}$$

$$\varepsilon = 4\frac{h\left[\ln\left(\frac{2M}{h}\right) + 1\right] - \ln\left(\frac{\eta}{4}\right)}{M}. \tag{3.7}$$

Here h is the VC dimension of a set of hyperplanes. The VC dimension is defined as the maximum number of samples that can be separated into any combination of two sets by the set of functions. Since the set of m-dimensional hyperplanes can separate at most $m + 1$ samples, the VC dimension of the set is $m + 1$ (see Fig. 3.2). From (3.7), the confidence level ϕ is decreased when the VC dimension is decreased.

Since the present problem is linearly separable, the empirical risk is zero, i.e., $R_{emp}(\mathbf{w}, b) = 0$. Thus, from (3.6) $\phi = \varepsilon$ and the generalization ability is maximized if the confidence level is minimized. Vapnik proved that the VC dimension of a set of constrained hyperplanes is lower than $m + 1$ and the confidence level is minimized when the optimal separating hyperplane is selected [17].

Here, we must bear in mind that the optimal separating hyperplane realizes the highest generalization ability from the standpoint of the VC dimension. The only assumption made is that the training and test data are generated by a single unknown distribution. Thus if the outliers are included in the training data or the training data are biased from the unknown distribution,

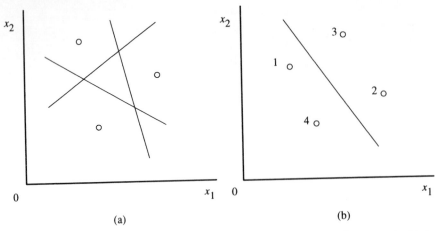

Fig. 3.2. VC dimension of a set of lines. (a) Any three data can be separated into any combination of two sets of data by a single line. (b) Sets of data {1, 4} and {2, 4} can be separated into the two sets by a line but sets of data {1, 2} and {3, 4} cannot. Thus the VC dimension of the set of lines is three

the optimal separating hyperplane may not realize the highest generalization ability.

Conversion of the n-class problem into the n two-class problems may generate unclassified regions as discussed in [49]. Consider the three-class classification problem with the two-dimensional input as shown in Fig. 3.3. Since the data in the region R_{123} are classified into indefinite classes, they are unclassifiable. In addition, regions R_{12}, R_{23}, and R_{31} in Fig. 3.3 are unclassified since for example in R_{12}, data are classified into classes 1 and 2 at the same time. This situation is the same even if the input space is mapped into the higher dimensional space. To avoid this, Kreßel [49] proposed to convert the n-class problem into $n(n-1)/2$ two-class problems which covers all pairs of classes. Another way of avoiding this situation is to introduce the distances from the separating hyperplanes in the unclassified regions.

Now consider determining the optimal separating hyperplane. The Euclidean distance from a training datum $\mathbf{x}$ to the separating hyperplane is given by $|D(\mathbf{x})|/\|\mathbf{w}\|$. Thus assuming the margin δ, all the training data must satisfy

$$\frac{y_k D(\mathbf{x}_k)}{\|\mathbf{w}\|} \geq \delta \quad \text{for} \quad k = 1, \ldots, M. \tag{3.8}$$

Now if $\mathbf{w}$ is a solution, $a\mathbf{w}$ is also a solution where a is a scalar. Thus we impose the following constraint:

$$\delta \|\mathbf{w}\| = 1. \tag{3.9}$$

From (3.8) and (3.9), to find the optimal separating hyperplane, we need to find $\mathbf{w}$ with the minimum Euclidean norm that satisfies (3.3).

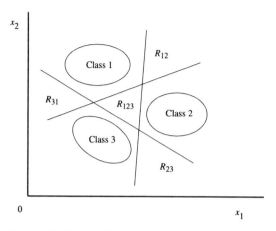

Fig. 3.3. Unclassified regions by the support vector machine. Each class is separated by the associated optimal line from the remaining classes. Then data in the regions R_{123}, R_{12}, R_{23}, and R_{31} are unclassified

The data that satisfy the equality in (3.3) are called support vectors. In Fig. 3.1, the data corresponding to the filled circles and the filled rectangle are support vectors. These data are nearest to the separating hyperplanes and thus are difficult to be classified.

Now the optimal separating hyperplane can be obtained by minimizing

$$\frac{1}{2}\|\mathbf{w}\|^2 \tag{3.10}$$

with respect to $\mathbf{w}$ and b subject to the constraints:

$$y_i\left(\mathbf{w}^t\,\mathbf{x}_i + b\right) \geq 1 \qquad \text{for} \qquad i = 1, \ldots, M. \tag{3.11}$$

The number of variables for the convex optimization problem given by (3.10) and (3.11) is the number of features plus 1: $m+1$. When the number of features is small, we can solve (3.10) and (3.11) by the quadratic programming technique. When the number of features is large, we can convert (3.10) and (3.11) into the equivalent dual problem whose number of variables is the number of training data.

First we convert the constrained problem given by (3.10) and (3.11) into the unconstrained problem:

$$Q(\mathbf{w}, b, \alpha) = \frac{1}{2}\mathbf{w}^t\,\mathbf{w} - \sum_{i=1}^{M} \alpha_i \left\{ y_i\left(\mathbf{w}^t\,\mathbf{x}_i + b\right) - 1\right\}, \tag{3.12}$$

where $\alpha = (\alpha_1, \ldots, \alpha_M)^t$ is the Lagrange multiplier. The optimal solution of (3.12) is given by the saddle point where (3.12) is minimized with respect to $\mathbf{w}$ and b and it is maximized with respect to α_i (≥ 0).

According to Kuhn-Tucker's theorem, the optimal solution $\mathbf{w}^*, b^*$, and α^* of (3.12) must satisfy

$$\frac{\partial Q(\mathbf{w}^*, b^*, \alpha^*)}{\partial b} = 0, \tag{3.13}$$

$$\frac{\partial Q(\mathbf{w}^*, b^*, \alpha^*)}{\partial \mathbf{w}} = 0. \tag{3.14}$$

Using (3.12), (3.13) and (3.14) reduce, respectively, to

$$\sum_{i=1}^{M} \alpha_i^* y_i = 0, \qquad \alpha_i^* \geq 0 \quad \text{for} \quad i = 1, \ldots, M, \tag{3.15}$$

$$\mathbf{w}^* = \sum_{i=1}^{M} \alpha_i^* y_i \mathbf{x}_i, \qquad \alpha_i^* \geq 0 \quad \text{for} \quad i = 1, \ldots, M. \tag{3.16}$$

According to the Kuhn-Tucker theorem, in (3.3) the equality holds for the training input-output pair $(\mathbf{x}_i, y_i)$ only if the associated α_i^* is not 0. In this case, the training data $\mathbf{x}_i$ are the support vectors.

Substituting (3.15) and (3.16) into (3.12), we obtain the following dual problem. Namely, maximize

$$Q(\alpha) = \sum_{i=1}^{M} \alpha_i - \frac{1}{2} \sum_{i,j=0}^{M} \alpha_i \alpha_j y_i y_j \mathbf{x}_i^t \mathbf{x}_j \tag{3.17}$$

with respect to α_i subject to the constraints

$$\sum_{i=1}^{M} y_i \alpha_i = 0, \alpha_i \geq 0 \quad \text{for} \quad i = 1, .., M. \tag{3.18}$$

Solving (3.17) and (3.18) for α_i $(i = 1, \ldots, M)$, we can obtain the support vectors for Classes 1 and 2. Then the optimal hyperplane is placed at the equal distances from the support vectors for Classes 1 and 2 and b^* is given by

$$b* = -\frac{1}{2}(\mathbf{w}^t \, \mathbf{s}_1 + \mathbf{w}^t \, \mathbf{s}_2), \tag{3.19}$$

where $\mathbf{s}_1$ and $\mathbf{s}_2$ are, respectively, arbitrary support vectors for Classes 1 and 2. From (3.16), (3.19) is rewritten as follows:

$$b^* = -\frac{1}{2} \sum_{k=1}^{M} y_k \alpha_k^* (\mathbf{s}_1^t \, \mathbf{x}_k + \mathbf{s}_2^t \, \mathbf{x}_k). \tag{3.20}$$

In the above discussion, we assumed that the training data are linearly separable. In the following we consider determining optimal hyperplane when the training data are not linearly separable. In this case, we want to determine the optimal hyperplane with both the maximum margin and the minimum classification error.

To allow the data that do not have the maximum margin to exist, we introduce the nonnegative slack variables ξ_i (> 0), $i = 1, \ldots, M$ into (3.3):

$$y_i \left(\mathbf{w}^t \, \mathbf{x}_i + b \right) \geq 1 - \xi_i \qquad \text{for} \quad i = 1, \ldots, M. \tag{3.21}$$

For the training data $\mathbf{x}_i$, if $0 < \xi_i < 1$, the data do not have the maximum margin but are still correctly classified. But if $\xi_i \geq 1$ the data are misclassified by the optimal hyperplane. To obtain the optimal hyperplane in which the number of training data that do not have the maximum margin is minimum, we need to minimize

$$Q(\mathbf{w}) = \sum_{i=1}^{n} \theta(\xi_i),$$

where

$$\theta(\xi_i) = \begin{cases} 1 & \text{for} \quad \xi_i > 0, \\ 0 & \text{for} \quad \xi_i = 0. \end{cases}$$

But this is a combinatorial optimization problem and is difficult to solve. Instead, we consider minimizing

$$\frac{1}{2} \|\mathbf{w}\|^2 + C \sum_{i=1}^{M} \xi_i \tag{3.22}$$

subject to the constraints

$$y_i \left(\mathbf{w}^t \, \mathbf{x}_i + b \right) \geq 1 - \xi_i \qquad \text{for} \qquad i = 1, \ldots, M, \tag{3.23}$$

where C is the upper bound that determines the tradeoff between the maximization of margin and minimization of classification error and is set to a large value. We call the obtained hyperplane soft margin hyperplane.

Similar to the linearly separable case, introducing the Lagrange multipliers α and β, we obtain

$$Q(\mathbf{w}, b, \xi, \alpha, \beta) = \frac{1}{2} \|\mathbf{w}\|^2 + C \sum_{i=1}^{M} \xi_i$$

$$- \sum_{i=1}^{M} \alpha_i \left(y_i \left(\mathbf{w}^t \, \mathbf{x} + b \right) - 1 + \xi_i \right) - \sum_{i=1}^{M} \beta_i \, \xi_i. \tag{3.24}$$

The conditions of optimality are given by

$$\frac{\partial Q(\mathbf{w}^*, b^*, \xi^*, \alpha^*, \beta^*)}{\partial b} = 0, \tag{3.25}$$

$$\frac{\partial Q(\mathbf{w}^*, b^*, \xi^*, \alpha^*, \beta^*)}{\partial \mathbf{w}} = 0, \tag{3.26}$$

$$\frac{\partial Q(\mathbf{w}^*, b^*, \xi^*, \alpha^*, \beta^*)}{\partial \xi} = 0. \tag{3.27}$$

Using (3.24), (3.25) to (3.27) reduce, respectively, to

$$\sum_{i=1}^{M} \alpha_i^* y_i = 0, \qquad \alpha_i^* \geq 0 \qquad \text{for} \quad i = 1, \ldots, M, \tag{3.28}$$

$$\mathbf{w}^* = \sum_{i=1}^{M} \alpha_i^* y_i \mathbf{x}_i, \qquad \alpha_i^* \geq 0 \qquad \text{for} \quad i = 1, \ldots, M, \tag{3.29}$$

$$\alpha_i + \beta_i = C, \quad \alpha_i^*, \beta_i^* \geq 0 \quad \text{for} \quad i = 1, \ldots, M. \tag{3.30}$$

Thus we obtain the following dual problem. Namely, find α_i $(i = 1, \ldots, M)$ that maximize

$$Q(\alpha) = \sum_{i=1}^{M} \alpha_i - \frac{1}{2} \sum_{i,j=0}^{M} \alpha_i \alpha_j y_i y_j \mathbf{x}_i^t \mathbf{x}_j \tag{3.31}$$

subject to the constraints

$$\sum_{i=1}^{M} y_i \alpha_i = 0, \qquad 0 \leq \alpha_i \leq C, \tag{3.32}$$

which is similar to the linearly separable case. The decision function is the same for separable and non-separable cases and is given by

$$D(\mathbf{x}) = \sum_{i=1}^{M} \alpha_i^* y_i \mathbf{x}_i^t \mathbf{x} + b^*. \tag{3.33}$$

Since α_i are nonzero for the support vectors, the summation in (3.33) is added only for the support vectors.

Then unknown datum $\mathbf{x}$ is classified as follows:

$$\mathbf{x} \in \begin{cases} \text{Class 1} & \text{if } D(\mathbf{x}) > 0, \\ \text{Class 2} & \text{otherwise.} \end{cases} \tag{3.34}$$

Thus, when training data are separable, the region $\{\mathbf{x} \,|\, 1 > D(\mathbf{x}) > -1\}$ is a generalization region.

3.1.3 Mapping to a High-dimensional Space

In support vector machines, an n-class problem is converted into n two-class problems and for each two-class problem the optimal hyperplane is determined to maximize the generalization ability. The reason why the hyperplane is used for decision function is that the calculation of the VC dimension is restricted to simple sets of functions such as hyperplanes. If the VC dimension is calculated, the optimal decision boundary that maximizes the generalization ability from the standpoint of the VC dimension is determined. Therefore, if the original input $\mathbf{x}$ are not sufficient to guarantee linear separability of the training data, the obtained classifier may not have high generalization ability although the hyperplanes are determined optimally. Thus to enhance linear separability, in the support vector machines, the original input space is mapped into a high-dimensional dot product space called feature space.

Now using the nonlinear vector function $\mathbf{g}(\mathbf{x}) = (g_1(\mathbf{x}), \ldots, g_l(\mathbf{x}))^t$ that map the m-dimensional input vector $\mathbf{x}$ into the l-dimensional feature space, the linear decision function in the feature space is given by

$$D(\mathbf{x}) = \mathbf{w}^t \mathbf{g}(\mathbf{x}), \tag{3.35}$$

where $\mathbf{w}$ is the l-dimensional vector and we drop the constant term assuming $g_1(\mathbf{x}) = 1$. Rewriting (3.35) in the dual form, we obtain

$$D(\mathbf{x}) = \sum_{i=1}^{M} \alpha_i\, y_i\, \mathbf{g}(\mathbf{x}_i)^t \mathbf{g}(\mathbf{x}). \tag{3.36}$$

According to the Hilbert-Schmidt theory the dot product in the feature space can be expressed by a symmetric kernel function $H(\mathbf{x}, \mathbf{x}')$:

$$H(\mathbf{x}, \mathbf{x}') = \sum_{j=1}^{l} g_j(\mathbf{x}) g_j(\mathbf{x}'), \tag{3.37}$$

if

$$\iint H(\mathbf{x}, \mathbf{x}')\, h(\mathbf{x})\, h(\mathbf{x}')\, d\mathbf{x}\, d\mathbf{x}' \geq 0 \tag{3.38}$$

is satisfied for all the square integrable functions $h(\mathbf{x})$ in the compact subset of the input space ($\int h^2(\mathbf{x})\, d\mathbf{x} < \infty$). This condition is called Mercer's condition.

Using the kernel function, without treating the high dimensional data explicitly, we can construct a nonlinear classifier using the method discussed in Section 3.1.2.

Then unknown data are classified using the kernel function as follows.

$$\mathbf{x} \in \begin{cases} \text{Class 1} & \text{if} \quad f(\mathbf{x}) = +1, \\ \text{Class 2} & \text{if} \quad f(\mathbf{x}) = -1, \end{cases} \tag{3.39}$$

where

$$f(\mathbf{x}) = \text{sign}\left(\sum_{\text{support vectors}} y_i\, \alpha_i^*\, H(\mathbf{x}, \mathbf{x}_i) \right). \tag{3.40}$$

Some of the kernel functions used in support vector machines are as follows:

- Polynomials with the degree of d

$$H(\mathbf{x}, \mathbf{x}') = (\mathbf{x}^t\, \mathbf{x}' + 1)^d.$$

- Radial basis functions

$$H(\mathbf{x}, \mathbf{x}') = \exp(-\gamma \|\mathbf{x} - \mathbf{x}'\|),$$

where γ is a positive parameter for slope control and the centers of the radial basis function neural network consist of the support vectors.

- Three-layer neural networks

$$H(\mathbf{x}, \mathbf{x}') = \frac{1}{1 + \exp(\nu \, \mathbf{x}^t \, \mathbf{x}' - a)},$$

where the values of ν and a need to be determined so that (3.38) is satisfied. The weights between the input and hidden neurons correspond to the support vectors. Since Mercer's condition is not always satisfied for three-layer neural networks, several approaches are made to overcome this problem [50].

3.2 Performance Evaluation

We evaluated performance of the support vector machine for the benchmark data sets listed in Table 1.1 on page 19 using the software developed by Royal Holloway, University of London [51]. We assumed that all the benchmark data sets were not linearly separable, namely the upper bound C was imposed on the variables in solving the optimization problems.

We used the following kernel functions:

- Polynomials: $H(\mathbf{x}, \mathbf{x}') = (\mathbf{x}^t \, \mathbf{x}' + 1)^d.$
- RBF: $H(\mathbf{x}, \mathbf{x}') = \exp(-\gamma \|\mathbf{x} - \mathbf{x}'\|).$

Since a small value of the upper bound C caused degradation of the recognition rate of the test data, we used $C = 5000$ throughout the simulations.

We used a SUN UltraSPARC-II (360MHz) workstation for all the data sets except for the three hiragana data sets. For the hiragana data sets we used a SUN UltraSPARC-IIi (335MHz) workstation. (For comparison, the calculation time listed in the tables in Chapter 10 was measured using a UltraSPARC-IIi (335MHz) workstation.)

3.2.1 Iris Data

Table 3.1 shows the results for the iris data. The recognition rates of the training data were all 100%. The recognition rates of the test data varied from 93.33% (5 misclassified data) to 96.00% (3 misclassified data). The maximum recognition rate was obtained for polynomial kernel functions with degrees higher than three and the radial basis function with $\gamma = 10$.

3.2.2 Numeral Data

Table 3.2 shows the results for the numeral data. The maximum recognition rate of 99.76% was obtained for the test data for the polynomial kernel function with the degree of three. For the polynomial kernel functions with degrees of four and five the recognition rates of the test data were very low.

Table 3.1. Performance for the iris data

Kernel	Parameter	Rates (%)	Time (s)
	$d = 1$	93.33 (100)	0.13
	$d = 2$	94.67 (100)	0.11
	$d = 3$	94.67 (100)	0.12
Polynomial	$d = 4$	<u>96.00</u> (100)	0.11
	$d = 5$	<u>96.00</u> (100)	0.12
	$d = 6$	<u>96.00</u> (100)	0.10
	$\gamma = 10$	<u>96.00</u> (100)	0.28
RBF	$\gamma = 1$	93.33 (100)	0.24
	$\gamma = 0.1$	93.33 (100)	0.24
	$\gamma = 0.01$	93.33 (100)	0.26

Since these kernel functions included all the features that were included in the polynomial kernel function with the degree of two, the associated classification problems were linearly separable. But the recognition rates of the training data were not 100%. This indicated that the optimization problems might have failed to converge properly.

Table 3.2. Performance for the numeral data

Kernel	Parameter	Rates (%)	Time (s)
	$d = 1$	99.63 (100)	26
	$d = 2$	99.63 (100)	25
Polynomial	$d = 3$	<u>99.76</u> (100)	24
	$d = 4$	89.75 (92.47)	30
	$d = 5$	89.51 (91.97)	30
	$\gamma = 10$	96.58 (100)	157
RBF	$\gamma = 1$	99.63 (100)	99
	$\gamma = 0.1$	99.63 (100)	83
	$\gamma = 0.01$	99.63 (100)	82

3.2.3 Thyroid Data

Table 3.3 shows the results for the thyroid data. The thyroid data are peculiar in that most of the thyroid data belong to one class and 16 inputs among 21 inputs are discrete. It is not clear whether these characteristics affected performance of the support vector machine, but recognition rates and training times varied greatly as kernel functions were changed. The best recognition rate of the test data was obtained for the polynomial kernel function with the degree of one. These results indicate that selection of a kernel function is important in realizing high generalization performance of the support vector machine.

Table 3.3. Performance for the thyroid data

Kernel	Parameter	Rates (%)	Time (s)
	$d = 1$	<u>97.43</u> (98.67)	609
	$d = 2$	97.40 (99.42)	2269
	$d = 3$	96.67 (99.87)	4003
Polynomial	$d = 4$	96.56 (100)	137
	$d = 5$	95.16 (96.80)	232
	$d = 6$	92.27 (96.21)	3121
	$\gamma = 10$	95.16 (100)	604
RBF	$\gamma = 1$	95.77 (100)	563
	$\gamma = 0.1$	93.33 (100)	873

3.2.4 Blood Cell Data

Table 3.4 shows the results for the blood cell data. Classes of blood cells were determined according to the growth stage of white blood cells and some of the class boundaries were very vague. The recognition rates of the test data were not changed very much for the different kernel functions. The best recognition rate of the test data of 92.19% was obtained by RBF with $\gamma = 0.01$.

3.2.5 Hiragana Data

Hiragana-50 Data. Table 3.5 shows the results for the hiragana-50 data. The recognition rates of the training data were all 100% for the kernel functions given in the table, and the recognition rates of the test data were good except for RBF with $\gamma = 1$. There were cases where the recognition rates of the training data were below 50% (not listed in the table).

Table 3.4. Performance for the blood cell data

Kernel	Parameter	Rates (%)	Time (s)
	$d = 1$	91.00 (96.71)	1060
	$d = 2$	92.10 (99.32)	1117
	$d = 3$	91.94 (99.93)	1134
Polynomial	$d = 4$	92.10 (100)	815
	$d = 5$	91.90 (100)	829
	$d = 6$	91.65 (100)	1046
	$\gamma = 10$	91.65 (100)	2678
RBF	$\gamma = 1$	92.13 (100)	2380
	$\gamma = 0.1$	92.16 (100)	2568
	$\gamma = 0.01$	92.19 (100)	2756

Table 3.5. Performance for the hiragana-50 data

Kernel	Parameter	Rates (%)	Time (s)
	$d = 1$	98.35 (100)	2657
	$d = 2$	98.91 (100)	2849
Polynomial	$d = 3$	98.89 (100)	3064
	$d = 6$	98.48 (100)	3301
	$\gamma = 1$	93.97 (100)	19624
RBF	$\gamma = 0.1$	99.07 (100)	7144
	$\gamma = 0.01$	98.89 (100)	6405

Hiragana-105 Data. Table 3.6 shows the results for the hiragana-105 data. The recognition rates for both of the training and test data were 100% except for the polynomial kernel function with the degree of one. The training time was from 3 to 9 hours.

Hiragana-13 Data. The hiragana-13 data were generated by calculating the central moments of the hiragana-105 data. Table 3.7 shows the results for the hiragna-13 data. Irrespective of kernel functions, the classification performance was good. But since the number of data was large, training took 2 to 6 hours.

Table 3.6. Performance for the hiragana-105 data

Kernel	Parameter	Rates (%)	Time (s)
	$d = 1$	99.87 (100)	9574
Polynomial	$d = 2$	100 (100)	10443
	$d = 3$	100 (100)	11008
RBF	$\gamma = 0.1$	100 (100)	32622
	$\gamma = 0.01$	100 (100)	25096

Table 3.7. Performance for the hiragana-13 data

Kernel	Parameter	Rates (%)	Time (s)
	$d = 1$	99.40 (99.95)	7848
	$d = 2$	99.57 (100)	8093
	$d = 3$	99.55 (100)	8034
Polynomial	$d = 4$	99.56 (100)	7686
	$d = 5$	99.55 (100)	7535
	$d = 6$	99.32 (100)	8041
	$\gamma = 1$	<u>99.77</u> (100)	23216
RBF	$\gamma = 0.1$	99.56 (100)	23212
	$\gamma = 0.01$	99.40 (100)	19525

3.2.6 Discussions

For most of the benchmark data sets, classification performance of the support vector machine was good. But since the quadratic optimization problem with the number of variables equal to that of training data needs to be solved, training time became prohibitively longer as the number of training data increased. Therefore, acceleration of training is one of the research topics [52]. The larger value of C makes the recognition rate of the training data higher, but it slows down training. Thus, by optimizing the value the training time may be shortened.

One of the basic concepts of support vector machines is to map the original input space into a high-dimensional feature space. But for the benchmark data the recognition rates of the test data using the polynomial kernel function with the degree of one were sufficiently good. For the thyroid data, the polynomial kernel function with the degree of one showed the best recognition rate. In [53], generalization performance of the support vector machine was experimentally shown to be suffered from irrelevant features. Thus, also for the support vector machine, selection of features is an important task in realizing high generalization ability [54].

In addition, for some benchmark data such as the thyroid and numeral data, the choice of kernel functions influenced the recognition rates of the test and training data immensely. Thus the proper choice of the kernel function is important [53].

In most benchmark data there was no overfitting, but for the thyroid data overfitting occurred. The recognition rate of the test data was best when the recognition rate of the training data was lower than 100%. One reason of overfitting might be the biased distributions of the data. Since most of the data belonged to one class, the optimum hyperplane might not be optimal; the hyperplane shifted from the optimal hyperplane towards the classes with a smaller number of data might be optimal.

4. Membership Functions

In conventional fuzzy classifiers, the range of each input variable is divided into several intervals. Then each interval is considered to be a fuzzy set and an associated membership function is defined. Thus, the input space is divided into several subregions that are parallel to input axes. For each subregion a fuzzy rule is defined; if the input is in the subregion, the input belongs to the class associated with the subregion. Then the degrees of membership of an unknown input for all the fuzzy sets are calculated and the input is classified into the class with the maximum degree of membership. Therefore, the membership functions directly influence the performance of the fuzzy classifier.

But since conventional fuzzy classifiers weigh heavily on linguistic interpretation, one-dimensional membership functions are essential. Then usually classification performance of the conventional fuzzy classifiers is below that of multilayer neural network classifiers especially when the input variables are correlated.

Therefore, in this chapter, we discuss multi-dimensional membership functions that are suited for pattern classification. First we discuss one-dimensional triangular, trapezoidal, and bell-shaped membership functions, and then extend them to multi-dimensional membership functions using the minimum, average, and product operators. We also clarify the relations between some membership functions.

4.1 One-dimensional Membership Functions

In this section we discuss three one-dimensional membership functions: triangular membership functions, trapezoidal membership functions, and bell-shaped membership functions. Usually, the degree of membership is in $[0, 1]$. But here, we allow the negative degree of membership so that any point in the input space is classified into one of the classes if the point is not on the class boundary.

4.1.1 Triangular Membership Functions

We consider a triangular membership function for the input variable x as shown in Fig. 4.1. At $x = c$, the degree of membership is 1 where c is the center of the membership function and it decreases as x moves away from c and reaches 0 at $c = v$ or $c = V$. When $x \geq V$ or $x \leq v$, the degree of membership is 0. The center c is not necessarily the middle point of v and V. The membership function $m(x)$ for the input variable x can be defined as follows:

$$m(x) = \begin{cases} 0 & \text{for} \quad x > V, \\ 1 - \dfrac{x-c}{V-c} & \text{for} \quad V \geq x \geq c, \\ 1 - \dfrac{c-x}{c-v} & \text{for} \quad c < x \leq v, \\ 0 & \text{for} \quad x < v. \end{cases} \tag{4.1}$$

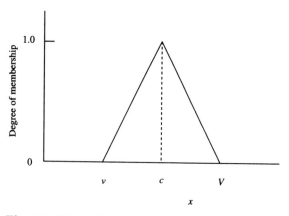

Fig. 4.1. Triangular membership function

To control the slope of the membership function given by (4.1), we introduce a positive parameter α as follows:

$$m(x) = \begin{cases} 0 & \text{for} \quad x > c + \alpha(V - c), \\ 1 - \dfrac{x-c}{\alpha(V-c)} & \text{for} \quad c + \alpha(V - c) \geq x \geq c, \\ 1 - \dfrac{c-x}{\alpha(c-v)} & \text{for} \quad c > x \geq c - \alpha(c - v), \\ 0 & \text{for} \quad x < c - \alpha(c - v). \end{cases} \tag{4.2}$$

Fig. 4.2 shows the triangular membership functions for different values of α. When the value of α is increased the slope decreases and when it is decreased, the slope increases.

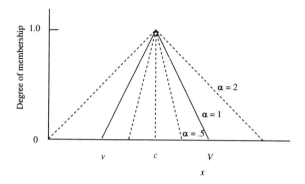

Fig. 4.2. Triangular membership functions for different values of α

Instead of using the membership function given by (4.2), we can use a one-dimensional tuned distance $h(x)$ from the center c given by:

$$h(x) = \frac{d(x)}{\alpha}, \tag{4.3}$$

where $d(x)$ is the one-dimensional weighted distance from c given by

$$d(x) = \begin{cases} \dfrac{x - c}{(V - c)} & \text{for} \quad x \geq c, \\[2mm] \dfrac{c - x}{(c - v)} & \text{for} \quad x < c. \end{cases} \tag{4.4}$$

Then the one-dimensional membership function $m(x)$ given by (4.2) is rewritten as follows:

$$m(x) = \begin{cases} 0 & \text{for} \quad x > c + \alpha(V - c), \\ 1 - h(x) & \text{for} \quad c + \alpha(V - c) \geq x \geq c - \alpha(c - v), \\ 0 & \text{for} \quad x < c - \alpha(c - v). \end{cases} \tag{4.5}$$

Equation (4.5) is a nonlinear function and tuning of α is difficult. Thus to make (4.5) a linear function in $x \geq c$ and $x < c$, we allow negative degrees of membership:

$$m(x) = 1 - h(x). \tag{4.6}$$

In this case, instead of using (4.6), we can use the tuned distance $h(x)$, in which the tuned minimum distance corresponds to the largest degree of membership.

The membership function given by (4.6) is piecewise linear. The membership can be quadratic by

$$m(x) = 1 - h^2(x), \tag{4.7}$$

where

$$h^2(x) = \frac{d^2(x)}{\alpha}. \tag{4.8}$$

Fig. 4.3 shows the piecewise linear and quadratic membership functions with the negative degrees of membership. Eqs. (4.6) and (4.7) can be written in a single form as follows:

$$m(x) = 1 - h^s(x), \tag{4.9}$$

where $s = 1$ or 2 and

$$h^s(x) = \frac{d^s(x)}{\alpha}. \tag{4.10}$$

If we do not want to use a negative degree of membership, we can use

$$m(x) = \exp\left(-h^s(x)\right). \tag{4.11}$$

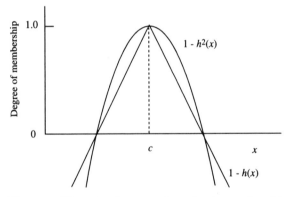

Fig. 4.3. Piecewise linear and quadratic membership functions with the negative degrees of membership

4.1.2 Trapezoidal Membership Functions

Fig. 4.4 shows a trapezoidal membership function $m(x)$. The degree of membership is 1 for $v \le x \le V$ and the degree of membership decreases, until it reaches 0, with the slope $-1/\alpha$ for $x > V$ where α is positive, and with the slope $1/\alpha$ for $x < v$. The membership function $m(x)$ is given by

$$m(x) = \begin{cases} 0 & \text{for} \quad x > V + \alpha, \\ 1 - \dfrac{x - V}{\alpha} & \text{for} \quad V + \alpha \geq x \geq V, \\ 1 & \text{for} \quad V > x > v, \\ 1 - \dfrac{v - x}{\alpha} & \text{for} \quad v > x \geq v - \alpha, \\ 0 & \text{for} \quad x < v - \alpha. \end{cases} \qquad (4.12)$$

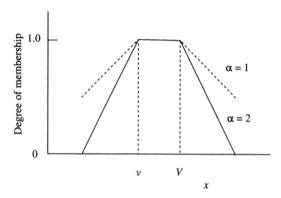

Fig. 4.4. Trapezoidal membership function

Similar to triangular membership functions, we can define a tuned distance $h(x)$:

$$h(x) = \frac{d(x)}{\alpha} = \begin{cases} \dfrac{x - V}{\alpha} & \text{for} \quad V + \alpha \geq x \geq V, \\ 0 & \text{for} \quad V > x > v, \\ \dfrac{v - x}{\alpha} & \text{for} \quad v > x \geq v - \alpha, \end{cases} \qquad (4.13)$$

where $d(x)$ is a one-dimensional distance. Using (4.13), (4.12) is rewritten as follows:

$$m(x) = \begin{cases} 0 & \text{for} \quad x > V + \alpha, \\ 1 - h(x) & \text{for} \quad V + \alpha \geq x \geq v - \alpha, \\ 0 & \text{for} \quad x < v - \alpha. \end{cases} \qquad (4.14)$$

If we allow the negative degree of membership:

$$m(x) = 1 - \frac{d(x)}{\alpha} = 1 - h(x), \qquad (4.15)$$

we can use $h(x)$ instead of $m(x)$.

To introduce a quadratic membership function, we generalize (4.15) as follows:

$$m(x) = 1 - h^s(x) = 1 - \frac{d^s(x)}{\alpha},$$ (4.16)

where $s = 1$ or 2. Equation (4.16) is equivalent to

$$m(x) = \exp\left(-h^s(x)\right).$$ (4.17)

4.1.3 Bell-shaped Membership Functions

Fig. 4.5 shows the bell-shaped membership functions for the input variable x where at the center c the degree of membership is 1 and it decreases as x moves away from the center. The membership function $m(x)$ is given by

$$m(x) = \exp\left(-\frac{(x-c)^2}{\alpha \sigma^2}\right),$$ (4.18)

where α is a positive tuning parameter to tune the membership function and σ^2 is a variance of x around c.

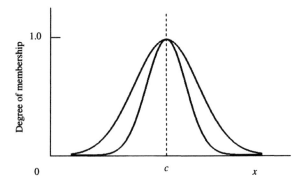

Fig. 4.5. Bell-shaped membership functions

Similar to the triangular membership function, instead of the bell-shaped membership function given by (4.18), we can consider the tuned distance $h(x)$ as follows:

$$h(x) = \frac{|x-c|}{\sqrt{\alpha}\,\sigma}.$$ (4.19)

By comparing (4.10) and (4.19), the tuned distances for the triangular membership function and the bell-shaped membership function are the same if $s = 2$, $\sigma = V - c$, and $c = (V + v)/2$.

Now consider the difference between the bell-shaped membership function given by (4.18) and the normal distribution. The one-dimensional normal distribution function $N(c, \sigma)$ is given by

$$N(c, \sigma) = \frac{1}{\sqrt{2\pi}\sigma} \exp\left(-\frac{(x - c)^2}{2\sigma^2}\right). \qquad (4.20)$$

Thus the difference between the bell-shaped membership function given by (4.18) and the normal distribution function given by (4.20) is that the value of the former function is 1 at $x = c$, while that for the latter is not restricted to 1. Then if the centers of two different bell-shaped membership functions are defined at c, the degree of membership of one membership function is larger than that of the other except for the center. But if the centers of two different normal distributions exist at c, the probability of the one normal distribution is larger than that of the other around the center, while the probability of the other is larger around the tails of the distributions (see Fig. 4.6). This means that if the centers of the bell-shaped membership functions are the same, for any datum the degree of membership of one membership function is always larger except for the center. But this is not a serious problem since the situation is rare in real problems.

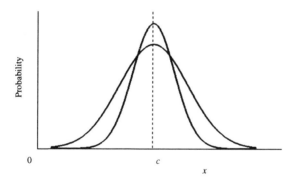

Fig. 4.6. Two normal distribution functions with the same centers

4.2 Multi-dimensional Membership Functions

As the extension of triangular membership functions, we consider rectangular pyramidal membership functions and polyhedral pyramidal membership functions. As the extension of trapezoidal membership functions, we consider truncated rectangular pyramidal membership functions and truncated polyhedral pyramidal membership functions. As the extension of one-dimensional bell-shaped membership functions, we consider multi-dimensional bell-shaped membership functions.

4.2.1 Extension to Multi-dimensional Membership Functions

Suppose l one-dimensional membership functions $m_k(\mathbf{x}), k = 1, \ldots, l$ are defined for the input vector $\mathbf{x}$ that forms an m-dimensional input space. Each one-dimensional membership function is defined on one of the input axes or a linear combination of the input axes.

We can define the membership function $m(\mathbf{x})$ for the input variable $\mathbf{x}$ applying several operators to the l one-dimensional membership functions. Here, we consider using the minimum, average, and product operators. Using the minimum operator, the membership function $m(\mathbf{x})$ is given by

$$m(\mathbf{x}) = \min_{k=1,\ldots,l} m_k(\mathbf{x}). \tag{4.21}$$

Using the average operator, the membership function $m(\mathbf{x})$ is given by

$$m(\mathbf{x}) = \frac{1}{l} \sum_{k=1}^{l} m_k(\mathbf{x}). \tag{4.22}$$

Using the product operator, the membership function $m(\mathbf{x})$ is given by

$$m(\mathbf{x}) = \prod_{k=1}^{l} m_k(\mathbf{x}). \tag{4.23}$$

When the triangular or trapezoidal membership functions with the negative degree of membership are used as one-dimensional membership functions, from (4.9) and (4.16), $m(\mathbf{x})$ is given by

$$m(\mathbf{x}) = 1 - h^s(\mathbf{x}), \tag{4.24}$$

where $h(\mathbf{x})$ is the tuned distance:

$$h^s(\mathbf{x}) = \min_{k=1,\ldots,l} h_k^s(\mathbf{x}) \tag{4.25}$$

for the minimum operator and $h(\mathbf{x})$ is the tuned distance:

$$h^s(\mathbf{x}) = \frac{1}{l} \sum_{k=1}^{l} h_k^s(\mathbf{x}) \tag{4.26}$$

for the average operator. When the minimum operator is used the membership functions for $s = 1$ and 2 are equivalent.

If the product operator is used for (4.11) and (4.17), $m(\mathbf{x})$ is given by

$$m(\mathbf{x}) = \exp\left(-h^s(\mathbf{x})\right)$$
$$= \exp\left(-\sum_{k=1}^{l} h_k^s(\mathbf{x})\right), \tag{4.27}$$

where $h^s(\mathbf{x})$ in (4.27) is given by multiplying l to that in (4.26).

4.2.2 Rectangular Pyramidal Membership Functions

Here we assume that the one-dimensional membership function $m(x)$ given by (4.5) is defined for the kth input variable of the m-dimensional input vector $\mathbf{x}$. Namely,

$$m_k(\mathbf{x}) = \begin{cases} 0 & \text{for} \quad x_k > c_k + \alpha(V_k - c_k), \\ 1 - h_k(\mathbf{x}) & \text{for} \quad c_k + \alpha(V_k - c_k) \geq x_k \geq c_k - \alpha(c_k - v_k), \\ 0 & \text{for} \quad x < c_k - \alpha(c_k - v_k), \end{cases} \quad (4.28)$$

where $h_k(\mathbf{x})$ is the one-dimensional tuned distance for x_k given by (4.3).

Since the value of α does not change the shape of the membership function, here we assume $\alpha = 1$ and first study the shape of the membership function $m(\mathbf{x})$ for the minimum and product operators. At $\mathbf{c} = (c_1, \ldots, c_m)^t$, the degree of membership is 1. Let the region A be defined by $A = \{\mathbf{x} \mid \mathbf{v} \leq \mathbf{x} \leq \mathbf{V}\}$ where $\mathbf{v} = (v_1, \ldots, v_m)^t$ and $\mathbf{V} = (V_1, \ldots, V_m)^t$. The region A is rectangular. Then the degree of membership is 0 when $x_k \leq v_k$ or $x_k \geq V_k$ ($k = 1, \ldots, m$). Thus on the surface and outside of A, the degree of membership is 0.

Inside of A, any $m_k(\mathbf{x})$ is positive and less than or equal to 1. Thus, in A

$$\prod_{k=1}^{m} m_k(\mathbf{x}) \leq \min_{k=1,\ldots,m} m_k(\mathbf{x}). \quad (4.29)$$

Namely, the degree of membership using the minimum operator is larger than or equal to that using the product operator inside of A.

The shape of the membership function using the minimum operator is a rectangular pyramid as shown in Fig. 4.7.

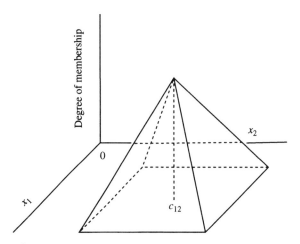

Fig. 4.7. Rectangular pyramidal membership function

Now we investigate the shape of the membership function $m(\mathbf{x})$ using the average operator assuming $\alpha = 1$. The degree of membership at $\mathbf{x} = \mathbf{c}$ is 1. Since $m(\mathbf{x}) = 0$ only when $x_k \leq v_k$ or $x_k \geq V_k$ for all $k, k = 1, \ldots, m$, there are regions where the degree of membership is positive outside of A. Figure 4.8 shows the contour lines of the membership function using the average operator for the two-dimensional input. The irregular contour lines occur outside of A.

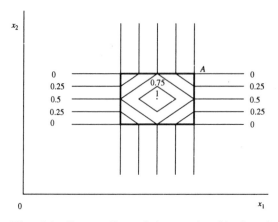

Fig. 4.8. Contour lines of the membership function with the average operator

Instead of (4.28), if the membership function with the negative degree of membership:

$$m_k(\mathbf{x}) = 1 - h_k(\mathbf{x}) \tag{4.30}$$

is used, we can avoid irregular contour lines as shown in Fig. 4.9. Here, we do not show the negative degree of membership for $m(\mathbf{x})$. Then the shape of the membership function with the average operator is a rectangular pyramid but the rectangle is not parallel to the input axes.

Instead of (4.30), we can use the quadratic membership function:

$$m_k(\mathbf{x}) = 1 - h_k^2(\mathbf{x}) = 1 - \frac{d_k^2(\mathbf{x})}{\alpha}. \tag{4.31}$$

If we use the product operator for the one-dimensional membership functions with the negative degrees of membership, a large degree of membership may be obtained by multiplying the even number of negative degrees of membership. Thus if we use the product operator, we should not use the one dimensional membership functions with the negative degrees of membership.

Hereafter we call the membership function with any of the three operators rectangular pyramidal membership function.

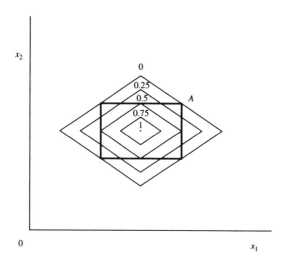

Fig. 4.9. Contour lines of the membership function with the average operator when the negative degree of membership is allowed

4.2.3 Truncated Rectangular Pyramidal Membership Functions

We assume that the one-dimensional membership function $m(x)$ given by (4.14) is defined for the kth input variable of the m-dimensional input vector **x**. Namely,

$$m_k(\mathbf{x}) = \begin{cases} 0 & \text{for} \quad x_k > V_k + \alpha, \\ 1 - h_k(\mathbf{x}) & \text{for} \quad V_k + \alpha \geq x_k \geq v_k - \alpha, \\ 0 & \text{for} \quad x_k < v_k - \alpha, \end{cases} \tag{4.32}$$

where $h_k(\mathbf{x})$ is the one-dimensional tuned distance for x_k given by (4.13)

First we discuss the shape of the membership function with the minimum and product operators. Let $A = \{\mathbf{x} \mid \mathbf{v} \leq \mathbf{x} \leq \mathbf{V}\}$, where $\mathbf{v} = (v_1, \ldots, v_m)^t$ and $\mathbf{V} = (V_1, \ldots, V_m)^t$, and $A' = \{\mathbf{x} \mid \mathbf{v} - \alpha\mathbf{1} \leq \mathbf{x} \leq \mathbf{V} + \alpha\mathbf{1}\}$, where $\mathbf{1} = (1, \ldots, 1)^t$. Then A and A' are rectangular. When **x** is outside of A', the degree of membership is zero and when **x** is inside of A, the degree of membership is 1. Then similar to the discussions for the rectangular pyramidal membership function, the shape of the membership function with the minimum operator is the truncated rectangular pyramidal membership function shown in Fig. 4.10, and the degree of membership with the product operator is smaller than or equal to that with the minimum operator.

Now we study the shape of the membership function with the average operator. To avoid irregular contour surfaces, we assume the membership function with the negative degree of membership:

$$m_k(\mathbf{x}) = 1 - h_k(\mathbf{x}), \tag{4.33}$$

where $h_k(\mathbf{x})$ is a tuned distance given by (4.13).

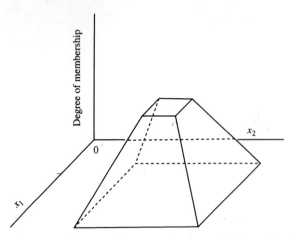

Fig. 4.10. Truncated rectangular pyramidal membership function

In A the degree of membership is 1. The degree of membership for the two inputs is shown in Fig. 4.11. Thus the shape of the membership function with the average operator is a truncated polyhedral pyramid.

Instead of (4.33), we can use a quadratic membership function:

$$m_k(\mathbf{x}) = 1 - h_k^2(\mathbf{x}) = 1 - \frac{d_k^2(\mathbf{x})}{\alpha}. \tag{4.34}$$

For the product operator, we do not use the one dimensional membership functions with the negative degrees of membership.

Hereafter we call the membership function with any of the three operators truncated rectangular pyramidal membership function.

4.2.4 Polyhedral Pyramidal Membership Functions

Assume that a convex polyhedron with l surfaces is in the input space and there is a vector $\mathbf{c} = (c_1, \ldots, c_m)^t$ in the polyhedron. Now the kth surface is given by

$$\mathbf{a}_k^t \, \mathbf{x} = b_k, \tag{4.35}$$

where $\mathbf{a}_k$ is the unit vector which is orthogonal to the kth surface and which is towards the outside of the polyhedron and b_k is a constant.

Let the distance from $\mathbf{c}$ to the kth surface be V_k. Then,

$$\mathbf{x} = \mathbf{c} + V_k \, \mathbf{a}_k \tag{4.36}$$

is on the kth surface. Substituting (4.36) into (4.35) and rearranging the equation give

$$V_k = b_k - \mathbf{a}_k^t \, \mathbf{c}. \tag{4.37}$$

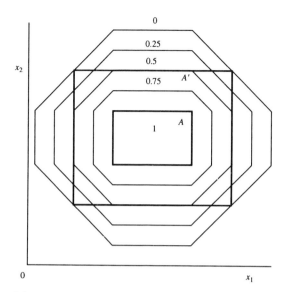

Fig. 4.11. Contour lines of the truncated rectangular pyramidal membership function with the average operator

Now we define a one-dimensional membership function for $\mathbf{x}$ in the $\mathbf{a}_k$ direction. Consider the line $\mathbf{c} + y_k \mathbf{a}_k$, where y_k takes on a real value, as the y_k axis. Then the component of $\mathbf{x}$ in the y_k axis is given by $y_k = \mathbf{a}^t(\mathbf{c} - \mathbf{x})$ as follows:

$$m_k(\mathbf{x}) = \begin{cases} 0 & \text{for} \quad y_k > \alpha V_k, \\ 1 - h_k(\mathbf{x}) & \text{for} \quad \alpha V_k \geq y_k, \\ 0 & \text{for} \quad y_k < 0, \end{cases} \tag{4.38}$$

where $y_k = \mathbf{a}^t(\mathbf{x} - \mathbf{c})$ and $h_k(\mathbf{x})$ is a one-dimensional tuned distance given by

$$h_k(\mathbf{x}) = \frac{d_k(\mathbf{x})}{\alpha} = \begin{cases} \dfrac{y_k}{\alpha V_k} & \text{for} \quad y_k \geq 0, \\ 0 & \text{for} \quad y_k < 0. \end{cases} \tag{4.39}$$

Here, $d_k(\mathbf{x})$ is a one-dimensional weighted distance and α is a tuning parameter.

When the minimum operator is used the shape of the membership function is a polyhedral pyramid as shown in Fig. 4.12.

When the product or average operator is used, discontinuity of the degree of membership occurs. Now consider a triangle shown in Fig. 4.13. Let the surfaces of the triangle be S_1, S_2, and S_3 and the associated regions that the degrees of one-dimensional membership are non-zero be H_1, H_2, and H_3, respectively. When the average operator is used, in the regions $H_1 \cap H_2$, $H_2 \cap H_3$, and $H_1 \cap H_3$, two one-dimensional membership functions are non-negative, and in each of the remaining regions only one one-dimensional membership

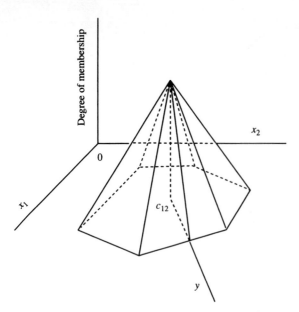

Fig. 4.12. Polyhedral pyramidal membership function

function is non-negative. Thus discontinuity of the multi-dimensional membership function occurs at the boundaries of these regions. This unfavorable discontinuity is avoided if each surface has a parallel surface as discussed in [55]. If this is not the case, we do not use the sum or product operator. Hereafter we call the membership function with the minimum operator (or with the average or product operator with parallel surfaces) the polyhedral pyramidal membership function.

4.2.5 Truncated Polyhedral Pyramidal Membership Functions

Assume that a convex polyhedron with l surfaces given by (4.35) is in the input space and there is a vector $\mathbf{c} = (c_1, \ldots, c_m)^t$ in the polyhedron.

Let the distance from $\mathbf{c}$ to the kth surface be V_k. Then, (4.37) holds.

Now we define a one-dimensional membership function for $\mathbf{x}$ in the $\mathbf{a}_k$ direction. Consider the line $\mathbf{c} + y_k \mathbf{a}_k$, where y_k takes a real value, as the y_k axis. Then the component of $\mathbf{x}$ in the y_k axis is given by $y_k = \mathbf{a}^t(\mathbf{x} - \mathbf{c})$ as follows:

$$
m_k(\mathbf{x}) = \begin{cases} 0 & \text{for} \quad y_k > V_k + \alpha, \\ 1 - h_k(\mathbf{x}) & \text{for} \quad V_k + \alpha \geq y_k \geq 0, \\ 0 & \text{for} \quad y_k < 0, \end{cases} \tag{4.40}
$$

where $y_k = \mathbf{a}^t(\mathbf{x} - \mathbf{c})$ and $h_k(\mathbf{x})$ is a one-dimensional tuned distance given by

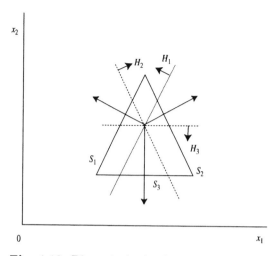

Fig. 4.13. Discontinuity in the polyhedral pyramidal membership function when the average operator is used

$$h_k(\mathbf{x}) = \begin{cases} \dfrac{y_k - V_k}{\alpha} & \text{for} \quad V_k + \alpha \geq y_k > V_k, \\ 0 & \text{for} \quad V_k \geq y_k. \end{cases} \tag{4.41}$$

Here, α controls the slope of the membership function.

When the minimum operator is used, the shape of the membership function is a truncated polyhedral pyramid as shown in Fig. 4.14. If we assume the negative degree of membership, (4.40) becomes

$$m_k(\mathbf{x}) = \begin{cases} 1 - h_k(\mathbf{x}) & \text{for} \quad y_k \geq 0, \\ 0 & \text{for} \quad y_k < 0. \end{cases} \tag{4.42}$$

If we use the average operator, the discontinuity of the multi-dimensional membership function occurs if each surface does not have a parallel surface. Thus if each surface has no parallel surface we use the minimum operator. Hereafter we call the membership function with the minimum operator the truncated polyhedral pyramidal membership function.

4.2.6 Bell-shaped Membership Functions

Assume that the one-dimensional bell-shaped membership function given by (4.18) is defined for the kth input variable of the m-dimensional input vector $\mathbf{x}$:

$$m_k(\mathbf{x}) = \exp\left(-\frac{(x_k - c_k)^2}{\alpha \sigma_k^2}\right), \tag{4.43}$$

where α is a positive tuning parameter, c_k is the kth element of the center vector $\mathbf{c}$, and σ_k^2 is a variance of x_k.

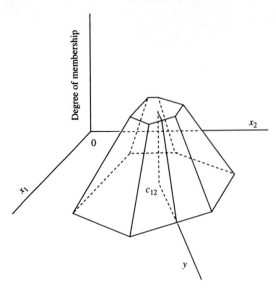

Fig. 4.14. Truncated polyhedral pyramidal membership function

The minimum of $m_k(\mathbf{x})$ for $k = 1, \ldots, m$ is equivalent to the maximum of the tuned distances $h_k(\mathbf{x})$:

$$h_k(\mathbf{x}) = \frac{|x_k - c_k|}{\sqrt{\alpha}\sigma}. \tag{4.44}$$

Thus using the minimum operator, $m(\mathbf{x})$ is given by

$$m(\mathbf{x}) = \min_{k=1,\ldots,m} m_k(\mathbf{x})$$

$$= \exp\left(-\max_{k=1,\ldots,m} h_k^2(\mathbf{x})\right). \tag{4.45}$$

Using the product operator, $m(\mathbf{x})$ is given by

$$m(\mathbf{x}) = \exp\left(-\sum_{k=1}^{m} \frac{(x_k - c_k)^2}{\alpha\sigma_k^2}\right)$$

$$= \exp\left(-\sum_{k=1}^{m} h_k^2(\mathbf{x})\right). \tag{4.46}$$

This is the special case of the following membership function:

$$m(\mathbf{x}) = \exp(-h(\mathbf{x})^2), \tag{4.47}$$

where $h(\mathbf{x})$ is a tuned distance from $\mathbf{c}$ to $\mathbf{x}$ and

$$h(\mathbf{x})^2 = \frac{d(\mathbf{x})^2}{\alpha}, \tag{4.48}$$

$$d(\mathbf{x})^2 = (\mathbf{x} - \mathbf{c})^t Q^{-1}(\mathbf{x} - \mathbf{c}). \tag{4.49}$$

Here $d(\mathbf{x})$ is a weighted distance known as the Mahalanobis distance and Q is a covariance matrix. Fig. 4.15 shows the contour curves of the bell-shaped membership function for the two-dimensional input vector $\mathbf{x}$. When Q is a diagonal matrix (4.47) reduces to (4.46).

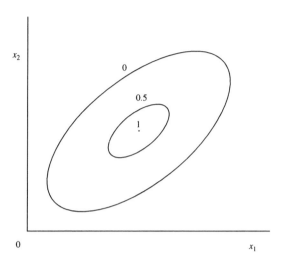

Fig. 4.15. Contour curves of the bell-shaped membership function

Let Q be non-diagonal and S be the orthogonal matrix that diagonalizes Q. Taking the orthogonal basis $\mathbf{y} = S\mathbf{x}$,

$$m(\mathbf{x}) = m(S^t\mathbf{y})$$
$$= \exp\left(-\frac{1}{\alpha}(\mathbf{y} - \mathbf{c}')^t(SQS^t)^{-1}(\mathbf{y} - \mathbf{c}')\right), \tag{4.50}$$

where $\mathbf{c}' = S\mathbf{c}$. Since SQS^t is a diagonal matrix, the membership function given by (4.47)–(4.49) is equivalent to the membership function with the product operator where the m one-dimensional bell-shaped membership functions are defined on the orthogonal basis $\mathbf{y}$ at center $\mathbf{c}'$. Hereafter, we use only the membership function given by (4.47)–(4.49) as the bell-shaped membership function.

4.2.7 Relations between Membership Functions

When the minimum operator is used, rectangular pyramidal membership functions are the special case of polyhedral pyramidal membership functions and truncated rectangular pyramidal membership functions are the special case of the truncated polyhedral pyramidal membership functions.

To study the relationship between bell-shaped membership functions and rectangular pyramidal membership functions, assume that the covariance matrix Q in (4.49) is diagonal, i.e., the membership function given by (4.46) is used and the quadratic one-dimensional membership functions with the negative degrees of membership are used for the rectangular pyramidal membership function. Then for the bell-shaped membership function, the tuned distance $h(\mathbf{x})$ is given by

$$h^2(\mathbf{x}) = \sum_{k=1}^{m} \frac{(x_k - c_k)^2}{\alpha \sigma_k^2}. \tag{4.51}$$

For the rectangular pyramidal membership function with the average operator and $s = 2$, the tuned distance is given by

$$h^2(\mathbf{x}) = \sum_{k=1}^{m} \frac{(x_k - c_k)^2}{\alpha(V_k - c_k)^2}, \tag{4.52}$$

where we assume $c_k = (V_k + v_k)/2$.

From (4.51) and (4.52), if $(V_k - c_k)^2 = \sigma_k^2$, the bell-shaped membership function with a diagonal covariance matrix is equivalent to the rectangular pyramidal membership function with the average operator and $s = 2$. Similarly, we can show that the bell-shaped membership function with a non-diagonal covariance matrix is expressed by the polyhedral pyramidal membership function with the average operator, $s = 2$, and parallel surfaces. Thus the polyhedral pyramidal membership function is the most general among rectangular pyramidal, polyhedral pyramidal, and bell-shaped membership functions.

5. Static Fuzzy Rule Generation

In generating fuzzy rules, overlaps between classes need to be resolved. There are two ways to resolve overlaps: one is to generate fuzzy rules without considering overlaps and then to resolve overlaps by tuning fuzzy rules, and the other is to resolve overlaps while generating fuzzy rules. We call the former static fuzzy rule generation and the latter dynamic fuzzy rule generation. In this chapter, we discuss static fuzzy rule generation.

In defining fuzzy rules it is usually necessary to define several fuzzy rules for each class to prevent underfitting. This means that we need to cluster class data. In static rule generation, clustering can be done before defining fuzzy rules or after defining and tuning fuzzy rules. We call the former preclustering, and the latter postclustering. Conventional clustering techniques that cluster unlabeled data can be used for preclustering. In addition, we can cluster class data considering overlaps between classes. In the following we discuss the classifier architecture for the static fuzzy rule generation, the definition of fuzzy rules, and rule generation by preclustering and postclustering.

5.1 Classifier Architecture

Consider classification of an m-dimensional input vector $\mathbf{x}$ into one of n classes. Assume that class i $(i = 1, \ldots, n)$ data are divided into several clusters ij $(j = 1, \ldots)$ where cluster ij denotes the jth cluster for class i. For each cluster ij, we define the following fuzzy rule using the training data included in the cluster:

$$R_{ij} : \quad \text{If } \mathbf{x} \text{ is } A_{ij} \text{ then } \mathbf{x} \text{ belongs to class } i, \tag{5.1}$$

where A_{ij} is a fuzzy region for cluster ij and on the region we define one of the membership functions $m_{ij}(\mathbf{x})$ discussed in Section 4.2. For the then-part, we do not define a membership function. Thus, for input $\mathbf{x}$ if the membership function $m_{kl}(\mathbf{x})$ is the largest, we classify input $\mathbf{x}$ into class k.

Fig. 5.1 shows the architecture of the fuzzy classifier. The output nodes, which correspond to the fuzzy rules, calculate the degrees of membership for the input vector $\mathbf{x}$ and $\mathbf{x}$ is classified into the class with the maximum degree of membership. This is the simplest architecture that is conceivable.

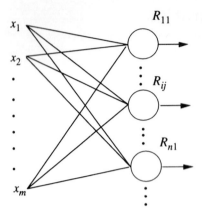

Fig. 5.1. Architecture of a fuzzy classifier [56, p.359 ©IEEE 1998]

According to the shape of the membership function for A_{ij}, the fuzzy rules that we discuss in this book are

1. the hyperbox fuzzy rules that have (truncated) rectangular pyramidal membership functions;
2. the polyhedral fuzzy rules that have polyhedral pyramidal membership functions; and
3. the ellipsoidal fuzzy rules that have bell-shaped membership functions.

In static fuzzy rule generation, after fuzzy rules are generated, we tune membership functions to improve classification performance. Therefore, if we use truncated membership functions, we need to avoid overlapping of the membership functions for different classes. Otherwise, data in the overlapping region are unclassifiable because the degrees of membership for different classes become 1. Thus to make tuning easy, in this chapter we do not use the truncated membership functions. They are used in Chapter 9. We call the hyperbox fuzzy rules that have rectangular pyramidal membership functions fuzzy rules with pyramidal membership functions.

5.2 Fuzzy Rules

Assuming that cluster ij is given, namely, the subset of the training data, X_{ij}, included in cluster ij is given, we discuss how to define the fuzzy rule given by (5.1). In the following we discuss fuzzy rules with pyramidal membership functions, polyhedral fuzzy rules, and ellipsoidal fuzzy rules.

5.2.1 Fuzzy Rules with Pyramidal Membership Functions

A hyperbox fuzzy rule is generated for cluster ij by calculating the minimum and maximum values in each input variable. Then the center of the hyperbox is taken as the apex of the rectangular pyramidal membership function. In the following we discuss the detailed procedure of rule generation.

For cluster ij we define the hyperbox A_{ij} that includes the training data in X_{ij} as follows:

$$A_{ij} = \{\mathbf{x} \,|\, v_{ijk} \leq x_k \leq V_{ijk}, k = 1, \ldots, m\}, \tag{5.2}$$

where x_k is the kth element of $\mathbf{x}$, and v_{ijk} and V_{ijk} are, respectively, the minimum and maximum values of the hyperbox A_{ij} with respect to x_k. Namely,

$$v_{ijk} = \min_{\mathbf{x} \in X_{ij}} x_k, \tag{5.3}$$

$$V_{ijk} = \max_{\mathbf{x} \in X_{ij}} x_k. \tag{5.4}$$

Then the center vector $\mathbf{c}_{ij}$ of the hyperbox A_{ij} is given by

$$\mathbf{c}_{ij} = \left(\frac{V_{ij1} + v_{ij1}}{2}, \ldots, \frac{V_{ijm} + v_{ijm}}{2} \right)^t. \tag{5.5}$$

Now we define, for A_{ij}, a rectangular pyramidal membership function with the average operator discussed in Section 4.2.2. Namely, first we define, for the input vector $\mathbf{x}$, a one-dimensional tuned distance $h_{ijk}(\mathbf{x})\ (k = 1, \ldots, m)$ as follows:

$$h^s_{ijk}(\mathbf{x}) = \frac{d^s_{ijk}(\mathbf{x})}{\alpha_{ij}}, \tag{5.6}$$

where $s = 1$ or 2 and α_{ij} is the tuning parameter that determines the slope of the membership function and the initial value is set to be 1, and $d_{ijk}(\mathbf{x})$ is the one-dimensional weighted distance given by

$$d_{ijk}(\mathbf{x}) = \frac{|c_{ijk} - x_k|}{w_{ijk}}, \tag{5.7}$$

where c_{ijk} is the kth element of $\mathbf{c}_{ij}$ and

$$w_{ijk} = \frac{V_{ijk} - v_{ijk}}{2}. \tag{5.8}$$

To avoid zero division, if $w_{ijk} < \varepsilon$ where ε is a minimum edge length, in (5.7) we replace w_{ijk} with ε.

According to the above definition, the weighted distance at any point on the surface of the hyperbox is the same. This is the same idea with that of the Mahalanobis distance discussed in Section 4.2.6 [56, 57].

Then the one-dimensional membership function for input variable x_k ($k = 1, \ldots, m$), $m_{ijk}(\mathbf{x})$, is defined as follows:

$$m_{ijk}(\mathbf{x}) = \begin{cases} 0 & \text{for} \quad x_k > c_{ijk} + \alpha_{ij} w_{ijk}, \\ 1 - h^s_{ijk}(\mathbf{x}) & \text{for} \quad c_{ijk} + \alpha_{ij} w_{ijk} \geq x_k \geq c_{ijk} - \alpha_{ij} w_{ijk}, \\ 0 & \text{for} \quad x_k < c_{ijk} - \alpha_{ij} w_{ijk}. \end{cases} \tag{5.9}$$

When $\mathbf{x}$ is far away from the center $\mathbf{c}_{ij}$, (5.9) becomes 0. Hence there may be cases where $\mathbf{x}$ is not classified into any class. To avoid this, we allow negative degrees of membership:

$$m_{ijk}(\mathbf{x}) = 1 - h_{ijk}^s(\mathbf{x}). \tag{5.10}$$

Hereafter we use (5.10) as the membership function. Instead of using (5.10), we can use $h_{ijk}^s(\mathbf{x})$, in which the tuned minimum distance corresponds to the largest degree of membership.

The degree of membership for the input variable $\mathbf{x}, m_{ij}(\mathbf{x})$, can be defined in two ways using (5.10). The first one uses the minimum operator:

$$m_{ij}(\mathbf{x}) = \min_{k=1,\ldots,m} m_{ijk}(\mathbf{x})$$
$$= 1 - \max_{k=1,\ldots,m} h_{ijk}^s(\mathbf{x}). \tag{5.11}$$

When we use the minimum operator, the input variable that gives the minimum $m_{ijk}(\mathbf{x})$ is the same irrespective of the value of s.

The second one uses the average operator:

$$m_{ij}(\mathbf{x}) = \frac{1}{m} \sum_{k=1}^m m_{ijk}(\mathbf{x})$$
$$= 1 - h_{ij}^s(\mathbf{x}), \tag{5.12}$$

where $h_{ij}(\mathbf{x})$ is the tuned distance:

$$h_{ij}^s(\mathbf{x}) = \frac{1}{m} \sum_{k=1}^m h_{ijk}^s(\mathbf{x}). \tag{5.13}$$

Since we use the membership functions with the negative degrees of membership, we do no use the product operator to generate the multi-dimensional membership functions (cf. Section 4.2.2 on page 72).

Now consider the sensitivity of the two membership functions to the slope or location change. First consider the slope change. Let the membership functions with the minimum and average operators be $m_{ij}'(\mathbf{x})$ for α_{ij}'. Then from (5.11) and (5.12),

$$m_{ij}'(\mathbf{x}) = 1 + \frac{\alpha_{ij}}{\alpha_{ij}'} \left(m_{ij}(\mathbf{x}) - 1 \right). \tag{5.14}$$

Thus the change of the slope is expressed in closed form for both membership functions.

Assume that $c_{ijk}(\mathbf{c}_{ij})$ is changed to $c_{ijk}'(\mathbf{c}_{ij}'), k \in \{1,\ldots,m\}$. For the membership function with the minimum operator, if $h_{ijk'}(\mathbf{x}), k' \in \{1,\ldots,m\}$ is the maximum for $\mathbf{c}_{ij}$, there is no guarantee that it is still the maximum for $\mathbf{c}_{ij}'$. Thus it is difficult to express the change of the location in closed form. For the membership function with the average operator, on the other hand,

$$m_{ij}'(\mathbf{x}) - m_{ij}(\mathbf{x}) = \frac{|c_{ijk} - x_{ij}|^s}{m\alpha_{ij} w_{ijk}^s} - \frac{|c_{ijk}' - x_{ij}|^s}{m\alpha_{ij} w_{ijk}^s}. \tag{5.15}$$

Thus the change of the location is expressed in closed form.

From the above discussions, for the membership functions with the minimum operator, tuning of centers is not easily achieved. Also since tuning of centers is essential to achieve a high recognition rate, in the following we use the membership function with the average operator given by (5.12) or the tuned distance given by (5.13). We call the fuzzy classifier with this membership function the fuzzy classifier with pyramidal membership functions.

5.2.2 Polyhedral Fuzzy Rules

We can approximate class regions by polyhedrons [55, 58]. In [55], first the multilayer neural network is trained and the separating hyperplanes are extracted from the trained network and class regions are approximated by shifting the separating hyperplanes. In [58], a class region is approximated by a convex hull that is the minimum convex region that includes the training data belonging to the class.

In this section we generate a convex hull, using the dynamic convex hull generation method developed for generating the Lyapunov function [59]. To use the dynamic convex hull generation method, we need to generate an initial convex hull whose dimension is the same with that of the convex hull including the training data for that class. Then starting from the initial convex hull, we expand the convex hull adding a training datum one at a time. If the training datum is in the convex hull, we do nothing. But if it is outside of the convex hull, we modify the facets, which are the surfaces of the convex hull, in front of the training datum.

In the following, first we explain a general procedure for generating a convex hull and then we discuss a procedure for generating an initial convex hull. Finally we explain the dynamic convex hull generation method.

Generation of a Convex Hull. We generate a convex hull using M m-dimensional input vectors $\mathbf{p}^0, \mathbf{p}^1, \ldots, \mathbf{p}^{M-1}$. To avoid confusion, we do not affix class labels to the input vectors. Let the convex hull generated by l input vectors $\mathbf{p}^0, \mathbf{p}^1, \ldots, \mathbf{p}^{l-1}$ be $P(l)$:

$$P(l) = \text{conv}\{\mathbf{p}^0, \mathbf{p}^1, \ldots, \mathbf{p}^{l-1}\}, \tag{5.16}$$

where $\text{conv}\{\cdot\}$ denotes the convex hull generated by the set of points in $\{\cdot\}$. Namely, we call the linear combination of the set consisting of l points $\{\mathbf{p}^0, \mathbf{p}^1, \ldots, \mathbf{p}^{l-1}\}$:

$$P(l) = \left\{ \mathbf{x} \mid \mathbf{x} = \lambda_0 \mathbf{p}^0 + \lambda_1 \mathbf{p}^1 + \cdots + \lambda_{l-1} \mathbf{p}^{l-1}, \right.$$

$$\left. \lambda_0 + \lambda_1 + \cdots + \lambda_{l-1} = 1, \ \lambda_k \geq 0, k \in \{0, 1, \ldots, l-1\} \right\} \tag{5.17}$$

a convex hull or a convex polyhedron. If the dimension of $P(l)$ is j, we denote the convex hull as $P^j(l)$. If we need to show neither the dimension nor the number of points of the convex hull, we simply denote it as P.

Let the dimension of the convex hull $P(M)$ generated by points $\mathbf{p}^0, \mathbf{p}^1, \ldots,$ $\mathbf{p}^{M-1}$ be m_c ($\leq m$). In generating $P(M)$ by dynamic convex hull generation, we first need to generate an m_c-dimensional initial convex hull P^{m_c} using $m_c + 1$ points in general positions. Here we say that $n+1$ points $\mathbf{a}^0, \mathbf{a}^1, \ldots, \mathbf{a}^n$ are in general positions when vectors $\mathbf{a}^1 - \mathbf{a}^0, \ldots, \mathbf{a}^n - \mathbf{a}^0$ are linearly independent. Then using the remaining $M - m_c - 1$ points one at a time, we modify the convex hull. In the following, we discuss the method for generating the initial convex hull P^{m_c}, and then we explain the generation of the convex hull $P^{m_c}(M)$ starting from the initial convex hull.

Generation of the Initial Convex Hull. Let the first $m + 1$ points $\mathbf{p}^0, \mathbf{p}^1, \ldots, \mathbf{p}^m$ among M points $\mathbf{p}^0, \mathbf{p}^1, \ldots, \mathbf{p}^{M-1}$ be in general positions. Then the initial convex hull can be generated by $P^m = \text{conv}\{\mathbf{p}^0, \mathbf{p}^1, \ldots, \mathbf{p}^m\}$. The $(m-1)$-dimensional convex hulls that form the surface of P^m are generated by deleting one point from $\mathbf{p}^0, \mathbf{p}^1, \ldots, \mathbf{p}^m$:

$$
\begin{aligned}
F_1 &= \text{conv}\{\mathbf{p}^1, \mathbf{p}^2, \ldots, \mathbf{p}^m\}, \\
F_2 &= \text{conv}\{\mathbf{p}^0, \mathbf{p}^2, \ldots, \mathbf{p}^m\}, \\
&\quad \cdots \\
F_{m+1} &= \text{conv}\{\mathbf{p}^0, \mathbf{p}^1, \ldots, \mathbf{p}^{m-1}\}.
\end{aligned} \tag{5.18}
$$

We call them the facets of P^m and we denote the set as $\mathcal{F}_{m-1}(P^m)$. The k-dimensional convex hulls that are intersections of P^m and the tangent hyperplane are called the k-dimensional faces and we denote the set as $\mathcal{F}_k(P^m)$.

If none of $m + 1$ points which are chosen from M points are in general positions, we cannot generate m-dimensional initial convex hull. As shown in Fig. 5.2, if four points in the three-dimensional input space are on a plane, the generated convex hull is of two dimensions. Thus to use the dynamic convex hull generation method, first we need to determine the rank of the M m-dimensional points. Instead of calculating the rank, we generate the initial convex hull reading points one at a time as follows.

1. **Generation of a one-dimensional convex hull.** Let $\mathbf{p}^0$ be a reference point, and read $\mathbf{p}^{i_1}$ ($i_1 \neq 0$). Calculate the one-dimensional distance between two points $\mathbf{p}^0$ and $\mathbf{p}^{i_1}$ and if the distance for the m_1-th ($1 \leq m_1 \leq m$) input axis is not zero, a one-dimensional convex hull is generated. Store $\mathbf{p}^{i_1}$ and m_1, and go to Step 3. Otherwise, go to Step 2.
2. If all the one-dimensional distances between $\mathbf{p}^0$ and $\mathbf{p}^{i_1}$ are zero, the two points are identical. Thus read another point $\mathbf{p}^{i_x}$ and repeat Step 1 until a one-dimensional convex hull is generated.
3. **Generation of an n-dimensional convex hull using the $(n-1)$-dimensional convex hull.** Assume that we have generated $(n-1)$-dimensional convex hull $\text{conv}(\mathbf{p}^0, \mathbf{p}^{i_1}, \ldots, \mathbf{p}^{i_{n-1}})$ in the $(x_{m_1}, \ldots, x_{m_{n-1}})$ subspace. Read $\mathbf{p}^{i_n} \neq \mathbf{p}^{i_k}$ ($k = 1, \ldots, n-1$), take $m_n \neq m_k$ ($k =$

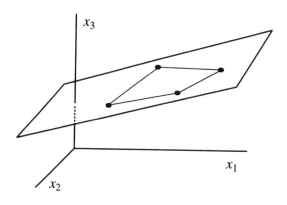

Fig. 5.2. A reduced initial convex hull

$1, \ldots, n - 1$), and let $\mathbf{p}_r^j = [p_{m_1}^j, \ldots, p_{m_{n-1}}^j, p_{m_n}^j]^t$. Namely, $\mathbf{p}_r^j$ is an n-dimensional vector whose n elements are chosen from $\mathbf{p}^j$.
Then if

$$\mathbf{p}_r^{i_n} \in H(\mathbf{p}^0, \mathbf{p}_r^{i_1}, \ldots, \mathbf{p}_r^{i_{n-1}}) \tag{5.19}$$

is satisfied, where $H(\mathbf{p}^0, \mathbf{p}_r^{i_1}, \ldots, \mathbf{p}_r^{i_{n-1}})$ is the hyperplane that includes $\mathbf{p}^0, \mathbf{p}_r^{i_1}, \ldots, \mathbf{p}_r^{i_{n-1}}$, go to Step 5. Otherwise, go to Step 4.

4. Store $\mathbf{p}^{i_n}$ and m_n to generate the n-dimensional convex hull and add 1 to n and go to Step 3.
5. Repeat Step 3 for the remaining input axes. If the convex hull is not expandable for all the dimensions, do Step 3 for the remaining points.
6. If for all the points and the remaining input axes the convex hull is not expandable, terminate the algorithm and store the final dimension as m_c ($m_c \leq m$).

Dynamic Convex Hull Generation. Let the dimension of the convex hull $P(M)$ generated by the M m-dimensional points $\mathbf{p}^0, \mathbf{p}^1, \ldots, \mathbf{p}^{M-1}$ be m_c, and the first $m_c + 1$ points $\mathbf{p}^0, \mathbf{p}^1, \ldots, \mathbf{p}^{m_c}$ be in general positions. Then the initial convex hull is given by $P^{m_c} = \mathrm{conv}\{\mathbf{p}^0, \mathbf{p}^1, \ldots, \mathbf{p}^{m_c}\}$. To generate $P(M)$ by the dynamic convex hull generation method starting from the initial convex hull P^{m_c}, first we generate the convex hull by adding $\mathbf{p}^{m_c+1}$ to P^{m_c}. Namely,

$$P(m_c + 2) = \mathrm{conv}\left\{ P^{m_c}, \mathbf{p}^{m_c+1} \right\}. \tag{5.20}$$

In general, for $i = 2, \ldots, M - m_c - i$ we modify the convex hull $P(m_c + i - 1)$ by

$$P(m_c + i) = \text{conv}\left\{ P(m_c + i - 1), \mathbf{p}^{m_c+i-1} \right\}. \tag{5.21}$$

Here if $\mathbf{p}^{m_c+i-1}$ is included in $P(m_c + i - 1)$,

$$P(m_c + i) = P(m_c + i - 1) \tag{5.22}$$

and thus we need not modify the convex hull. But if $\mathbf{p}^{m_c+i-1}$ is not included in $P(m_c + i - 1)$, we need to modify $P(m_c + i - 1)$.

In the following, according to [59], we explain how to generate $P(n+1) = \text{conv}\{P(n), \mathbf{p}^n\}$ by adding $\mathbf{p}^n$ to the m_c-dimensional convex hull $P(n)$.

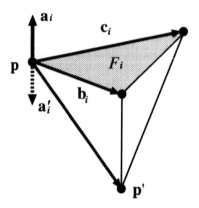

Fig. 5.3. Orthogonal vector in the outer direction

1. We calculate the vector orthogonal to facet $F_i \in \mathcal{F}_{m_c-1}(P(n))$, $\mathbf{a}_i$, by the outer product of two linearly independent vectors on the facet. Fig. 5.3 shows an example for the three-dimensional case. In the figure, calculating the outer product of $\mathbf{b}_i$ and $\mathbf{c}_i$ in facet F_i, we get

$$\mathbf{a}_i = \mathbf{b}_i \times \mathbf{c}_i. \tag{5.23}$$

In the figure $\mathbf{a}_i$ is in the outer direction and $\mathbf{a}_i'$ is in the inner direction. To determine the direction of the orthogonal vector, we use the point that is not on F_i, e.g. $\mathbf{p}'$ in the figure, and if

$$\mathbf{a}_i^t (\mathbf{p}' - \mathbf{p}) < 0, \tag{5.24}$$

$\mathbf{a}_i$ is in the outer direction. If

$$\mathbf{a}_i^t (\mathbf{p}' - \mathbf{p}) > 0, \tag{5.25}$$

$\mathbf{a}_i$ is in the inner direction and we replace $\mathbf{a}_i$ with $-\mathbf{a}_i$.

2. We color the facet F_i red, yellow, or blue according to where the given point $\mathbf{p}^n$ lies against the facet. Namely, we calculate the dot product of $\mathbf{a}_i$ and $\mathbf{q}_i^n$ that is a vector from an arbitrary point on the facet F_i to $\mathbf{p}^n$, and we paint the facet F_i the color C_i:

$$C_i = \begin{cases} \text{Red} & \text{for } \mathbf{a}_i^t\,\mathbf{q}_i^n > 0, \\ \text{Yellow} & \text{for } \mathbf{a}_i^t\,\mathbf{q}_i^n = 0, \\ \text{Blue} & \text{for } \mathbf{a}_i^t\,\mathbf{q}_i^n < 0. \end{cases} \tag{5.26}$$

Using the two-dimensional example shown in Fig. 5.4, we explain the coloring procedure. Let $\mathbf{p}^n$ be given as in the figure. Let the vectors from the three facets to $\mathbf{p}^n$ be $\mathbf{q}_1^n$, $\mathbf{q}_2^n$, and $\mathbf{q}_3^n$ and the associated outer orthogonal vectors be $\mathbf{a}_1$, $\mathbf{a}_2$, and $\mathbf{a}_3$, respectively. Then from (5.26), the colors of the facets are $C_1 = \text{Red}$, $C_2 = \text{Yellow}$, and $C_3 = \text{Blue}$.

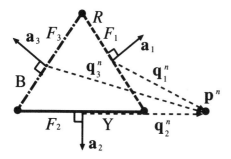

Fig. 5.4. Coloring the facets of a two-dimensional convex hull

3. After coloring facets in Step 2, we color the faces, included in the set $\mathcal{F}_{m_c-2}(P(n))$, that are intersections of facets. The color is determined by mingling the colors of the facets as listed in Table 5.1.

Table 5.1. Face coloring according to facet colors

Facet color	Facet color		
	Red	Yellow	Blue
Red	Red	Orange	Purple
Yellow	Orange	Yellow	Green
Blue	Purple	Green	Blue

4. From Step 2, according to the location of $\mathbf{p}^n$ against the facets, the set of facets of the m_c-dimensional convex hull $P(n)$, $\mathcal{F}_{m_c-1}(P(n))$, are divided into

$$\mathcal{F}_{m_c-1}(P(n)) = \mathcal{F}_{m_c-1}^{B}(P(n)) \cup \mathcal{F}_{m_c-1}^{R}(P(n))$$
$$\cup\, \mathcal{F}_{m_c-1}^{Y}(P(n)), \tag{5.27}$$

where the superscripts B, R, and Y denote that the associated facets are painted blue, red, and yellow, respectively. Now consider how to modify the facets. The facets included in $\mathcal{F}_{m_c-1}^{B}(P(n))$ are not seen from $\mathbf{p}^n$. Thus, we need not modify the facets. Namely,

$$\mathcal{F}_{m_c-1}^{B}(P(n+1)) = \mathcal{F}_{m_c-1}^{B}(P(n)). \tag{5.28}$$

Next, the facets included in $\mathcal{F}_{m_c-1}^{R}(P(n))$ are seen from $\mathbf{p}^n$. Thus, these facets need to be modified to be the convex hull $\mathcal{F}_{m_c-1}^{R}(P(n+1))$ including $\mathbf{p}^n$ and the faces that connect blue and red facets:

$$\mathcal{F}_{m_c-1}^{R}(P(n+1)) = \{\mathrm{conv}\{E \cup \mathbf{p}^n\} \mid E \in \mathcal{F}_{m_c-2}^{P}(P(n))\}, \tag{5.29}$$

where the superscript P denotes purple. Finally, $\mathbf{p}^n$ is on the extension of the facets included in $\mathcal{F}_{m_c-1}^{Y}(P(n))$. Thus the modified set of facets, $\mathcal{F}_{m_c-1}^{Y}(P(n+1))$, are given by

$$\mathcal{F}_{m_c-1}^{Y}(P(n+1)) = \{\mathrm{conv}\{E \cup \mathbf{p}^n\} \mid E \in \mathcal{F}_{m_c-2}^{G}(P(n))\}, \tag{5.30}$$

where the superscript G denotes green.

Using (5.27) to (5.30), we modify the convex hull.

To accelerate convex hull generation, we maximize the initial convex hull. In generating the initial convex hull, we check whether the point is included in the hyperplane (cf. (5.19)). This is done by calculating the distance between the point and the hyperplane. If the distance is not zero, we expand the convex hull so that it includes the point. To maximize the volume of the initial convex hull, we search the point with the maximum distance.

For the convex hull we define a polyhedral pyramidal membership function with the minimum operator discussed in Section 4.2.4 and we call the classifier the fuzzy classifier with polyhedral regions.

5.2.3 Ellipsoidal Fuzzy Rules

Extraction of Fuzzy Rules. In ellipsoidal fuzzy rules, for cluster ij, we define the ellipsoidal region A_{ij} given by the center vector $\mathbf{c}_{ij}$ of the training data belonging to cluster ij, X_{ij}, and the covariance matrix Q_{ij} around the center vector. Here, the center vector $\mathbf{c}_{ij} = (c_{ij1}, \ldots, c_{ijm})^t$ is calculated by

$$\mathbf{c}_{ij} = \frac{1}{|X_{ij}|} \sum_{\mathbf{x} \in X_{ij}} \mathbf{x}, \tag{5.31}$$

where $|X_{ij}|$ is the number of data belonging to cluster ij.

The membership function $m_{ij}(\mathbf{x})$ of A_{ij} for input $\mathbf{x}$ is given by

$$m_{ij}(\mathbf{x}) = \exp(-h_{ij}^2(\mathbf{x})), \tag{5.32}$$

$$h_{ij}^2(\mathbf{x}) = \frac{d_{ij}^2(\mathbf{x})}{\alpha_{ij}}, \tag{5.33}$$

$$d_{ij}^2(\mathbf{x}) = (\mathbf{x} - \mathbf{c}_{ij})^t Q_{ij}^{-1}(\mathbf{x} - \mathbf{c}_{ij}), \tag{5.34}$$

where $d_{ij}(\mathbf{x})$ is the weighted distance between $\mathbf{x}$ and $\mathbf{c}_{ij}$, $h_{ij}(\mathbf{x})$ is the tuned distance, α_{ij} (> 0) is the tuning parameter for cluster ij, and Q_{ij} is the $m \times m$ covariance matrix of cluster ij.

We calculate the covariance matrix Q_{ij} by

$$Q_{ij} = \frac{1}{|X_{ij}|} \sum_{\mathbf{x} \in X_{ij}} (\mathbf{x} - \mathbf{c}_{ij})(\mathbf{x} - \mathbf{c}_{ij})^t. \tag{5.35}$$

For input $\mathbf{x}$ if the membership function $m_{kl}(\mathbf{x})$ is the largest, input $\mathbf{x}$ is classified into class k. The exponential function in (5.32) makes the output range of (5.32) lie in $[0,1]$. Thus, if we classify input $\mathbf{x}$ using the input of the exponential function in (5.32), we need to find the smallest $h_{ij}(\mathbf{x})$.

If we add the linear layer to the outputs of the above architecture, we can obtain the radial basis function neural network. The centers and the covariance matrices of the radial basis function network are determined by the steepest descent method [60] or estimated by the Gram-Schmidt orthogonalization [61]. Here, we determine them using (5.31) and (5.35). These are good estimates if the training data obey the normal distributions. To improve generalization ability when the training data do not obey the normal distributions, we introduce the tuning parameter α_{ij} that resolves overlaps of ellipsoidal regions belonging to different classes. An increase of α_{ij} decreases the slope of the membership function $m_{ij}(\mathbf{x})$ or increases the value of $m_{ij}(\mathbf{x})$. A decrease of α_{ij} increases the slope of $m_{ij}(\mathbf{x})$ or decreases the value of $m_{ij}(\mathbf{x})$. The sophisticated tuning algorithm of the tuning parameter α_{ij} is discussed in Section 7.2.1. We can also improve the generalization ability by tuning the center vectors as discussed in Section 7.2.2. We call the classifier the fuzzy classifier with ellipsoidal regions.

If the covariance matrix Q_{ij} is singular, one way to calculate the inverse matrix is to make all the off-diagonal elements of Q_{ij} to zero. By making the covariance matrix diagonal, the principal axes of the associated ellipsoidal region become parallel to the input axes. This may worsen the generalization ability of the classifier. Thus in the following we discuss to improve the generalization ability when Q_{ij} becomes singular.

Restricting Singular Values of Q_{ij}. The covariance matrix Q_{ij} is guaranteed to be positive semi-definite, and to be positive definite, the number of the training data belonging to class i needs to be at least larger than the number of input variables (see Appendix B.3). Namely,

$$|X_{ij}| \geq m + 1. \tag{5.36}$$

When the covariance matrix Q_{ij} is singular, we decompose Q_{ij} into singular values (see Appendix B.2) [47]. Decomposing the $m \times m$ symmetric matrix Q_{ij} into singular values, we have $Q_{ij} = USU^t$, where U is the orthogonal matrices satisfying $UU^t = I$ and $U^tU = I$. Here, I is the unit matrix and S satisfies

$$S = \text{diag}(\lambda_1, \lambda_2, \ldots, \lambda_m), \tag{5.37}$$

where diag denotes a diagonal matrix and $\lambda_1 \geq \lambda_2 \geq \cdots \geq \lambda_m \geq 0$.

If Q_{ij} is singular and its rank is r ($m > r \geq 0$), $\lambda_{r+1} = \lambda_{r+2} = \cdots = \lambda_m = 0$. Then the pseudo-inverse of Q_{ij}, Q_{ij}^+, is defined by

$$Q_{ij}^+ = U\text{diag}(\lambda_1^{-1}, \lambda_2^{-1}, \ldots, \lambda_r^{-1}, 0, \ldots, 0)U^t. \tag{5.38}$$

The singular value is judged as 0, when

$$\lambda < \eta, \tag{5.39}$$

where η is a small positive value and is called a precision parameter.

When Q_{ij} is positive semi-definite, by a small value of λ_i, λ_i^{-1} becomes large and the principal vectors in Q_{ij}^+ are imbedded by this component. To avoid this, we set a value to η in (5.39) larger than the value specified by the precision of computation.

Speedup by Symmetric Cholesky Factorization. When Q_{ij} is positive definite, each diagonal element is positive and the value is the maximum among the column elements. Namely,

$$q_{ii} > |q_{ij}|, \; i, j = 1, 2, \cdots, m, \qquad i \neq j. \tag{5.40}$$

Thus Q_{ij} can be decomposed into the two triangular matrices by the symmetric Cholesky factorization without pivot exchanges as follows [47]:

$$Q_{ij} = L_{ij}L_{ij}^t, \tag{5.41}$$

where L_{ij} is the real-valued regular lower triangular matrix and each element of L_{ij} is given by

$$l_{op} = \frac{q_{op} - \sum_{n=1}^{p-1} l_{pn}l_{on}}{l_{pp}} \quad \text{for} \quad o = 1, \cdots, m, \quad p = 1, \cdots, o-1, \tag{5.42}$$

$$l_{aa} = \sqrt{q_{aa} - \sum_{n=1}^{a-1} l_{an}^2} \quad \text{for} \quad a = 1, 2, \cdots, m. \tag{5.43}$$

Using L_{ij}, (5.34) is written as follows:

$$d_{ij}^2(\mathbf{x}) = (L_{ij}^{-1}(\mathbf{x} - \mathbf{c}_{ij}))^t L_{ij}^{-1}(\mathbf{x} - \mathbf{c}_{ij}). \tag{5.44}$$

Now we define the vector $\mathbf{y}_{ij}(= (y_{ij1}, \cdots, y_{ijm})^t)$ by

$$\mathbf{y}_{ij} = L_{ij}^{-1}(\mathbf{x} - \mathbf{c}_{ij}). \tag{5.45}$$

Then solving the following equation for $\mathbf{y}_{ij}$

$$L_{ij}\mathbf{y}_{ij} = \mathbf{x} - \mathbf{c}_{ij}, \tag{5.46}$$

we calculate $d_{ij}^2(\mathbf{x})$ without calculating the inverse of Q_{ij}:

$$d_{ij}^2(\mathbf{x}) = \mathbf{y}_{ij}^t \mathbf{y}_{ij}. \tag{5.47}$$

When the number of training data is small, we cannot calculate L_{ij}. In this case, the value in the square root of (5.43) is non-positive. If this happens we stop the decomposition and calculate the pseudo-inverse.

5.3 Class Boundaries

In this section we discuss how class boundaries change according to the changes of membership functions [62]. To facilitate illustrations of class boundaries, we restrict our discussions to a classification problem with two input variables and with one fuzzy rule for each class. First we discuss the general characteristics of class boundaries for the fuzzy rules with pyramidal membership functions and ellipsoidal fuzzy rules, and then we show the class boundaries for the iris data with the third and fourth input variables.

5.3.1 Fuzzy Rules with Pyramidal Membership Functions

Assuming that one rule is defined for each class and dropping the subscripts for the cluster numbers, the membership functions with the average operator for Classes 1 and 2 are given, respectively, by

$$m_1(\mathbf{x}) = 1 - \frac{1}{2}\left(\frac{|c_{11} - x_1|}{w_{11}}\right)^s - \frac{1}{2}\left(\frac{|c_{12} - x_2|}{w_{12}}\right)^s, \tag{5.48}$$

$$m_2(\mathbf{x}) = 1 - \frac{1}{2\alpha}\left(\frac{|c_{21} - x_1|}{w_{21}}\right)^s - \frac{1}{2\alpha}\left(\frac{|c_{22} - x_2|}{w_{22}}\right)^s, \tag{5.49}$$

where $s = 1$ or 2. Here we assume that the tuning parameter for Class 1 is 1. Then the class boundary is given by

$$\left(\frac{|c_{11} - x_1|}{w_{11}}\right)^s + \left(\frac{|c_{12} - x_2|}{w_{12}}\right)^s$$
$$= \frac{1}{\alpha}\left(\frac{|c_{21} - x_1|}{w_{21}}\right)^s + \frac{1}{\alpha}\left(\frac{|c_{22} - x_2|}{w_{22}}\right)^s. \tag{5.50}$$

When $s = 1$, (5.50) becomes

$$\frac{|c_{11} - x_1|}{w_{11}} + \frac{|c_{12} - x_2|}{w_{12}} = \frac{1}{\alpha}\frac{|c_{21} - x_1|}{w_{21}} + \frac{1}{\alpha}\frac{|c_{22} - x_2|}{w_{22}}. \tag{5.51}$$

Thus the class boundary is piecewise linear. Fig. 5.5 shows the class boundary for $\mathbf{c}_1 = (0,0)^t$, $\mathbf{c}_2 = (1,1)^t$, $w_{11} = w_{21} = 1$, and $w_{12} = w_{22} = 2$. When $\alpha = 1$, the class boundary is piecewise linear. When $\alpha > 1$, Class 1 is surrounded by a polygon. As α becomes larger, the polygon that surrounds Class 1 becomes smaller. Since we assume negative degree of membership, any point, except on the class boundary, in the input space is classified into one of the two classes. Thus the point which is far away from c_1 and c_2 are classified into Class 2. This is avoided if we do not allow negative degrees of membership but there exists an unclassified region. When $\alpha < 1$, Class 2 is surrounded by a polygon.

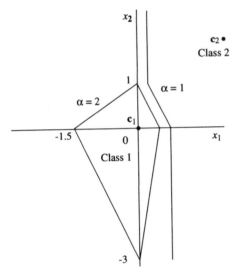

Fig. 5.5. Class boundary for the average operator ($s = 1$)

When $s = 2$, the class boundary given by (5.50) is either liner or quadratic. Assuming $w_{11}^2 = \alpha w_{21}^2$ and $w_{12}^2 = \alpha w_{22}^2$, (5.50) becomes linear:

$$\frac{2(c_{21} - c_{11})}{w_{11}^2} x_1 + \frac{2(c_{22} - c_{12})}{w_{12}^2} x_2 = \frac{c_{21}^2 - c_{11}^2}{w_{11}^2} + \frac{c_{22}^2 - c_{12}^2}{w_{12}^2}. \tag{5.52}$$

If

$$\frac{\alpha w_{21}^2 - w_{11}^2}{w_{11}^2 w_{21}^2} = \frac{\alpha w_{22}^2 - w_{12}^2}{w_{12}^2 w_{22}^2}, \tag{5.53}$$

(5.50) is a circle, and if (5.53) does not hold (5.50) is an ellipsoid. Fig. 5.6 shows the class boundaries for $c_1 = (0,0)^t$, $c_2 = (1,1)^t$, $w_{11} = w_{21} = 1$, and $w_{12} = w_{22} = 2$. When $\alpha = 1$, the class boundary is linear. When $\alpha > 1$, Class 1 is surrounded by an ellipsoid whose principal axes are parallel to the input variables.

When the minimum operator is used, the membership functions are given by

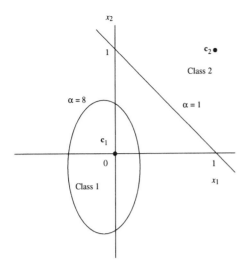

Fig. 5.6. Class boundary for the average operator $(s = 2)$

$$m_1(\mathbf{x}) = 1 - \max\left\{\frac{|c_{11} - x_1|}{w_{11}}, \frac{|c_{12} - x_2|}{w_{12}}\right\}, \tag{5.54}$$

$$m_2(\mathbf{x}) = 1 - \max\left\{\frac{|c_{21} - x_1|}{\alpha w_{21}}, \frac{|c_{22} - x_2|}{\alpha w_{22}}\right\}. \tag{5.55}$$

Thus the class boundary is given by

$$\max\left\{\frac{|c_{11} - x_1|}{w_{11}}, \frac{|c_{12} - x_2|}{w_{12}}\right\}$$
$$= \max\left\{\frac{|c_{21} - x_1|}{\alpha w_{21}}, \frac{|c_{22} - x_2|}{\alpha w_{22}}\right\}. \tag{5.56}$$

Fig. 5.7 shows the class boundary for $\mathbf{c}_1 = (0,0)^t$, $\mathbf{c}_2 = (1,1)^t$, $w_{11} = w_{21} = 1$, and $w_{12} = w_{22} = 2$. When $\alpha = 1$, the class boundary is piecewise linear. When $\alpha > 1$, the Class 1 is surrounded by a polygon.

5.3.2 Ellipsoidal Fuzzy Rules

Assuming that one rule is defined for each class and dropping the subscripts for the cluster numbers, the ellipsoidal membership functions for Classes 1 and 2 are given, respectively, by

$$m_1(\mathbf{x}) = \exp\left(-(\mathbf{x} - \mathbf{c}_1)^t Q_1^{-1}(\mathbf{x} - \mathbf{c}_1)\right), \tag{5.57}$$

$$m_2(\mathbf{x}) = \exp\left(-\frac{1}{\alpha}(\mathbf{x} - \mathbf{c}_2)^t Q_2^{-1}(\mathbf{x} - \mathbf{c}_2)\right). \tag{5.58}$$

Here we assume that the tuning parameter for Class 1 is 1. Thus the class boundary is given by

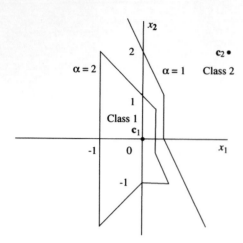

Fig. 5.7. Class boundary for the minimum operator

$$(\mathbf{x} - \mathbf{c}_1)^t Q_1^{-1}(\mathbf{x} - \mathbf{c}_1) = \frac{1}{\alpha}(\mathbf{x} - \mathbf{c}_2)^t Q_2^{-1}(\mathbf{x} - \mathbf{c}_2). \tag{5.59}$$

Let the covariance matrices Q_1 and Q_2 be nonsingular. Then the class boundary is linear when $\alpha Q_2 = Q_1$. Fig. 5.8 shows the class boundary when

$$Q_1 = Q_2 = \begin{pmatrix} 1 & \frac{1}{2} \\ \frac{1}{2} & 1 \end{pmatrix}. \tag{5.60}$$

If $\alpha = 1$, the class boundary is linear and if $\alpha > 1$, Class 1 is surrounded by an ellipsoid. If $\alpha < 1$, Class 2 is surrounded by an ellipsoid.

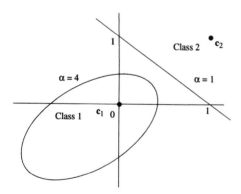

Fig. 5.8. Class boundary for the ellipsoidal fuzzy rules

5.3.3 Class Boundaries for the Iris Data

Using the iris data listed in Table 1.1 on page 19, we show the class boundaries for different kinds of fuzzy rules. The iris data consist of three classes; each class has 25 training data and 25 test data. Among the four features, we used the third and fourth features that most influence classification performance. Using the training data we defined the fuzzy rules and tuned only the slopes of the membership functions using the method discussed in Section 7.2.1.

Fuzzy Rules with Pyramidal Membership Functions. Table 5.2 shows the classification results for the training and test data when pyramidal membership functions with $s = 1$ were used. The numerals in the brackets show the numbers of classifications for the training data. For the training data, a Class 2 datum was misclassified into Class 3 and a Class 3 datum was misclassified into Class 2. For the test data, two Class 2 data were misclassified into Class 3.

Table 5.2. Classification performance using pyramidal membership functions ($s = 1$)

Class	Class 1	Class 2	Class 3
1	25 (25)	0 (0)	0 (0)
2	0 (0)	23 (24)	2 (1)
3	0 (0)	0 (1)	25 (24)

Figs. 5.9 and 5.10 show the class boundaries between Classes 1 and 2, and Classes 1 and 3, respectively. In the figures only the training data are plotted. For both cases, data of different classes were well separated and there were no misclassifications both for the training and test data.

Figs. 5.11 and 5.12 show the class boundaries between Classes 2 and 3 with the training data and test data, respectively. Two classes overlapped and in Fig. 5.11, one datum belonging to Class 2 was misclassified into Class 3 and one datum belonging to Class 3 was misclassified into Class 2. In Fig. 5.12, two test data belonging to Class 2 were misclassified into Class 3.

Table 5.3 shows the classification performance for the training and test data when pyramidal membership functions ($s = 2$) were used. The numerals in the brackets show the numbers of classifications for the training data. Two Class 2 test data were misclassified into Class 3, and one Class 2 training datum was misclassified into Class 3 and one Class 3 training datum was misclassified into Class 2.

Figs. 5.13 and 5.14 show the class boundaries between Classes 2 and 3 when training data and test data were plotted, respectively.

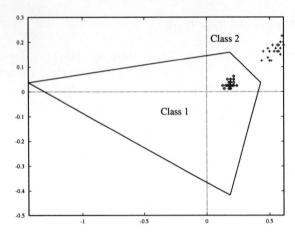

Fig. 5.9. Class boundary between Classes 1 and 2 by pyramidal membership functions $(s = 1)$ with the training data

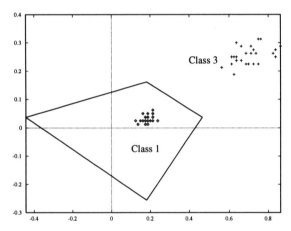

Fig. 5.10. Class boundary for Classes 1 and 3 by pyramidal membership functions $(s = 1)$ with the training data

Table 5.3. Classification performance of pyramidal membership functions $(s = 2)$

Class	Class 1	Class 2	Class 3
1	25 (25)	0 (0)	0 (0)
2	0 (0)	23 (24)	2 (1)
3	0 (0)	0 (1)	25 (24)

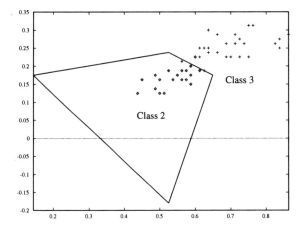

Fig. 5.11. Class boundary between Classes 2 and 3 by the pyramidal membership functions ($s = 1$) with the training data

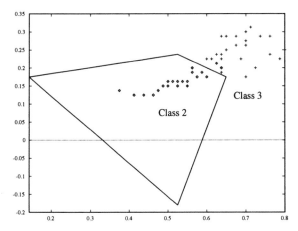

Fig. 5.12. Class boundary between Classes 2 and 3 by the pyramidal membership functions ($s = 1$) with the test data

Class Boundary for the Polyhedral Fuzzy Rules. Fig. 5.15 shows the three polygons generated by the method discussed in Section 5.2.2 and Table 5.4 shows the classification performance when polyhedral fuzzy rules were used. Numerals in the brackets show the numbers of classifications for the training data. From the table, two Class 2 test data were misclassified into Class 3 and one Class 2 training datum was misclassified into Class 3. As seen from the figure, since the polyhedrons for Classes 2 and 3 overlap, resolution of misclassification of the training datum is impossible even if the slope of the membership function is tuned.

The class boundary generated by the polyhedral fuzzy rules are shown in Figs. 5.16 and 5.17 for the training and test data, respectively. As for the

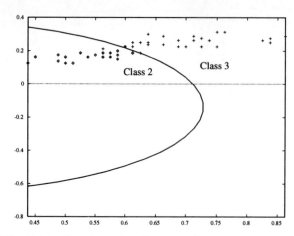

Fig. 5.13. Class boundary between Classes 2 and 3 by the pyramidal membership functions ($s = 2$) with the training data

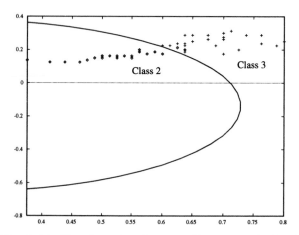

Fig. 5.14. Class boundary between Classes 2 and 3 by the pyramidal membership functions ($s = 2$) with the test data

class boundaries generated by the pyramidal membership functions and the polyhedral membership functions, the polyhedral membership functions well separate the overlapping regions for Classes 2 and 3.

Class Boundary for the Ellipsoidal Fuzzy Rules. Table 5.5 shows the classification performance when ellipsoidal fuzzy rules were used. Numerals in the brackets show the numbers of classifications for the training data. From the table, three Class 2 test data were misclassified into Class 3 and one Class 2 training datum was misclassified into Class 3.

The class boundary generated by the ellipsoidal fuzzy rules are shown in Figs. 5.18 and 5.19 for the training and test data, respectively. As for the class boundaries generated by the pyramidal membership functions and

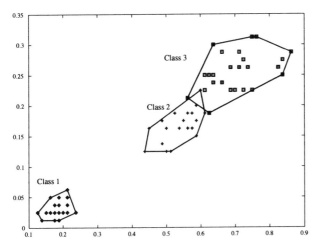

Fig. 5.15. Generated polygons

Table 5.4. Classification performance of polyhedral fuzzy rules

Class	Class 1	Class 2	Class 3
1	25 (25)	0 (0)	0 (0)
2	0 (0)	23 (24)	2 (1)
3	0 (0)	0 (0)	25 (25)

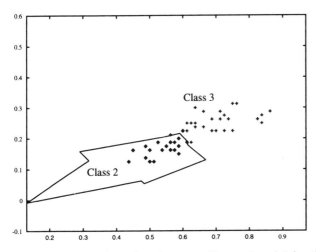

Fig. 5.16. Class boundary between Classes 2 and 3 by the polyhedral rules with the training data

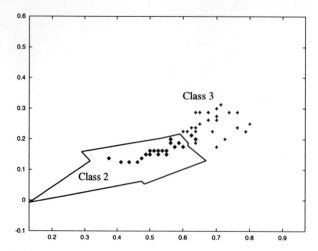

Fig. 5.17. Class boundary between Classes 2 and 3 by the polyhedral rules with the test data

the bell-shaped membership functions, the bell-shaped membership functions well represent the distribution of training and test data. Comparing the class boundaries by the polyhedral and bell-shaped membership functions, the former represents the class boundary in the overlapping region better, but the boundary is not smooth.

Table 5.5. Classification performance of ellipsoidal fuzzy rules

Class	Class 1	Class 2	Class 3
1	25 (25)	0 (0)	0 (0)
2	0 (0)	22 (24)	3 (1)
3	0 (0)	0 (0)	25 (25)

5.4 Training Architecture

In this section we discuss two ways to generate fuzzy rules: fuzzy rule generation by preclustering and fuzzy rule generation by postclustering. By preclustering, the training data for each class are clustered in advance and a fuzzy rule is generated for each cluster. By postclustering, on the other hand, clustering of the training data is postponed until after fuzzy rule generation.

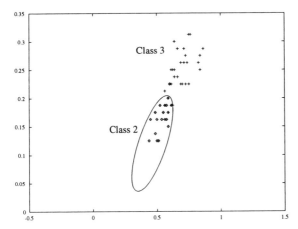

Fig. 5.18. Class boundary between Classes 2 and 3 by the ellipsoidal rules with the training data

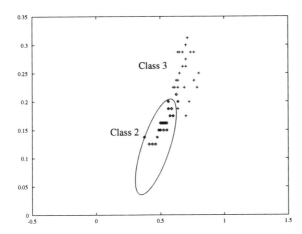

Fig. 5.19. Class boundary between Classes 2 and 3 by the ellipsoidal rules with the test data

5.4.1 Fuzzy Rule Generation by Preclustering

The process of fuzzy rule generation by preclustering is as follows. First we divide the training data for each class into several clusters using one of the clustering methods discussed in Chapter 6, and using the data belonging to each cluster we define a fuzzy rule. Then we tune the center vectors and the slopes of the membership functions until the recognition rate of the training data is not improved or the recognition rate reaches 100% using the direct methods discussed in Section 7.2.

Using the two-dimensional training data for classes i and j shown in Fig. 5.20, we explain the concept of fuzzy rule generation by preclustering. Here we use ellipsoidal fuzzy rules. First we divide the training data into clusters and suppose that class i is divided into two clusters and class j into one. Then we approximate each cluster by the ellipsoid with the center and the covariance matrix calculated using (5.31) and (5.35), respectively. Let Fig. 5.20 show the result and the ellipsoids in the figure have the same degree of membership.

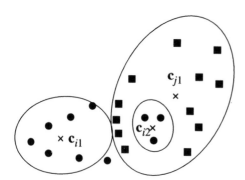

Fig. 5.20. Define a rule for each cluster

Since we determine the centers and the covariance matrices without considering the overlap of class regions, we can improve the recognition rate of the training data by tuning the tuning parameters α_{i1}, α_{i2}, and α_{j1} and moving locations of the centers. Here we explain the improvement by changing the tuning parameters. In this case, the three data belonging to cluster $i2$ may not be correctly classified. But if we increase α_{i2}, the degree of membership for cluster $i2$ increases. Thus we increase α_{i2} so that the three data are correctly classified and no data belonging to class i are misclassified.

The general flow of fuzzy rule generation is as follows.

1. Using a clustering method we divide the training data for each class into several clusters. For cluster ij, we define the hyperbox, polyhedral, or ellipsoidal fuzzy rule given by (5.1).
2. Let the maximum allowable number of new misclassifications in tuning α_{ij} be $l_M - 1$. Increase or decrease the values of α_{ij} ($i = 1, \ldots, n, j = 1, \ldots$) successively so that the recognition rate of the training data is maximized, allowing at most $l_M - 1$ new misclassifications. (cf. Section 7.2.1.)
3. Let the maximum allowable number of new misclassifications in tuning the center c_{ijk} be $l_{M_c} - 1$. For each fuzzy rule, tune c_{ijk} for $k = 1, \ldots, m$ successively so that the recognition rate of the training data is maximized, allowing at most $l_{M_c} - 1$ new misclassifications. (cf. Section 7.2.2.)
4. Iterate Steps 2 and 3 until there is no improvement in the recognition rate of the training data.

We call the execution of Steps 2 and 3 one epoch of tuning. For polyhedral fuzzy rules we tune only the slopes.

5.4.2 Fuzzy Rule Generation by Postclustering

In postclustering, clustering is postponed after the training is completed. Namely, first for the data belonging to a class we define one fuzzy rule. Then we tune the slopes and locations of the membership functions successively until there is no improvement in the recognition rate of the training data. Then if the number of the data belonging to a class that are misclassified into another class exceeds a prescribed number, we define a new cluster to which those data belong. Then we tune the newly defined fuzzy rules in the similar way as stated above, fixing the already obtained fuzzy rules. We iterate cluster generation and tuning of newly generated fuzzy rules until the number of the data belonging to a class that are misclassified into another class does not exceed the prescribed number.

Using the two-dimensional training data for classes i and j used in Fig. 5.20, we explain the concept of fuzzy rule generation by postclustering. Here we use ellipsoidal fuzzy rules. First we approximate each class region by an ellipsoid with the center and the covariance matrix calculated using (5.31) and (5.35), respectively. Let Fig. 5.21 show the result and the ellipsoids in the figure have the same degree of membership. Then the class boundary of classes i and j becomes as shown in the figure and the seven data in the overlapping region of the two ellipsoids are misclassified.

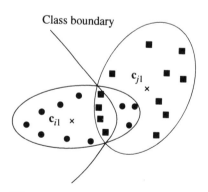

Fig. 5.21. Define a rule for each class [57, p.870 ©IEEE 1998]

Since we determine the centers and the covariance matrices without considering the overlap of class regions, we can improve the recognition rate of the training data by tuning the tuning parameters α_{i1} and α_{j1} and tuning the locations of the centers. In the following we explain by tuning the tuning parameters. If we increase α_{i1}, the degree of membership for class i increases.

Thus we can make the three data belonging to class i in the overlapping region be correctly classified, while the four misclassified data remain misclassified. Or if we decrease α_{i1}, the degrees of membership for class i decreases. Thus we can make the four data belonging to class j in the overlapping region be correctly classified, while the three misclassified data remain misclassified. We can do the same thing by increasing or decreasing α_{j1}. Thus, the maximum recognition rate is obtained if we decrease α_{i1} or increase α_{j1}. Fig. 5.22 shows the case when α_{i1} is decreased.

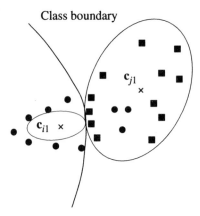

Fig. 5.22. Decrease α_{i1} to improve the recognition rate [57, p. 870 ©IEEE 1998]

After tuning, the three data belonging to class i that are in the ellipsoid for class j are misclassified. These data can be correctly classified if we define a cluster for these data as shown in Fig. 5.23. But if we define a cluster for a small number of training data, the generalization ability becomes poor because of overfitting. Therefore, we introduce a minimum number of the training data belonging to a class that are misclassified into another class, N_c. If the number of the data belonging to a class that are misclassified into another class is larger than or equal to N_c, we define a cluster and calculate the center and the covariance matrix using (5.31) and (5.35), respectively. In the following we describe the general flow of fuzzy rule generation by postclustering.

1. Assuming that each class consists of one cluster, define fuzzy rules given by (5.1).
2. Let the maximum allowable number of misclassified data that are previously correctly classified be $l_M - 1$. Increase or decrease α_{i1} ($i = 1, \ldots, n$) successively so that the recognition rate of the training data is maximized, allowing $l_M - 1$ correctly classified data to be misclassified. (cf. Section 7.2.1.)

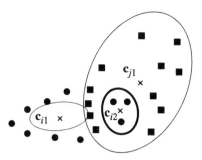

Fig. 5.23. Generate a cluster for class i [57, p. 870 ©IEEE 1998]

3. Let the maximum allowable number of new misclassifications in tuning the center c_{ijk} be $l_{M_c} - 1$. For each fuzzy rule, tune c_{ijk} for $k = 1, \ldots, m$ successively so that the recognition rate of the training data is maximized, allowing at most $l_{M_c} - 1$ new misclassifications. (cf. Section 7.2.2.)
4. If the number of the data belonging to a class that are misclassified into another class is larger than or equal to N_c, and these misclassified data are not previously defined as a cluster, define a new cluster that includes only these misclassified data. Then define the fuzzy rules given by (5.1). Go to Step 5. Otherwise, go to Step 6.
5. Tune the tuning parameters α_{ij} and the centers c_{ijk} of the newly defined fuzzy rules successively, fixing the already tuned parameters. The tuning procedure is the same as in Steps 2 and 3. Go to Step 4.
6. Tune all the centers and the slopes until there is no improvement in the recognition rate and terminate the algorithm.

In Step 4, we check whether the training data belonging to a class that are misclassified into another class are previously defined as a cluster. The reason is as follows. If the recognition rate is not improved by the newly defined cluster, the same misclassified data remain. Thus, if we do not check this situation, a new cluster will be defined indefinitely.

Tuning of α_{ij} and c_{ijk} in Steps 2, 3, and 5 is done if the recognition rate of the training data is improved. In addition, in Step 5, we fix the parameters that are already tuned. Thus by fuzzy rule generation by postclustering, the recognition rate of the training data is monotonically improved. But this does not guarantee that the recognition rate of the test data is also monotonically improved.

In Step 5, we fix the parameters of the already tuned fuzzy rules. Thus in Step 6, we tune all the fuzzy rules to further improve the recognition rate.

6. Clustering

Clustering is to divide a set of unlabeled data into several subsets called clusters according to the data distribution. Batch clustering clusters gathered data usually by iterative computations, while on-line clustering clusters incoming data according to the relative relations to the already clustered data.

In this chapter, our discussion focuses on preclustering the training data belonging to a class into several clusters so that the recognition rate of the classifier is improved. Conventional clustering techniques can be used for preclustering. In addition, we can cluster class data considering overlaps between classes [63].

The conventional clustering methods cluster unlabeled data. Using one of the conventional methods, we can cluster the training data belonging to the same class. But this is not a good strategy because clusters are determined without considering the overlap between classes. One way to overcome this problem is to cluster all the training data at the same time setting the number of clusters larger than that of classes. If a cluster includes data with different classes, we split the cluster so that one cluster consists of the data belonging to the same class. If this happens the training data in the overlapping regions are clustered and there is a possibility of improving the recognition rate.

In the following we discuss the fuzzy c-means clustering algorithm [64], [65], the Kohonen network [66], the minimum volume clustering algorithm [56], the fuzzy min-max clustering algorithm [67], and the overlap resolving clustering algorithm [63].

6.1 Fuzzy c-means Clustering Algorithm

The fuzzy c-means clustering algorithm is an extension of the c-means clustering algorithm, which is based on a crisp clustering criterion. The c-means algorithm divides data $\mathbf{x}$ included in a set X into c cluster sets $S_1, S_2, \ldots, S_c$, by minimizing

$$\sum_{i=1}^{c} \sum_{\mathbf{x} \in S_i} \|\mathbf{x} - \mathbf{v}_i\|^2, \tag{6.1}$$

where $\|\cdot\|$ is the Euclidean distance between $\mathbf{x}$ and $\mathbf{v}_i$ and $\mathbf{v}_i$ is the average of data belonging to S_i. In this method the data which are near $\mathbf{v}_i$ are clustered as cluster i.

The fuzzy c-means clustering algorithm introduces m_{ij} which is the degree of membership of $\mathbf{x}_j$ for cluster i. Thus we need not explicitly define cluster sets. Namely, instead of minimizing (6.1), we minimize

$$\sum_{i=1}^{c}\sum_{j}(m_{ij})^{l}\|\mathbf{x}_j - \mathbf{v}_i\|^2 \tag{6.2}$$

with respect to m_{ij} and $\mathbf{v}_i$ with the constraint that the sum of the degrees of membership of datum $\mathbf{x}_j$ for all the clusters is 1:

$$\sum_{i=1}^{c} m_{ij} = 1, \tag{6.3}$$

where l is an integer larger than 1. Since we do not define cluster sets, unlike $\mathbf{v}_i$ in (6.1), $\mathbf{v}_i$ in (6.2) is not the average. We call $\mathbf{v}_i$ the cluster center of cluster i. In [68], the constraint given by (6.3) is excluded.

Solving (6.2) and (6.3), we want to get a solution in which if datum $\mathbf{x}_j$ is nearest to the cluster center $\mathbf{v}_i$ among $\mathbf{v}_k(k = 1,\ldots,c)$, m_{ij} is the largest among $m_{ik}(k = 1,\ldots,c)$. The integer l works to eliminate noises and as l becomes larger, more data with small degrees of membership are neglected.

Taking the partial derivatives of (6.2) with respect to m_{ij} and $\mathbf{v}_i$, and equating the resulting equations to zero, under the constraint of (6.3), we get:

$$\mathbf{v}_i = \frac{\sum_j (m_{ij})^l \mathbf{x}_j}{\sum_j (m_{ij})^l}, \tag{6.4}$$

$$m_{ij} = \left(\sum_{k=1}^{c} \left(\frac{\|\mathbf{x}_j - \mathbf{v}_i\|^2}{\|\mathbf{x}_j - \mathbf{v}_k\|^2} \right)^{1/(l-1)} \right)^{-1}. \tag{6.5}$$

Substituting some proper initial values of m_{ij} and $\mathbf{v}_i$ into (6.4) and (6.5), respectively, we obtain the next m_{ij} and $\mathbf{v}_i$. Then with the newly obtained m_{ij} and $\mathbf{v}_i$ as the initial values we iterate the above procedure until m_{ij} and $\mathbf{v}_i$ converge. If datum $\mathbf{x}_j$ coincides with $\mathbf{v}_i$, namely

$$\mathbf{x}_j = \mathbf{v}_i, \tag{6.6}$$

m_{ij} cannot be determined by using (6.5). In this case, since $m_{kj} = 0$ for $\mathbf{x}_j \neq \mathbf{v}_k$, we need to determine m_{ij} using (6.3) and

$$m_{kj} = 0 \quad \text{for} \quad \mathbf{x}_j \neq \mathbf{v}_k. \tag{6.7}$$

If only one $\mathbf{v}_i$ satisfies (6.6) for datum $\mathbf{x}_j$, $m_{ij} = 1$, but if plural $\mathbf{v}_i$'s satisfy (6.6), the solution is non-unique. We need to consider this situation if we program the algorithm [69].

6.2 The Kohonen Network

The Kohonen network [66], [70] provides both unsupervised and supervised learning models. Here, we discuss only the unsupervised model. Since the network weights are autonomously trained according to the distribution of input data in the input space, the network is also called a self-organizing network. Fig. 6.1 (a) shows the structure of the Kohonen network which is similar to a two-layer neural network. The differences are that neighborhood relations are defined in the output neurons and that unsupervised learning is adopted. In the following we call the output neuron simply the neuron.

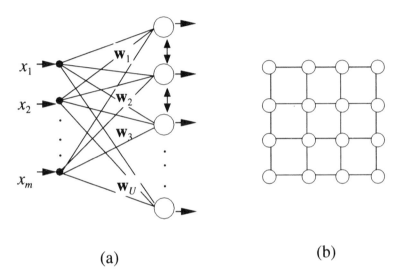

(a)   (b)

Fig. 6.1. The Kohonen network. (a) Structure. (b) Neighborhood relation

Fig. 6.1 (b) shows neighborhood relations represented by a two-dimensional lattice. Neurons that are connected to a neuron with one branch are the nearest to that neuron. We assume that the distances of any two neurons that are connected with one branch are the same.

The training is done by showing one input datum at a time. When an input is presented to the network, the neuron whose weight vector is the nearest in the input space is selected. This neuron is called the firing neuron. The weight vector of the firing neuron is modified so that the weight vector becomes closer to the input vector. In addition, the weight vectors of the neurons which are near the firing neuron are modified so that the weight vectors become closer to the input vector. The learning method that selects one neuron from the competing neurons is called competitive learning. In the trained network, neurons are assigned to the points where many data are gathered. Thus the Kohonen network has an ability to compress data or select cluster centers. In addition, since the weights of the neurons which are near in the neighborhood relations are trained to be near in the input space, the network is considered to be a mapping that preserves neighborhood relations in the input space.

Now we explain the training procedure using the following definitions. Let the m-dimensional input vector be $\mathbf{x} = (x_1, \ldots, x_m)^t$, the number of training data be M, the number of neurons be U, the weight vector between the input and the ith neuron be $\mathbf{w}_i = (w_{i1}, \ldots, w_{im})^t$, where w_{ij} is the weight between the jth input and the ith neuron. The detailed training procedure is as follows.

1. Initialize weight vectors $\mathbf{w}_j$ for $j = 1, \ldots, U$.
2. Iterate Steps 3 and 4 for $l = 1, \ldots, M$. We count this as one epoch number. The training terminates when all the corrections of weights are less than the specified value or the network is trained for a specified number of epochs.
3. For the lth training data find the firing neuron, i.e., the neuron with minimum $\|\mathbf{x} - \mathbf{w}_j\|$ for $j = 1, \ldots, U$. Namely, find the neuron i that satisfies

$$i = \arg \min_{j=1,\ldots,U} \|\mathbf{x} - \mathbf{w}_j\|, \tag{6.8}$$

where arg min returns the subscript that gives the minimum value of $\|\mathbf{x} - \mathbf{w}_j\|$ and $\|\mathbf{x} - \mathbf{w}_j\|$ is the Euclidean distance between $\mathbf{x}$ and $\mathbf{w}_j$:

$$\|\mathbf{x} - \mathbf{w}_j\| = \sqrt{\sum_{k=1}^{m} (x_k - w_{jk})^2}. \tag{6.9}$$

4. Modify the weight vectors $\mathbf{w}_j$ of the neurons near the firing neuron (including the weight vector of the firing neuron) so that $\mathbf{w}_j$ becomes nearer to $\mathbf{x}$:

$$\mathbf{w}_j^{new} = \mathbf{w}_j^{old} + \varepsilon h_{ij}(\mathbf{x} - \mathbf{w}_j^{old}), \tag{6.10}$$

where ε is a positive constant and satisfies $0 < \varepsilon < 1$ and h_{ij} is a neighborhood function determined by the distance between neurons i and j. The neighborhood function h_{ij} may be given by the step function in which h_{ij} is constant within some specified distance from the neuron j and zero outside of the specified distance or the Gaussian function:

$$h_{ij} = \exp\left(-\frac{d_{ij}}{\sigma^2}\right), \tag{6.11}$$

where d_{ij} is a distance determined by the neighborhood relation between neurons i and j and σ^2 is an adjusting parameter. Equation (6.11) takes the maximum value of 1 when $i = j$, namely for the firing neuron and decreases when the distance becomes large. Parameters ε and σ are decreased during training to prevent the weight vectors from oscillations.

Consider simplifying the calculation of (6.8) [71]. Since the square of (6.9) is positive or zero,

$$\frac{1}{2} \sum_{k=1}^{m} x_k^2 \geq \sum_{k=1}^{m} x_k w_{jk} - \frac{1}{2} \sum_{k=1}^{m} w_{jk}^2, \tag{6.12}$$

where the equality holds when $\mathbf{x} = \mathbf{w}_j$. Similar to the multilayer neural network, assign a bias neuron which always outputs 1. Let the weight between the bias neuron and the jth neuron be

$$w_{j,m+1} = -\frac{1}{2} \sum_{k=1}^{m} w_{jk}^2. \tag{6.13}$$

Now similar to the multilayer neural network, the right hand side of (6.12) is obtained by taking the dot product of $\mathbf{x}$ and $\mathbf{w}_j$ and adding $w_{j,m+1}$. Therefore, by assuming this as the output of the jth neuron, the neuron whose weight vector is the closest to the input vector has the maximum output value. Thus the firing neuron corresponds to the neuron with the maximum output.

If we associate each training datum with the firing neuron after training, we can reduce the dimension of the training data to the dimension of the weights, namely, data compaction is performed. Then the weight vector of the firing neuron is the representative value of the training data. Since the weight vectors of the neighboring neurons are corrected as well as that of the firing neuron, the data compaction is performed while retaining the neighborhood relation of the training data.

To use the Kohonen network for clustering the training data, we train the network setting the number of neurons more than the number of classes and train the network using all the training data. After training, we input a class datum into the Kohonen network and associate the datum with the firing neuron. The set of data associated with a neuron forms a cluster. If a cluster includes data with different classes, we split the cluster into plural clusters in which only the data with the same class are included.

6.3 Minimum Volume Clustering Algorithm

Here we discuss a simple clustering algorithm that divides the training data by a line parallel to an input axis at a time. As a measure of the degree of data gathering, we use the size of the hyperboxes that include the training data. Consider dividing the training data included in cluster ij as shown in Fig. 6.2. In Figs. 6.2 (a) and (b) the training data are divided by $x_1 = c_{ij1}$ and $x_2 = c_{ij2}$, respectively, where $\mathbf{c}_{ij} = (c_{ij1}, c_{ij2})^t$ is the center vector of the data included in cluster ij. The total size of the hyperboxes that include the training data in Fig. 6.2 (b) is smaller than that in Fig. 6.2 (a). Thus we consider that dividing the training data by $x_2 = c_{ij2}$ is more favorable. But if the number of the data included in either class is smaller than the specified number we do not choose this axis for division. It may be better to divide the data by a hyperplane expressed by a linear combination of the input variables. But here, for simplicity, we do not consider this. In the following, we show the more detailed algorithm.

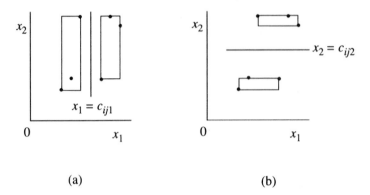

(a)   (b)

Fig. 6.2. Concept of clustering. (a) Division by $x_1 = c_{ij1}$. (b) Division by $x_2 = c_{ij2}$ [56, p. 360 ©IEEE 1998]

1. Let N_{max} and N_{min} be the upper and lower bounds of the number of data belonging to each cluster. First assume that each class has one cluster.
2. Select a cluster, e.g. cluster ij, whose number of data exceeds N_{max}. Then the center of the hyperbox, $\mathbf{c}_{ij}$, is given by

$$\mathbf{c}_{ij} = \frac{1}{|X_{ij}|} \sum_{\mathbf{x} \in X_{ij}} \mathbf{x}, \tag{6.14}$$

where X_{ij} is the set of training data included in cluster ij and $|X_{ij}|$ is the number of data included in X_{ij}. The size of the two hyperboxes that include the training data belonging to cluster ij, when the training data are divided by $x_k = c_{ijk}$, is given by

$$S_{ijk} = \prod_{l=1}^{m} \left(\max_{\substack{\mathbf{x} \in X_{ij} \\ x_k \geq c_{ijk}}} x_l - \min_{\substack{\mathbf{x} \in X_{ij} \\ x_k \geq c_{ijk}}} x_l \right)$$

$$+ \prod_{l=1}^{m} \left(\max_{\substack{\mathbf{x} \in X_{ij} \\ x_k < c_{ijk}}} x_l - \min_{\substack{\mathbf{x} \in X_{ij} \\ x_k < c_{ijk}}} x_l \right). \tag{6.15}$$

We select such k that minimizes

$$\min_{\substack{k=1,\ldots,m \\ N_{ij1} > N_{min}, N_{ij2} > N_{min}}} S_{ijk}, \tag{6.16}$$

where N_{ij1} and N_{ij2} are the numbers of data satisfying $x_k \geq c_{ijk}$ and $x_k < c_{ijk}$, respectively. If there is no such k that minimizes (6.16) we do not divide cluster ij.

3. Iterate Step 2 until there is no cluster whose number of data exceeds N_{max} and which is dividable.

6.4 Fuzzy Min-max Clustering Algorithm

Here we discuss the clustering method [67] based on the fuzzy min-max classifier discussed in Section 9.1. In clustering, the training data are usually unlabeled. But here we assume that the training data are labeled and we consider clustering the training data belonging to the same class.

Unlike the fuzzy min-max classifier, we only generate or expand hyperboxes by a single scan of the training data. Namely when a datum is inputted, a new hyperbox including only this datum is generated if there is no hyperbox of the same class that includes this datum or that is expandable. If the datum is included in a hyperbox, nothing is done. Otherwise, the hyperbox is expanded to include that datum. When the hyperbox generation is over, the training data included in the same hyperbox forms a cluster.

Now we explain the clustering procedure more in detail. The jth hyperbox A_{ij} for class i in the m-dimensional input space is defined as follows:

$$A_{ij} = \{\mathbf{x} \mid v_{ijk} \leq x_k \leq V_{ijk}, k = 1, \ldots, m\} \qquad \text{for} \quad j = 1, \ldots, \qquad (6.17)$$

where x_k is the kth element of input vector $\mathbf{x}$ and v_{ijk} and V_{ijk} are, respectively, the minimum and maximum values of hyperbox A_{ij} with respect to x_k. We assume that the maximum size of a hyperbox, i.e., the maximum sum length of m edges emanating from a vertex, is bounded by $m\theta$ where θ is a positive parameter.

On the hyperbox A_{ij}, we define the truncated rectangular pyramidal membership functions with the average operator. Namely, for the kth input, the one-dimensional membership function is given by

$$m_{A_{ij}}(x_k) = \begin{cases} 1 - (x_k - V_{ij})^s & \text{for} \quad x_k \geq V_{ij}, \\ 1 & \text{for} \quad V_{ij} > x_k \geq v_{ij}, \\ 1 - (v_{ij} - x_k)^s & \text{for} \quad v_{ij} > x_k, \end{cases} \qquad (6.18)$$

where $s = 1$ for linear membership functions and $s = 2$ for quadratic membership functions. In the following we assume the linear membership functions. The m-dimensional membership function with the average operator is given by

$$m_{A_{ij}}(\mathbf{x}) = \frac{1}{m} \sum_{k=1}^{m} m_{A_{ij}}(x_k). \qquad (6.19)$$

For the first datum of the training data, we generate the hyperbox A_{i1} that include the datum. For a datum $\mathbf{x}$ except for the first datum, we check if the datum is included in one of the hyperboxes. If $\mathbf{x}$ is included in a hyperbox, we do nothing and go to the next datum. If it is not included, we check if some hyperboxes A_{ik} $(k = 1, \ldots)$ are expandable, namely if the following inequality holds:

$$m\theta \geq \sum_{k=1}^{m} (\max(V_{ijk}, x_k) - \min(v_{ijk}, x_k)). \tag{6.20}$$

If there are expandable hyperboxes, we calculate the degrees of membership for these hyperboxes using (6.19). Let A_{ij} have the maximum degree of membership. Then we expand A_{ij} changing the minimum values of hyperbox A_{ij} by

$$v_{ijk} \leftarrow \min(v_{ijk}, x_k) \qquad \text{for } k = 1, \ldots, m \tag{6.21}$$

and the maximum values by

$$V_{ijk} \leftarrow \max(V_{ijk}, x_k) \qquad \text{for } k = 1, \ldots, m. \tag{6.22}$$

If there is no expandable hyperbox, we define a new hyperbox:

$$v_{ijk} = V_{ijk} = x_k \qquad \text{for } k = 1, \ldots, m. \tag{6.23}$$

The above method can create hyperboxes by one scan of the training data. The problem with this method is that clustering depends on the processing order of the training data. One way to avoid this is to randomize the processing order.

6.5 Overlap Resolving Clustering Algorithm

Since performance of pattern classification is determined by how the overlaps between classes are resolved, we approximate the class region by an ellipsoid, analyze the overlaps of the ellipsoids, and cluster the data in the overlapping regions. In the following first we define the overlapping regions, explain how to extract the data in the overlapping regions, and discuss the clustering algorithm.

6.5.1 Approximation of Overlapping Regions

Let the set of m-dimensional data in class i be X_i. Then the Mahalanobis distance of a datum $\mathbf{x}$ from the center of class i, $d_i(\mathbf{x})$, is given by

$$d_i^2(\mathbf{x}) = (\mathbf{x} - \mathbf{c}_i)^t Q_i^{-1} (\mathbf{x} - \mathbf{c}_i), \tag{6.24}$$

where $\mathbf{c}_i$ is the center of class i and Q_i is the covariance matrix of class i, and they are given, respectively, by

$$\mathbf{x}_i = \frac{1}{|X_i|} \sum_{\mathbf{x} \in X_i} \mathbf{x}, \tag{6.25}$$

$$Q_i = \frac{1}{|X_i|} \sum_{\mathbf{x} \in X_i} (\mathbf{x} - \mathbf{c}_i)(\mathbf{x} - \mathbf{c}_i)^t. \tag{6.26}$$

Using the Mahalanobis distance, we classify an input vector $\mathbf{x}$ into the class:

$$\arg \min_{i=1,\dots,n} d_i(\mathbf{x}). \tag{6.27}$$

6.5.2 Extraction of Data from the Overlapping Regions

Suppose two-dimensional inputs of classes j and k exist as shown in Fig. 6.3. Let each ellipsoid in the figure be at an equal Mahalanobis distance from the class center. Then we want to extract data in the overlapping regions denoted as A and B.

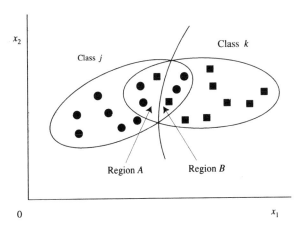

Fig. 6.3. Overlapping regions of two classes

In the following we discuss how to extract class k data in the overlapping regions A and B. Let B_{jk} be the set of class j data included in region B. Then

$$B_{jk} = \{\mathbf{x} \mid d_j(\mathbf{x}) \geq d_k(\mathbf{x}), \mathbf{x} \in X_j\}, \tag{6.28}$$

where X_j is the set of data belonging to class j. Next we calculate the maximum Mahalanobis distance from the center of class j among the data in B_{jk}:

$$d_{j\,\max} = \max_{\mathbf{x} \in B_{jk}} d_j(\mathbf{x}). \tag{6.29}$$

Let B'_{kj} be the set of class k data whose Mahalanobis distances from the center of class j are less than $d_{j\,\text{max}}$:

$$B'_{kj} = \{\mathbf{x} \mid d_j(\mathbf{x}) < d_{j\,\text{max}}, \mathbf{x} \in X_k\}. \tag{6.30}$$

When $B_{jk} = \phi$, namely there are no class j data in region B, we set $B'_{kj} = \phi$. This means that if $B_{jk} = \phi$, we can consider that there are no overlaps between classes j and k in region B.

Let A_{kj} be the set of class k data in region A. Then A_{kj} is given by

$$A_{kj} = \{\mathbf{x} \mid d_j(\mathbf{x}) \le d_k(\mathbf{x}), \mathbf{x} \in X_k\}. \tag{6.31}$$

Thus $D_{kj} = A_{kj} \cup B'_{kj}$ is the set of class k data included in the overlapping region for classes k and j. Then we can cluster X_k into $X_k - D_{kj}$ and D_{kj}. Similarly, defining $D_{jk} = A_{jk} \cup B'_{jk}$ as the set that includes class j data included in the overlapping region for classes k and j, we can cluster X_j into $X_j - D_{jk}$ and D_{jk}.

If three classes j, k, and l overlap, we calculate $D_{jkl} = D_{jk} \cap D_{jl}$. If $D_{jkl} \ne \phi$, we cluster X_j into four clusters: $X_j - (D_{jk} \cup D_{jl})$, $D_{jk} - D_{jkl}$, $D_{jl} - D_{jkl}$, and D_{jkl}. Similarly, we can cluster X_k and X_l.

6.5.3 Clustering Algorithm

Let N_{min} be the minimum number of data included in a cluster when the cluster is divided. Namely, we allow cluster division only when any of the divided clusters has at least N_{min} data. Then, assuming that only two classes overlap, i.e., $D_{jkl} = \phi$, the clustering algorithm to resolve overlaps between two classes is as follows.

1. Let $j_1 = 1$ and X_j be the set of class j data.
2. For all combinations of $(j, k), j = 1, \ldots, n, k \ne j, k = 1, \ldots, n$, if

$$|X_j - D_{jk}| \ge N_{\text{min}}, |D_{jk}| \ge N_{\text{min}}, \tag{6.32}$$

do the following:

$$X_{jj_1} = X_j - D_{jk},$$
$$X_{j(j_1+1)} = D_{jk},$$
$$X_j \leftarrow X_j - D_{jk},$$
$$j_1 \leftarrow j_1 + 2,$$

where X_{jk} is the set of class j data included in cluster jk.

In the above algorithm, data are deleted from X_j. Thus, if $D_{ijk} \ne \phi$, the generated clusters depend on the processing order of classes.

7. Tuning of Membership Functions

In the previous chapter, we discuss extraction of the three types of fuzzy rules: those with pyramidal membership functions, those with polyhedral regions, and those with ellipsoidal regions using the data included in the associated clusters. Since these fuzzy rules are generated without considering the overlap between classes, their classification performance may not be good. Therefore to improve their classification performance, in this chapter we discuss tuning the slopes and the locations of the membership functions. The direct methods directly maximize the recognition rate of the training data by counting the net increase in the recognition rate when the slope or location of the membership function is changed [56, 72]. The indirect methods maximize the continuous objective function that leads to improving the recognition rate. Finally, we evaluate the performance of the tuning methods for the benchmark data.

7.1 Problem Formulation

Assuming that we have defined the fuzzy rules R_{ij} given by (5.1) in classifying the m-dimensional input vector $\mathbf{x}$ into one of n classes, we consider improving classification performance by tuning membership functions. The membership functions used in this section are

- rectangular pyramidal membership functions with the minimum or average operator;
- polyhedral pyramidal membership functions with the minimum operator;
- truncated rectangular pyramidal membership functions with the minimum or average operator;
- truncated polyhedral pyramidal membership functions with the minimum operator; and
- bell-shaped membership functions.

From the discussion in Section 4.2, if the negative degree of membership is assumed, these membership functions $m_{ij}(\mathbf{x})$ are equivalent to the tuned distance:

$$h_{ij}(\mathbf{x}) = \frac{d_{ij}(\mathbf{x})}{\alpha_{ij}} \qquad (7.1)$$

for the piecewise linear membership function and the squared tuned distance:

$$h_{ij}^2(\mathbf{x}) = \frac{d_{ij}^2(\mathbf{x})}{\alpha_{ij}} \tag{7.2}$$

for the quadratic membership function where $d_{ij}(\mathbf{x})$ and α_{ij} are the (weighted) distance and the tuning parameter for the jth fuzzy rule for class i, respectively. Thus hereafter we consider the tuned distance or the squared tuned distance instead of the membership function. Notice that even if we change α_{ij}, the (weighted) distance $d_{ij}(\mathbf{x})$ does not change for given $\mathbf{x}$.

If, for the input vector $\mathbf{x}$, $h_{ij}^s(\mathbf{x})$ ($s \in \{1,2\}$) is the smallest, $\mathbf{x}$ is classified into class i. Thus the problem of tuning slopes and locations is as follows.

Let the M training data pairs be $(\mathbf{x}_i, y_i)$ ($i = 1, \ldots, M$) where y_i is the class label for $\mathbf{x}_i$. Then maximize the recognition rate of the training data:

$$
\begin{aligned}
R_{tr} &= \frac{100}{M} \sum_{k=1}^{M} I \left(\arg\min_{i} {}_{i,j} h_{ij}^s(\mathbf{x}_k) = y_k \right) \\
&= \frac{100}{M} \sum_{k=1}^{M} I \left(\arg\min_{i} {}_{i,j} \frac{d_{ij}^s(\mathbf{x}_k)}{\alpha_{ij}} = y_k \right)
\end{aligned} \tag{7.3}
$$

by tuning α_{ij} and locations of the membership functions where $s = 1$ for the piecewise linear functions and $s = 2$ for the quadratic membership functions, and

$$I(a = b) = \begin{cases} 1 & \text{for } a = b, \\ 0 & \text{for } a \neq b. \end{cases} \tag{7.4}$$

Equation (7.3) is a discrete function and usually it is difficult to optimize the function. In the following, we discuss the direct methods in which (7.3) is maximized by tuning α_{ij} and locations of the membership functions, and the indirect method in which the continuous version of (7.3) is minimized.

7.2 Direct Methods

Now consider how misclassification is resolved by tuning slopes and locations of the membership functions using an example of classifying a one-dimensional input into one of two classes. Assume that trapezoidal membership functions are defined as shown in Fig. 7.1. Then for Datum 1, the degree of membership for Class 1 is 1. Thus Datum 1 is classified into Class 1 irrespective of the slope of the membership function for Class 2. Suppose that Datum 1 belongs to Class 2. Then by tuning slopes we cannot classify this datum correctly, but by moving the location of the Class 1 membership function to the left as shown in the figure, this datum is correctly classified.

If we use the triangular membership functions as shown in Fig. 7.2, Datum 1 can be classified into Class 2 by shifting the membership function for Class

1 or by decreasing the slope of the membership function for Class 2 as shown in Fig 7.3.

While resolving misclassification, new misclassification may occur. Thus in tuning slopes and locations of the membership functions, we need to consider resolution of misclassification as well as prevention of misclassification. In the following we discuss tuning of slopes and locations of the membership functions so that (7.3) is maximized.

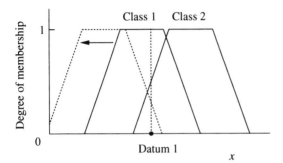

Fig. 7.1. Trapezoidal membership functions. Datum 1 is classified into Class 1 irrespective of the slope of the Class 2 membership function. To classify this datum into Class 2, the Class 1 membership function needs to be shifted as in the dotted line

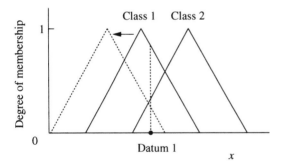

Fig. 7.2. Triangular membership functions. Datum 1 is classified into Class 1. To classify this datum into Class 2, the Class 1 membership function needs to be shifted as in the dotted line

7.2.1 Tuning of Slopes

Concept. In this section we consider maximizing (7.3) by changing α_{ij} ($i = 1, \ldots, n, j = 1, \ldots$). If we increase α_{ij}, the degree of membership $m_{ij}(\mathbf{x})$

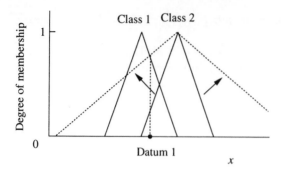

Fig. 7.3. Triangular membership functions. Datum 1 that is classified into Class 1 can be classified into Class 2 if the slopes of the membership function are changed as in the dotted lines

(the tuned distance $h_{ij}(\mathbf{x})$) increases (decreases), and if we decrease it, the degree of membership (the tuned distance) decreases (increases). Thus we can improve the recognition rate by iterate tuning α_{ij} one at a time without allowing new misclassification. To explain this, we consider a two-class case with one rule for each class as shown in Fig. 7.4. (In the figures we use the triangular function as the membership function.) In this case, Datum 1, belonging to Class 1, is misclassified into Class 2. This datum can be correctly classified if we increase α_{11} so that the membership function lies between the shaded regions, without causing Datum 2 to be misclassified. This can also be achieved when we decrease α_{21}. By tuning membership functions successively without allowing new misclassification, the recognition rate is improved monotonically and reaches a maximum. This is, however, a local maximum. Thus to avoid being trapped into a local maximum, we allow new misclassification during tuning.

Figure 7.5 shows tuning that allows misclassification. Datum 1 is correctly classified into Class 2, while Data 2, 3, and 4 are misclassified into Class 2. If we increase α_{11} or decrease α_{21}, Datum 1 is first misclassified, but if we allow Datum 1 to be misclassified we can make Data 2, 3, and 4 be correctly classified. Figure 7.5 shows this when α_{21} is decreased so that the degree of membership for Class 2 lies between the shaded regions. Then by allowing one datum to be misclassified, three data are correctly classified, i.e., the recognition rate is improved by two data.

Now suppose we tune the tuning parameter α_{ij}. Up to some value we can increase or decrease α_{ij} without making the correctly classified datum $\mathbf{x}$ be misclassified. Thus we can calculate the upper bound $U_{ij}(\mathbf{x})$ or lower bound $L_{ij}(\mathbf{x})$ of α_{ij} that causes no misclassification of $\mathbf{x}$ as shown in Fig. 7.6. Now let $U_{ij}(1)$ and $L_{ij}(1)$ denote the upper and lower bounds that do not make the correctly classified data be misclassified, respectively. Likewise, $U_{ij}(l)$ and $L_{ij}(l)$ denote the upper and lower bounds in which $l - 1$ correctly classified data are misclassified, respectively. Then, for instance, if we set a value in

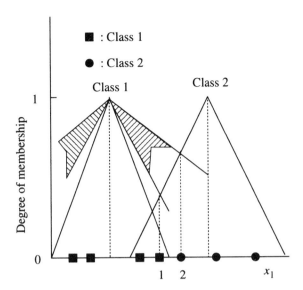

Fig. 7.4. Concept of tuning. If the slope of the membership function for Class 1 is decreased so that the resulting function lies between the shaded regions, Datum 1 can be correctly classified [56, p. 361 ©IEEE 1998]

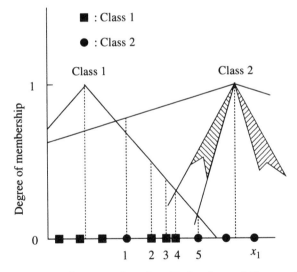

Fig. 7.5. Concept of tuning. If the slope of the membership function for Class 2 is increased so that the resulting function lies between the shaded regions, Datum 1 is misclassified but Data 2, 3, and 4 are correctly classified [56, p. 361 ©IEEE 1998]

the interval $[U_{ij}(1), U_{ij}(2))$ to α_{ij}, one correctly classified datum belonging to class i is misclassified, where $[a, b]$ and (a, b) denote the closed and open intervals, respectively.

Similarly, if we increase or decrease α_{ij}, misclassified data may be correctly classified. Let $\beta_{ij}(l)$ denote the upper bound of α_{ij} that is smaller than $U_{ij}(l)$ and that resolves misclassification (see Fig. 7.7 (a)). Let $\gamma_{ij}(l)$ denote the lower bound of α_{ij} that is larger than $L_{ij}(l)$ and that resolves misclassification (see Fig. 7.7 (b)). Figure 7.8 shows an example. If we change the current α_{ij} to the tuned α_{ij} in $(\beta_{ij}(2), U_{ij}(2))$, one correctly classified datum is misclassified but four misclassified data are correctly classified.

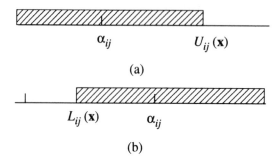

(a)

(b)

Fig. 7.6. Upper and lower bounds of the tuning parameter α_{ij} that cause no misclassification of $\mathbf{x}$. (**a**) For a datum belonging to class i. (**b**) For a datum not belonging to class i

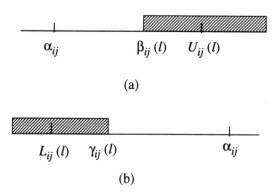

(a)

(b)

Fig. 7.7. Intervals of the tuning parameter α_{ij} that resolve $l-1$ misclassifications. (**a**) For a datum belonging to class i. (**b**) For a datum not belonging to class i

Then the next task is to find which interval among $(L_{ij}(l), \gamma_{ij}(l))$ and $(\beta_{ij}(l), U_{ij}(l))$ $(l = 1, \ldots)$ gives the maximum recognition rate. To limit the search space, we introduce the maximum l, i.e., l_M. Let $(L_{ij}(l), \gamma_{ij}(l))$ be the

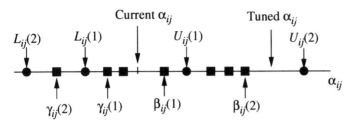

● : Correctly classified

■ : Misclassified

Fig. 7.8. Determination of tuned α_{ij}. If the current α_{ij} is modified to the tuned α_{ij} in $(\beta_{ij}(2), U_{ij}(2))$, one correctly classified datum is misclassified but four misclassified data are correctly classified [56, p. 364 ©IEEE 1998]

interval that gives the maximum recognition rate of the training data among $(L_{ij}(k), \gamma_{ij}(k))$ and $(\beta_{ij}(k), U_{ij}(k))$ for $k = 1, \ldots, l_M$. Then even if we set any value in the interval to α_{ij}, the recognition rate of the training data does not change but the recognition rate of the test data may change. To control the generalization ability, we set α_{ij} as follows:

$$\alpha_{ij} = \beta_{ij}(l) + \delta(U_{ij}(l) - \beta_{ij}(l)) \tag{7.5}$$

for $(\beta_{ij}(l), U_{ij}(l))$, where δ satisfies $0 < \delta < 1$ and

$$\alpha_{ij} = \gamma_{ij}(l) - \delta(\gamma_{ij}(l) - L_{ij}(l)) \tag{7.6}$$

for $(L_{ij}(l), \gamma_{ij}(l))$.

According to the above discussion, the tuning algorithm becomes as follows.

1. Set a positive number to parameter l_M, where $l_M - 1$ is the maximum number of misclassifications allowed for tuning α_{ij}, set a value in (0, 1) to δ in (7.5) and (7.6), and set the same positive initial value (usually 1) to α_{ij}.
2. For α_{ij} $(i = 1, \ldots, n, j = 1, \ldots)$, calculate $L_{ij}(l), U_{ij}(l), \beta_{ij}(l)$, and $\gamma_{ij}(l)$ for $l = 1, \ldots, l_M$. Find the interval $(L_{ij}(l), \gamma_{ij}(l))$ or $(\beta_{ij}(l), U_{ij}(l))$ that realizes the maximum recognition rate of the training data, and change α_{ij} using (7.5) or (7.6).
3. Iterate Step 2 until there is no improvement in the recognition rate of the training data.

We call the update of all α_{ij} $(i = 1, \ldots, n, j = 1, \ldots)$ one epoch of tuning slopes, and if there is no improvement in the recognition rate for the two consecutive iterations, or the recognition rate of the training data reaches 100%, we stop tuning. Our tuning algorithm determines, for each fuzzy rule R_{ij}, the optimum tuning parameter α_{ij}, allowing the data that are correctly classified before tuning R_{ij} to become misclassified after tuning R_{ij} as long as the recognition rate of the training data is improved.

The special feature of the fuzzy rule tuning is that outliers (Datum 1 in Figure 7.5 are automatically eliminated by allowing the data that are correctly classified before tuning to be misclassified after tuning.

In the following, we calculate the upper bound and the lower bound of α_{ij} that allow the $l - 1 (\geq 0)$ data that are correctly classified to become misclassified. Then, we check how many data that are misclassified are correctly classified if α_{ij} is changed within the bounds. Finally, α_{ij} is determined so that the recognition rate of the training data is maximized.

Upper and Lower Bounds of the Tuning Parameters. We calculate the upper bound $U_{ij}(l)$ and the lower bound $L_{ij}(l)$ of α_{ij} allowing the $l - 1$ data that are correctly classified to be misclassified. We divide a set of input data into X and Y, where X consists of the data correctly classified using the set of fuzzy rules $\{R_{ij}\}$ and Y consists of the misclassified data. Then, for $\mathbf{x} (\in X)$ that belongs to class i we find the cluster ij that satisfies

$$h_{ij}(\mathbf{x}) \leq \min_{k \neq j} h_{ik}(\mathbf{x}). \tag{7.7}$$

If (7.7) does not hold, $\mathbf{x}$ remains to be correctly classified even if we change α_{ij}. If $\mathbf{x}$ further satisfies

$$h_{ij}^s(\mathbf{x}) = \frac{d_{ij}^s(\mathbf{x})}{\alpha_{ij}} < \min_{o \neq i} h_{op}^s(\mathbf{x}) < \min_{k \neq j} h_{ik}^s(\mathbf{x}), \tag{7.8}$$

there is a lower bound $L_{ij}(\mathbf{x})$ to keep $\mathbf{x}$ correctly classified:

$$L_{ij}(\mathbf{x}) = \frac{d_{ij}^s(\mathbf{x})}{\min_{o \neq i} h_{op}^s(\mathbf{x})} < \alpha_{ij}. \tag{7.9}$$

If (7.8) is not satisfied, namely

$$h_{ij}^s(\mathbf{x}) = \frac{d_{ij}^s(\mathbf{x})}{\alpha_{ij}} < \min_{k \neq j} h_{ik}^s(\mathbf{x}) < \min_{o \neq i} h_{op}^s(\mathbf{x}), \tag{7.10}$$

α_{ij} can be decreased without making $\mathbf{x}$ misclassified.

Now the lower bound $L_{ij}(1)$, which is defined as the lower bound that does not make any correctly classified data be misclassified, is

$$L_{ij}(1) = \max_{\mathbf{x} \in X} L_{ij}(\mathbf{x}). \tag{7.11}$$

To clarify the discussion, we assume that $L_{ij}(\mathbf{x})$ is different for different $\mathbf{x}$. Then (7.11) is satisfied by one $\mathbf{x}$. Similarly, $L_{ij}(2)$, which is defined as the lower bound that allows one correctly classified datum to be misclassified, is the second maximum among $L_{ij}(\mathbf{x})$ and is given by

$$L_{ij}(2) = \max_{\mathbf{x} \in X, L_{ij}(\mathbf{x}) \neq L_{ij}(1)} L_{ij}(\mathbf{x}). \tag{7.12}$$

In general,

$$L_{ij}(l) = \max_{\mathbf{x} \in X, L_{ij}(\mathbf{x}) \neq L_{ij}(1),\dots,L_{ij}(l-1)} L_{ij}(\mathbf{x}). \qquad (7.13)$$

In the similar manner that we determined the lower bound $L_{ij}(l)$, we can determine the upper bound $U_{ij}(l)$. We choose $\mathbf{x}\,(\in X)$ which belongs to class $o\,(\neq i)$. Let cluster op have the minimum tuned distance $h_{op}(\mathbf{x})$:

$$h_{op} = \min_q h_{op}(\mathbf{x}). \qquad (7.14)$$

Since the tuned distance $h_{ij}(\mathbf{x})$ is larger than $h_{op}(\mathbf{x})$, the upper bound $U_{ij}(\mathbf{x})$ of α_{ij} in which $\mathbf{x}$ remains correctly classified is:

$$U_{ij}(\mathbf{x}) = \frac{d_{ij}^s(\mathbf{x})}{\min\limits_q h_{oq}^s(\mathbf{x})}. \qquad (7.15)$$

Now the upper bound $U_{ij}(1)$, which is defined as the upper bound that does not make any correctly classified data be misclassified, is

$$U_{ij}(1) = \min_{\mathbf{x} \in X} U_{ij}(\mathbf{x}). \qquad (7.16)$$

Here we also assume that $U_{ij}(\mathbf{x})$ is different for different $\mathbf{x}$. Then (7.16) is satisfied by one $\mathbf{x}$. Similarly, $U_{ij}(2)$, which is defined as the upper bound that allows one correctly classified datum to be misclassified, is the second minimum among $U_{ij}(\mathbf{x})$ and is given by

$$U_{ij}(2) = \min_{\mathbf{x} \in X, U_{ij}(\mathbf{x}) \neq U_{ij}(1)} U_{ij}(\mathbf{x}). \qquad (7.17)$$

In general,

$$U_{ij}(l) = \min_{\mathbf{x} \in X, U_{ij}(\mathbf{x}) \neq U_{ij}(1),\dots,U_{ij}(l-1)} U_{ij}(\mathbf{x}). \qquad (7.18)$$

Thus α_{ij} is bounded by

$$\cdots < L_{ij}(l) < L_{ij}(l-1) < \cdots < L_{ij}(1) < \alpha_{ij} < U_{ij}(1)$$
$$< \cdots < U_{ij}(l-1) < U_{ij}(l) < \cdots. \qquad (7.19)$$

If we change α_{ij} in the range of $(L_{ij}(1), U_{ij}(1))$, the correctly classified data $\mathbf{x}\,(\in X)$ remain to be correctly classified where (a, b) denotes an open interval. If we change α_{ij} in the range of $[U_{ij}(l-1), U_{ij}(l))$, or $(L_{ij}(l), L_{ij}(l-1)]$, the $l-1$ correctly classified data $\mathbf{x}\,(\in X)$ are misclassified where $[a, b]$ denotes a closed interval.

Resolution of Misclassification by Changing the Tuning Parameters. For $\mathbf{x}\,(\in Y)$ which is misclassified into class i or which belongs to class i but is misclassified into class $o\,(\neq i)$, we check whether it can be correctly classified by changing α_{ij}. Let $\mathbf{x}$, which belongs to class i, be misclassified into class o. This datum can be correctly classified if

$$\alpha_{ij} > V_{ij}(\mathbf{x}) = \frac{d_{ij}^s(\mathbf{x})}{\min\limits_p h_{op}^s(\mathbf{x})}, \tag{7.20}$$

irrespective of the values of $h_{ik}(\mathbf{x})$ $(k \neq i)$ where $V_{ij}(\mathbf{x})$ is the lower bound.

Let Inc(l) denote the number of misclassified data that are correctly classified if we set the value of α_{ij} in $[U_{ij}(l-1), U_{ij}(l))$. We increase Inc(l) by one if $V_{ij}(\mathbf{x})$ is included in $(\alpha_{ij}, U_{ij}(l))$ and we define

$$\beta_{ij}(l) = \max_{V_{ij}(\mathbf{x}) < U_{ij}(l)} V_{ij}(\mathbf{x}). \tag{7.21}$$

If α_{ij} is set to be larger, in $[U_{ij}(l-1), U_{ij}(l))$, than $\max(\beta_{ij}(l), U_{ij}(l-1))$, Inc($l$) data are correctly classified although the $l-1$ correctly classified data are misclassified.

Similarly, let $\mathbf{x}$, which belongs to class o, be misclassified into class i. Then we check whether $\mathbf{x}$ can be correctly classified by decreasing α_{ij}. First, the minimum tuned distance for class o should be the second minimum among n classes, namely q in the following equation needs to be o:

$$\min_k h_{ik}(\mathbf{x}) < \min_{q \neq i} h_{qr}(\mathbf{x}). \tag{7.22}$$

Second, $h_{ij}(\mathbf{x})$ needs to be the minimum in class i, and the second minimum in class i is larger than the minimum tuned distance in class o:

$$h_{ij}(\mathbf{x}) < \min_p h_{op}(\mathbf{x}) < \min_{k \neq j} h_{ik}(\mathbf{x}). \tag{7.23}$$

Then, the datum can be correctly classified if

$$\alpha_{ij} < K_{ij}(\mathbf{x}) = \frac{d_{ij}^s(\mathbf{x})}{\min\limits_p h_{op}^s(\mathbf{x})}, \tag{7.24}$$

where $K_{ij}(\mathbf{x})$ is the upper bound.

Let Dec(l) denote the number of misclassified data that are correctly classified if we set the value of α_{ij} in $(L_{ij}(l), L_{ij}(l-1)]$. We increase Dec(l) by one if $K_{ij}(\mathbf{x})$ is included in $(L_{ij}(l), \alpha_{ij})$. We define

$$\gamma_{ij}(l) = \min_{K_{ij}(\mathbf{x}) > L_{ij}(l)} K_{ij}(\mathbf{x}). \tag{7.25}$$

If α_{ij} is set to be smaller, in $(L_{ij}(l), L_{ij}(l-1)]$, than $\min(\gamma_{ij}(l), L_{ij}(l-1))$, Dec($l$) data are correctly classified although the $l-1$ correctly classified data are misclassified.

Modification of the Tuning Parameters. For Inc(l), $l = 1, \ldots, l_M$, where l_M is a positive integer, we find l that satisfies

$$\max_l (\text{Inc}(l) - l + 1). \tag{7.26}$$

Similarly, for Dec(l), $l = 1, \ldots l_M$, we find l that satisfies

$$\max_l (\text{Dec}(l) - l + 1). \tag{7.27}$$

If there are plural l's that satisfy (7.26) or (7.27), we chose the smallest l. First we consider the case where (7.26) is larger than or equal to (7.27). If we increase α_{ij} so that it is larger than $\beta_{ij}(l)$ in $(\alpha_{ij}, U_{ij}(l))$, the net increase of the correctly classified data is $\text{Inc}(l) - l + 1$. Thus we set α_{ij} in $[\beta_{ij}(l), U_{ij}(l))$ as follows:

$$\alpha_{ij} = \beta_{ij}(l) + \delta(U_{ij}(l) - \beta_{ij}(l)), \tag{7.28}$$

where δ satisfies $0 < \delta < 1$. Here, $\beta_{ij}(l) \geq U_{ij}(l-1)$ holds, otherwise l cannot satisfy (7.26).

Likewise, if (7.26) is smaller than (7.27), we decrease α_{ij} so that it is smaller than $\gamma_{ij}(l)$ in $(L_{ij}(l), \gamma_{ij}(l)]$ as follows:

$$\alpha_{ij} = \gamma_{ij}(l) - \delta(\gamma_{ij}(l) - L_{ij}(l)). \tag{7.29}$$

Equations (7.28) and (7.29) are the same with (7.5) and (7.6), respectively.

7.2.2 Tuning of Locations

Concept. Similar to tuning of slopes, we can tune locations of the membership functions but the procedure is more complicated and there is a limitation on the membership functions that are tunable by the direct method.

Allowing misclassification, we move the fuzzy region A_{ij} so that the recognition rate of the training data is maximized. To simplify matters, we move A_{ij} along one axis at a time. If we use the minimum operator for the pyramidal membership functions, the axis where the degree of one-dimensional membership is the minimum changes when the location of A_{ij} changes and thus it is difficult to tune locations directly. Thus we only consider the average operator for the pyramidal membership functions.

If we tune the locations of truncated rectangular pyramidal functions, the regions where the degrees of membership for more than one class are 1 may appear. Thus we also need to tune the size of the truncated rectangles to avoid this which complicates tuning. Thus we exclude the truncated rectangular pyramidal membership functions from tuning locations.

For the (truncated) polyhedral pyramidal membership functions any two of the surfaces given by (4.35) are not, in general, perpendicular or parallel to each other. Thus if the fuzzy region A_{ij} is moved in the direction parallel to one surface, all the degrees of membership of the one-dimensional membership functions are affected. Thus we exclude these membership functions from consideration. When we move the location of a rectangular pyramidal membership function along one input axis, only the one-dimensional membership function for this axis changes. Similarly when we move the location of a bell-shaped membership function along one principal axis, only the one-dimensional membership function for this axis changes. Thus we can tune the membership functions along these axes one at a time [73].

According to the above discussions, the membership functions for A_{ij} that we consider are as follows:

- Rectangular pyramidal membership functions with the average operator:

$$m_{ij}(\mathbf{x}) = 1 - h_{ij}^s(\mathbf{x})$$

$$= 1 - \frac{1}{m} \sum_{k=1}^{m} h_{ijk}^s(\mathbf{x}), \tag{7.30}$$

where $s = 1$ or 2 and $h_{ijk}(\mathbf{x})$ is the one-dimensional tuned distance for the kth axis and

$$h_{ijk}^s(\mathbf{x}) = \frac{d_{ijk}^s(\mathbf{x})}{\alpha_{ij}}$$

$$= \frac{|x_k - c_{ijk}|^s}{\alpha_{ij} w_{ijk}^s}. \tag{7.31}$$

Here we assume that the center $\mathbf{c}_{ij}$ is the middle point of A_{ij} and $w_{ijk} = (V_{ijk} + v_{ijk})/2$.

- Bell-shaped membership functions:

$$m_{ij}(\mathbf{x}) = \exp(-h_{ij}^2(\mathbf{x}))$$

$$= \exp\left(-\frac{d_{ij}^2(\mathbf{x})}{\alpha_{ij}}\right), \tag{7.32}$$

$$d_{ij}^2(\mathbf{x}) = (\mathbf{x} - \mathbf{c}_{ij})^t Q_{ij}^{-1} (\mathbf{x} - \mathbf{c}_{ij}). \tag{7.33}$$

First, we explain the concept of tuning locations, using simple one-dimensional cases with two classes shown in Fig. 7.9. In the figure, one fuzzy rule is defined for each class. Datum 1 that belongs to Class 1 is misclassified into Class 2, because the degree of membership for Class 2 is larger than that for Class 1. But this misclassification can be resolved with no misclassification if we move the center of the Class 1 fuzzy rule into the shaded region. This resolution of misclassification can also be achieved by moving the center of the Class 2 fuzzy rule to the right. Fig. 7.10 shows a more complicated situation. Datum 1 is correctly classified into Class 2 but Data 2 and 3 are misclassified into Class 2. Data 2 and 3 are correctly classified if the center of the Class 1 fuzzy rule is moved into the shaded region, although Datum 1 is now misclassified. In this case, since the recognition rate is improved by one datum, we allow misclassification.

When we tune a bell-shaped membership function $m_{ij}(\mathbf{x})$, we transform the input vector $\mathbf{x}$ into $\mathbf{x}'$ by $\mathbf{x}' = S\mathbf{x}$ where S is the orthogonal matrix that diagonalizes Q_{ij}. Then the squared tuned distance becomes

$$h_{ij}^2(S^t \mathbf{x}') = (\mathbf{x}' - \mathbf{c}_{ij}')^t (SQ_{ij}S^t)^{-1} (\mathbf{x}' - \mathbf{c}_{ij}')$$

$$= \sum_{k=1}^{m} \frac{(x_k' - c_{ijk}')^2}{\alpha_{ij} \sigma_{ijk}^2}, \tag{7.34}$$

where $\mathbf{c}_{ij}' = S\mathbf{c}_{ij}$ and σ_{ijk}^2 is the kth diagonal element of $SQ_{ij}S^t$. Thus if $m w_{ijk}^2 = \sigma_{ijk}^2$, the bell-shaped membership function is equivalent to the

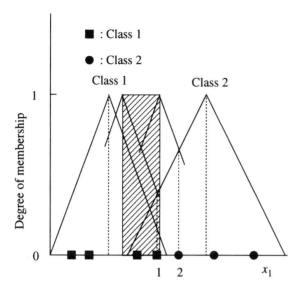

Fig. 7.9. Concept of tuning. If the center of the membership function for Class 1 is moved in the shaded region, Datum 1 can be correctly classified

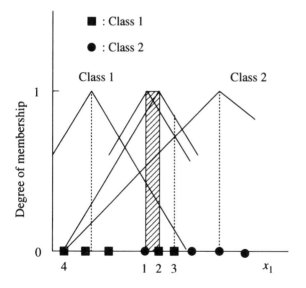

Fig. 7.10. Concept of tuning. If the center of the membership function for Class 1 is moved in the shaded region, Datum 1 is misclassified but Data 2 and 3 are correctly classified

rectangular pyramidal membership function with $s = 2$. Therefore, in the following we discuss only the tuning method for the rectangular pyramidal membership functions.

Now suppose we tune center c_{ij} on the kth input axis. To some value we can move c_{ijk} to the right or left without increasing misclassification of class i data. If c_{ijk} moves to x_k, $m_{ij}(\mathbf{x})$ increases until c_{ijk} reaches x_k. If c_{ijk} moves away from x_k, $m_{ij}(\mathbf{x})$ decreases. Thus for the correctly classified input vector $\mathbf{x}$ that belongs to class i, there is an interval $(I_{ijk}(\mathbf{x}), D_{ijk}(\mathbf{x}))$ (see Fig. 7.11 (a)), in which $\mathbf{x}$ is correctly classified when c_{ijk} is moved. For the correctly classified input vector $\mathbf{x}$ that does not belong to class i, under some condition that is discussed later, there are intervals $(-\infty, D_{ijk}(\mathbf{x}))$ and $(I_{ijk}(\mathbf{x}), \infty)$ (see Figs. 7.11 (b) and (c)) in which $\mathbf{x}$ is correctly classified. Thus there are an upper bound $D_{ijk}(1) = \min D_{ijk}(\mathbf{x}) (\geq c_{ijk})$ and a lower bound $I_{ijk}(1) = \max I_{ijk}(\mathbf{x}) (< c_{ijk})$ that result in no increase in misclassification. Likewise, we have the upper bound $D_{ijk}(l)$ and lower bound $I_{ijk}(l)$ in which $l - 1$ correctly classified data are misclassified. Here, suppose that some correctly classified datum that belongs to a class other than class i is misclassified when $c_{ijk} = D_{ijk}(o), 1 \leq o < l$ (see Fig. 7.11 (b)). This datum is correctly classified when $c_{ijk} = D_{ijk}(l)$ and $c_{ijk} > I_{ijk}(o)$. This may occur for $I_{ijk}(l)$. But, in calculating $D_{ijk}(l)$ and $I_{ijk}(l)$ we do not consider this to make the calculation simple.

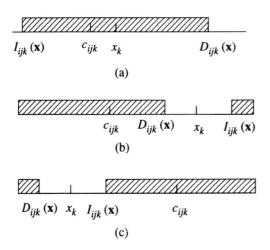

(a)

(b)

(c)

Fig. 7.11. Upper and lower bounds of the center c_{ijk} that cause no new misclassification. (a) For a datum belonging to class i. (b) For a datum not belonging to class i $(c_{ijk} \leq x_k)$. (c) For a datum not belonging to class i $(c_{ijk} > x_k)$

Similarly, if we move c_{ijk} to the right or left, misclassified data may be correctly classified. For the misclassified input vector $\mathbf{x}$ that belongs to class i, there is an interval $(I_{ijk}(\mathbf{x}), D_{ijk}(\mathbf{x}))$ where it is correctly classified (see

Figs. 7.12 (a) and (b)). For the misclassified input vector $\mathbf{x}$ that does not belong to class i, under some condition, there are intervals $(I_{ijk}(\mathbf{x}), \infty)$ and $(-\infty, D_{ijk}(\mathbf{x}))$ where it is correctly classified (see Fig. 7.12 (c)).

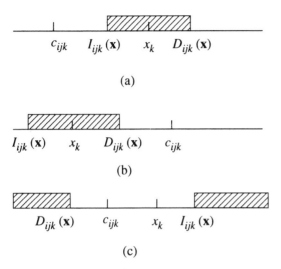

(a)

(b)

(c)

Fig. 7.12. Intervals of the center c_{ijk} that resolve misclassification. (a) For a datum belonging to class i $(c_{ijk} > x_k)$. (b) For a datum belonging to class i $(c_{ijk} \le x_k)$. (c) For a datum not belonging to class i

Then using the upper bounds $D_{ijk}(l)$ and the lower bounds $I_{ijk}(l)$ of c_{ijk} that allow $l - 1$ new misclassifications and the intervals that resolve misclassification, we can determined the interval of c_{ijk} that makes the highest recognition rate. Fig. 7.13 shows an example. In the upper part of the figure, the six intervals indicate that the misclassified data are correctly classified if c_{ijk} is moved into those intervals. The lower part of the figure shows the net increase of the number of correctly classified data for different c_{ijk}. Suppose c_{ijk} moves from the current position to the right. When c_{ijk} enters into the interval 3, one misclassification is resolved. If c_{ijk} enters into the interval 5, three misclassifications are resolved. But if c_{ijk} passes over $D_{ijk}(1)$, one misclassification occurs and the net increase of the correctly classified data becomes two. Thus in this example, if c_{ijk} is moved in the shaded interval, the number of correctly classified data increases by three.

In the following we discuss the determination of the upper bound $D_{ijk}(l)$ and the lower bound $I_{ijk}(l)$ that allow $l - 1$ new misclassifications, the determination of the intervals that resolve misclassification, and the determination of the center c_{ijk} that achieves the maximum recognition rate allowing at most $l_{M_c} - 1$ new misclassifications.

Upper and Lower Bounds of c_{ijk}. We calculate the upper bound $D_{ijk}(l)$ and the lower bound $I_{ijk}(l)$ of c_{ijk} in which $l - 1 (\ge 0)$ data are newly

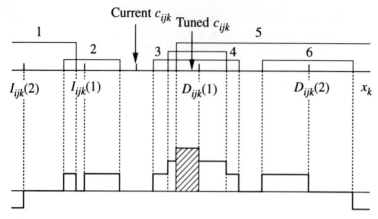

Fig. 7.13. Movement of c_{ijk}. If the current c_{ijk} is moved to the tuned c_{ijk} in the shaded interval, three misclassified data are correctly classified without causing new misclassification

misclassified. We divide a set of input data into X and Y, where X consists of the data correctly classified using the set of fuzzy rules $\{R_{ij}\}$ and Y consists of the misclassified data. Then, for $\mathbf{x}\,(\in X)$ that belongs to class i we choose cluster ij that satisfies

$$h_{ij}(\mathbf{x}) \le \min_{k \ne j} h_{ik}(\mathbf{x}). \tag{7.35}$$

If $\mathbf{x}$ further satisfies

$$h_{ij}(\mathbf{x}) < h_{op}(\mathbf{x}) = \min_{o \ne i, q = 1,\dots} h_{oq}(\mathbf{x}) < \min_{k \ne j} h_{ik}(\mathbf{x}), \tag{7.36}$$

there are a lower bound $I_{ijk}(\mathbf{x})$ and an upper bound $D_{ijk}(\mathbf{x})$ to keep $\mathbf{x}$ correctly classified (see Fig. 7.11 (a)).

To calculate these bounds, first we calculate the change in the tuned distance when c_{ijk} is moved. Among the one-dimensional tuned distances $h_{ijl}(\mathbf{x})\,(l = 1,\dots,m)$, only $h_{ijk}(\mathbf{x})$ changes. Let $h_{ij}^k(\mathbf{x})$ denote the tuned distance of the fuzzy rule R_{ij} for the input vector $\mathbf{x}$, when the center vector c_{ij} is moved in the kth input axis to c'_{ijk}. Then, the difference of the (squared) tuned distance $\Delta h_{ij}^{ks}(\mathbf{x}) = h_{ij}^{ks}(\mathbf{x}) - h_{ij}^s(\mathbf{x})$ is given by

$$\begin{aligned}
\Delta h_{ij}^{ks}(\mathbf{x}) &= h_{ij}^{ks}(\mathbf{x}) - h_{ij}^s(\mathbf{x}) \\
&= \frac{|c'_{ijk} - x_k|^s}{m\,\alpha_{ij}\,w_{ijk}^s} - \frac{|c_{ijk} - x_k|^s}{m\,\alpha_{ij}\,w_{ijk}^s}.
\end{aligned} \tag{7.37}$$

Thus

$$h_{ij}^{ks}(\mathbf{x}) - h_{ij}^s(\mathbf{x}) \ge -\frac{|c_{ijk} - x_k|^s}{m\,\alpha_{ij}\,w_{ijk}^s}, \tag{7.38}$$

where the equality holds for $c'_{ijk} = x_k$.

At the upper or lower bound,

$$h_{ij}^{ks}(\mathbf{x}) = \Delta h_{ij}^{ks}(\mathbf{x}) + h_{ij}^{s}(\mathbf{x})$$
$$= h_{op}^{s}(\mathbf{x}). \tag{7.39}$$

From (7.38) and (7.39),

$$h_{op}^{s}(\mathbf{x}) - h_{ij}^{s}(\mathbf{x}) \geq -\frac{|c_{ijk} - x_k|^s}{m\,\alpha_{ij}\,w_{ijk}^{s}}, \tag{7.40}$$

and from $h_{op}(\mathbf{x}) - h_{ij}(\mathbf{x}) > 0$, the upper and lower bounds always exist and are given by (see Fig. 7.11 (a))

$$D_{ijk}(\mathbf{x}) = x_k + \mathrm{sign}(h_{op}(\mathbf{x}) - h_{ij}(\mathbf{x}))a_k, \tag{7.41}$$
$$I_{ijk}(\mathbf{x}) = x_k - \mathrm{sign}(h_{op}(\mathbf{x}) - h_{ij}(\mathbf{x}))a_k, \tag{7.42}$$

where

$$\mathrm{sign}(x) = \begin{cases} 1 & \text{for } x \geq 0, \\ 0 & \text{for } x < 0, \end{cases} \tag{7.43}$$

$$a_k = \left(m\,\alpha_{ij}\,w_{ijk}^{s}\,(h_{op}^{s}(\mathbf{x}) - h_{ij}^{s}(\mathbf{x})) + |c_{ijk} - x_k|^s\right)^{1/s}. \tag{7.44}$$

If (7.36) is not satisfied, namely

$$h_{ij}(\mathbf{x}) < \min_{k \neq j} h_{ik}(\mathbf{x}) < \min_{o \neq i, p = 1, \dots} h_{op}(\mathbf{x}), \tag{7.45}$$

c_{ijk} can be moved without making $\mathbf{x}$ become misclassified.

Now we choose $\mathbf{x}\,(\in X)$ which belongs to class $o\,(\neq i)$. Let cluster op have the minimum tuned distance $h_{op}(\mathbf{x})$:

$$h_{op} = \min_{q} h_{oq}(\mathbf{x}). \tag{7.46}$$

Since the tuned distance $h_{ij}(\mathbf{x})$ is larger than $h_{op}(\mathbf{x})$, if (7.40) is satisfied, there are $D_{ijk}(\mathbf{x})$ given by (7.41) and $I_{ijk}(\mathbf{x})$ given by (7.42) (see Figs. 7.11 (b) and (c)).

Now the lower bound $I_{ijk}(1)$, which is defined as the lower bound that does not cause new misclassification, is

$$I_{ijk}(1) = \max_{\mathbf{x} \in X} I_{ijk}(\mathbf{x}). \tag{7.47}$$

To clarify the discussion, we assume that $I_{ijk}(\mathbf{x})$ is different for different $\mathbf{x}$. Then (7.47) is satisfied by one $\mathbf{x}$. Similarly, $I_{ijk}(2)$, which is defined as the lower bound that allows one new misclassification, is the second maximum among $I_{ijk}(\mathbf{x})$ and is given by

$$I_{ijk}(2) = \max_{\mathbf{x} \in X,\, I_{ijk}(\mathbf{x}) \neq I_{ijk}(1)} I_{ijk}(\mathbf{x}). \tag{7.48}$$

In general,

$$I_{ijk}(l) = \max_{\mathbf{x} \in X,\, I_{ijk}(\mathbf{x}) \neq I_{ijk}(1),\dots,I_{ijk}(l-1)} I_{ijk}(\mathbf{x}). \tag{7.49}$$

As we discussed earlier, when c_{ijk} is moved to $I_{ijk}(l)$, the data which are misclassified when c_{ijk} is moved to $I_{ijk}(o)\,(1 \leq o < l)$ may become correctly classified. Even when this happens, we do not consider these data in calculation of $I_{ijk}(l)$.

Now the upper bound $D_{ijk}(1)$, which is defined as the upper bound that does not cause any new misclassification, is

$$D_{ijk}(1) = \min_{\mathbf{x} \in X} D_{ijk}(\mathbf{x}). \tag{7.50}$$

Here we also assume that $D_{ijk}(\mathbf{x})$ is different for different $\mathbf{x}$. Then (7.50) is satisfied by one $\mathbf{x}$. Similarly, $D_{ijk}(2)$, which is defined as the upper bound that allows one new misclassification, is the second minimum among $D_{ijk}(\mathbf{x})$ and is given by

$$D_{ijk}(2) = \min_{\mathbf{x} \in X,\, D_{ijk}(\mathbf{x}) \neq D_{ijk}(1)} D_{ijk}(\mathbf{x}). \tag{7.51}$$

In general,

$$D_{ijk}(l) = \min_{\mathbf{x} \in X,\, D_{ijk}(\mathbf{x}) \neq D_{ijk}(1),\ldots,D_{ijk}(l-1)} D_{ijk}(\mathbf{x}). \tag{7.52}$$

As we discussed earlier, when c_{ijk} is moved to $D_{ijk}(l)$, the data which are misclassified when c_{ijk} is moved to $D_{ijk}(o)\,(1 \leq o < l)$ may become correctly classified. Even when this happens, we do not consider these data in calculation of $D_{ijk}(l)$.

Thus c_{ijk} is bounded by

$$\cdots < I_{ijk}(l) < I_{ijk}(l-1) < \cdots < I_{ijk}(1) < c_{ijk} < D_{ijk}(1)$$
$$< \cdots < D_{ijk}(l-1) < D_{ijk}(l) < \cdots. \tag{7.53}$$

If we change c_{ijk} in the range of $(I_{ijk}(1), D_{ijk}(1))$, there is no new misclassification. If we change c_{ijk} in the range of $[D_{ijk}(l-1), D_{ijk}(l))$, or $(I_{ijk}(l), I_{ijk}(l-1)]$, the $l-1\,(l \geq 2)$ correctly classified data are misclassified.

Resolution of Misclassification by Changing c_{ijk}. For $\mathbf{x}\,(\in Y)$ which is misclassified into class i or which belongs to class i but is misclassified into class $o\,(\neq i)$, we check whether it can be correctly classified by moving c_{ijk}. For the datum $\mathbf{x}$ that belongs to class i but that is misclassified into class o with the minimum tuned distance of $h_{op}(\mathbf{x})$, if (7.40) is satisfied, there is an interval $(I_{ijk}(\mathbf{x}), D_{ijk}(\mathbf{x}))$ in which $\mathbf{x}$ is correctly classified, where $I_{ijk}(\mathbf{x})$ and $D_{ijk}(\mathbf{x})$ are given by (7.41) and (7.42), respectively (see Figs. 7.12 (a) and (b)).

Let $\mathbf{x}$, which belongs to class o, be misclassified into class i. Then similar to the above discussions, we calculate the intervals $(-\infty, D_{ijk}(\mathbf{x}))$ and $(I_{ijk}(\mathbf{x}), \infty)$ that make $\mathbf{x}$ be correctly classified by moving c_{ijk}. First, the minimum tuned distance for class o should be the second minimum among n classes:

$$\min_{k} h_{ik}(\mathbf{x}) < h_{op}(\mathbf{x}) = \min_{q \neq i,\, r=1,\ldots} h_{qr}(\mathbf{x}). \tag{7.54}$$

Second, $h_{ij}(\mathbf{x})$ needs to be the minimum in class i, and the second minimum in class i is larger than the minimum tuned distance in class o:

$$h_{ij}(\mathbf{x}) < h_{op}(\mathbf{x}) < \min_{k \neq j} h_{ik}(\mathbf{x}). \tag{7.55}$$

Then, since $h_{op}(\mathbf{x}) > h_{ij}(\mathbf{x})$, $D_{ijk}(\mathbf{x})$ and $I_{ijk}(\mathbf{x})$ exist and are given by (7.41) and (7.42), respectively (see Fig. 7.12 (c)).

Calculation of the Net Increase of the Recognition Rate. To calculate the net increase of the recognition rate by moving c_{ijk}, we determine the stepwise function that shows the net increase/decrease of the number of correctly classified data when c_{ijk} is changed as shown in Fig. 7.13. New misclassification occurs when c_{ijk} is moved in the left direction to $I_{ijk}(1), \ldots, I_{ijk}(l_{M_c} - 1)$, or to $I_{ijk}(l_{M_c})$, and it moves in the right direction to $D_{ijk}(1), \ldots, D_{ijk}(l_{M_c} - 1)$, or $D_{ijk}(l_{M_c})$, and the value of the stepwise function decreases by 1 at these points. Since misclassification is resolved when c_{ijk} is moved in the right direction to $I_{ijk}(\mathbf{x})$ and misclassified again when moved to $D_{ijk}(\mathbf{x})$, and moved in the left direction to $D_{ijk}(\mathbf{x})$ and misclassified again when moved to $I_{ijk}(\mathbf{x})$, the stepwise function increases (or decreases) at these points. Thus to generate the stepwise function we need to find points at which the stepwise function increases or decreases in the increasing order in the interval $[c_{ijk}, D_{ijk}(l_{M_c})]$ and in the decreasing order in the interval $[I_{ijk}(l_{M_c}), c_{ijk}]$.

Now we sort, in the increasing order, $I_{ijk}(\mathbf{x})$, $D_{ijk}(\mathbf{x})$ and $D_{ijk}(l)$ that belong to the interval $[c_{ijk}, D_{ijk}(l_{M_c})]$ and define the sorted points as $\beta_{ijk}^u(o)$ $(o = 1, \ldots, o_{M_u})$ where o_{M_u} is the number of points included in $[c_{ijk}, D_{ijk}(l_{M_c})]$. Associated with $\beta_{ijk}^u(o)$, we define $\text{Inc}^u(o)$ as follows:

$$\text{Inc}^u(o) = \begin{cases} 1 & \text{for} \quad \beta_{ijk}^u(o) = I_{ijk}(\mathbf{x}), \\ -1 & \text{for} \quad \beta_{ijk}^u(o) = D_{ijk}(l) \text{ or } D_{ijk}(\mathbf{x}). \end{cases} \tag{7.56}$$

We sort, in the decreasing order, $D_{ijk}(\mathbf{x}), I_{ijk}(\mathbf{x})$, and $I_{ijk}(l)$ that belong to the interval $[I_{ijk}(l_{M_c}), c_{ijk})$ and define the sorted points as $\beta_{ijk}^l(o)$ $(o = 1, \ldots, o_{M_l})$ where o_{M_l} is the number of points included in $[I_{ijk}(l_{M_c}), c_{ijk})$. Associated with $\beta_{ijk}^l(o)$ we define $\text{Inc}^l(o)$ as follows:

$$\text{Inc}^l(o) = \begin{cases} 1 & \text{for} \quad \beta_{ijk}^l(o) = D_{ijk}(\mathbf{x}), \\ -1 & \text{for} \quad \beta_{ijk}^l(o) = I_{ijk}(l) \text{ or } I_{ijk}(\mathbf{x}). \end{cases} \tag{7.57}$$

Now we can determine the values of the stepwise function $f^u(o)$ at $\beta_{ijk}^u(o)$ as follows:

begin
    $f^u(1) = \text{Inc}^u(1)$
    $o = 1$
    while $o \leq o_{M_u}$ do
        $f^u(o) = f^u(o - 1) + \text{Inc}^u(o)$
    end
end

Similarly, we can determine the values of the stepwise function $f^l(o)$ at $\beta^l_{ijk}(o)$ as follows:

begin
$\quad f^l(1) = \text{Inc}^l(1)$
$\quad o = 1$
$\quad \text{while } o \leq o_{M_l} \text{ do}$
$\quad\quad f^l(o) = f^l(o - 1) + \text{Inc}^l(o)$
$\quad \text{end}$
end

Modification of c_{ijk}. For $f^u(o), o = 1, \ldots, o_{M_u}$, we find o^u that satisfies

$$f^u(o^u) = \max_{o=1,\ldots,o_{M_u}} f^u(o). \tag{7.58}$$

Similarly, for $f^l(o), l = 1, \ldots, o_{M_l}$, we find o^l that satisfies

$$f^l(o^l) = \max_{o=1,\ldots,o_{M_l}} f^l(o). \tag{7.59}$$

If there are plural o's that satisfy (7.58) or (7.59), we choose the smallest o. If $f^u(o^u) \geq f^l(o^l)$, we move c_{ijk} to the right. If we move c_{ijk} in the interval $(\beta^u_{ijk}(o^u), \beta^u_{ijk}(o^u + 1))$ the recognition rate is maximized. Thus we set c_{ijk} in $(\beta^u_{ijk}(o^u), \beta^u_{ijk}(o^u + 1))$ as follows:

$$c_{ijk} = \beta^u_{ijk}(o^u) + \delta^c(\beta^u_{ijk}(o^u + 1) - \beta^u_{ijk}(o^u)), \tag{7.60}$$

where δ^c satisfies $0 < \delta^c < 1$.

Likewise, if $f^u(o^u) < f^l(o^l)$, we move c_{ijk} to the left. If we move c_{ijk} in the interval $(\beta^l_{ijk}(o^l + 1), \beta^l_{ijk}(o^l))$ the recognition rate is maximized. Thus we set c_{ijk} in $(\beta^l_{ijk}(o^l + 1), \beta^l_{ijk}(o^l))$ as follows:

$$c_{ijk} = \beta^l_{ijk}(o^l) - \delta^c(\beta^l_{ijk}(o^l) - \beta^l_{ijk}(o^l + 1)). \tag{7.61}$$

According to the above discussion, the tuning of c_{ijk} becomes as follows.

1. Set a positive number to parameter l_{M_c}, where $l_{M_c} - 1$ is the maximum number of new misclassifications allowed for tuning c_{ijk}, set a value in $(0, 1)$ to δ^c in (7.60) and (7.61).
2. For c_{ijk} $(i = 1, \ldots, n, j = 1, \ldots, k = 1, \ldots, m)$, calculate $\beta^l_{ijk}(o), f^l(o)$, $\beta^u_{ijk}(o)$, and $f^u(o)$. Find the interval $(\beta^l_{ijk}(o^l + 1), \beta^l(o^l))$ or $(\beta^u(o^u), \beta^u_{ijk}(o^u + 1))$ that realizes the maximum recognition rate of the training data, and change c_{ijk} using (7.60) or (7.61).
3. Iterate Step 2 until there is no improvement in the recognition rate of the training data.

We call the update of all c_{ijk} one epoch of tuning centers. The above method is a local optimization method and guarantees a monotonic increase in the recognition rate during tuning. By allowing new misclassification during tuning, a relatively good local minimum can be obtained as shown in Section 7.4.

According to our experiments, the value of δ^c did not affect the recognition rate of the test data significantly, but a small value of δ^c sometimes gave a better recognition rate of the test data. Thus in the experiments in Section 7.4, we used 0.1.

7.2.3 Order of Tuning

Since the direct methods for tuning slopes and locations of the membership functions are the local optimization methods, classification performance may differ according to the order of tuning. One way to avoid this is to randomize the tuning order. For one epoch of tuning slopes or locations, instead of starting from the first rule and end in the last, we may shuffle the order. In addition to rule shuffling, we may shuffle the order of axes when we tune locations. Further, rule order and axis order may be shuffled, e.g. we may start tuning the location of the third axis of the second rule and then the location of the fifth axis of the fourth rule, etc.

7.3 Indirect Methods

Instead of maximizing discrete objective function given by (7.3), we can minimize its continuous counterparts:

1. For the pyramidal membership functions,

$$
\begin{aligned}
E &= \frac{1}{2} \sum_{k=1}^{M} \sum_{i} \sum_{j} \left(1 - h_{ij}^s(\mathbf{x}_k) - I(y_k = i)\right)^2 \\
&= \frac{1}{2} \sum_{k=1}^{M} \sum_{i} \sum_{j} \left(1 - \gamma_{ij} d_{ij}^s(\mathbf{x}_k) - I(y_k = i)\right)^2,
\end{aligned}
\tag{7.62}
$$

where $(\mathbf{x}_i, y_i)$ for $i = 1, \ldots, M$ are the M training data pairs, y_i is the class label for $\mathbf{x}_i$, $s = 1$ for the piecewise linear functions and $s = 2$ for the quadratic membership functions, $\gamma_{ij} = 1/\alpha_{ij}$, and

$$
I(a = b) = \begin{cases} 1 & \text{for } a = b, \\ 0 & \text{for } a \neq b. \end{cases}
\tag{7.63}
$$

2. For the bell-shaped membership functions,

$$
E = \frac{1}{2} \sum_{k=1}^{M} \sum_{i} \sum_{j} \left(\exp\left(-\gamma_{ij} d_{ij}^s(\mathbf{x}_k)\right) - I(y_k = i)\right)^2.
\tag{7.64}
$$

7.3.1 Tuning of Slopes Using the Least-squares Method

Since (7.62) is a square function of γ_{ij}, γ_{ij} can be determined by the least-squares method. Namely, (7.62) is minimized when the following equation holds:

$$\frac{\partial E}{\partial \gamma_{ij}} = \sum_{k=1}^{M} \sum_{i} \sum_{j} \left(1 - \gamma_{ij} d_{ij}^{s}(\mathbf{x}_k) - I(y_k = i)\right) d_{ij}^{s}(\mathbf{x}_k) = 0. \qquad (7.65)$$

Therefore the optimum value for γ_{ij} can be obtained by

$$\gamma_{ij} = \frac{\displaystyle\sum_{k=1}^{M} \sum_{i} \sum_{j} (1 - I(y_k = i)) \, d_{ij}^{s}(\mathbf{x}_k)}{\displaystyle\sum_{k=1}^{M} \sum_{i} \sum_{j} \left(d_{ij}^{2s}(\mathbf{x}_k)\right)}. \qquad (7.66)$$

Although we can determine γ_{ij} using (7.66) without any iteration, we must bear in mind that γ_{ij} determined by (7.66) is optimum in minimizing (7.62) not maximizing (7.3).

7.3.2 Tuning by the Steepest Descent Method

Similar to the back-propagation algorithm, we can tune slopes and centers by applying the steepest descent method to (7.62) or (7.64). We tune centers and slopes for each pair of data $(\mathbf{x}_k, y_k)$ that is misclassified. For tuning slopes, for a misclassified data pair $(\mathbf{x}_k, y_k)$, we modify γ_{ij} by

$$\gamma_{ij}^{new} = \gamma_{ij}^{old} - \alpha_s \frac{\partial E_k}{\partial \gamma_{ij}}, \qquad (7.67)$$

where α_s is a learning rate for tuning slopes and

$$E_k = \frac{1}{2} \sum_{i} \sum_{j} \left(1 - \gamma_{ij} d_{ij}^{s}(\mathbf{x}_k) - I(y_k = i)\right)^2. \qquad (7.68)$$

Thus,

$$\frac{\partial E_k}{\partial \gamma_{ij}} = -\left(1 - \gamma_{ij} d_{ij}^{s}(\mathbf{x}_k) - I(y_k = i)\right) d_{ij}^{s}(\mathbf{x}_k). \qquad (7.69)$$

Let $I(y_k = i) = 1$, namely, $\mathbf{x}_k$ belongs to class i. Then corrections should be made for γ_{ij} and $\gamma_{i'j'}$ that satisfy

$$\max_{j}(1 - \gamma_{ij} d_{ij}^{s}(\mathbf{x}_k)) \leq (1 - \gamma_{i'j'} d_{i'j'}^{s}(\mathbf{x}_k)). \qquad (7.70)$$

Similarly, for tuning centers, we modify $\mathbf{c}_{ij}$ by

$$\mathbf{c}_{ij}^{new} = \mathbf{c}_{ij}^{old} - \alpha_c \frac{\partial E_k}{\partial \mathbf{c}_{ij}}, \qquad (7.71)$$

where α_c is a learning rate for tuning centers, and

$$\frac{\partial E_k}{\partial \mathbf{c}_{ij}} = -s \left(1 - \gamma_{ij} d_{ij}^s(\mathbf{x}_k) - I(y_k = i)\right) d_{ij}^{s-1}(\mathbf{x}_k) \frac{\partial d_{ij}(\mathbf{x}_k)}{\partial \mathbf{c}_{ij}}. \tag{7.72}$$

Again, if $\mathbf{x}_k$ belongs to class i, corrections should be made for $\mathbf{c}_{ij}$ and $\mathbf{c}_{i'j'}$ that satisfy

$$\max_j \exp\left(-\gamma_{ij} d_{ij}^s(\mathbf{x}_k)\right) \leq \exp\left(-\gamma_{i'j'} d_{i'j'}^s(\mathbf{x}_k)\right). \tag{7.73}$$

Tuning is slow and ineffective compared to the direct methods that directly maximize (7.3) since by the steepest descent method, the corrections may not resolve misclassification or may incur other misclassifications and they are not checked in calculating corrections.

7.4 Performance Evaluation

The fuzzy classifier with pyramidal membership functions, the fuzzy classifier with polyhedral regions, and the fuzzy classifiers with ellipsoidal regions discussed in Sections 5.2.1, 5.2.2, and 5.2.3, respectively, were evaluated when fuzzy rules were tuned by the direct methods discussed in Section 7.2. For evaluation we used the data sets listed in Table 1.1 on page 19. (For the fuzzy classifier with polyhedral regions, we did not tune the membership functions and we used only the iris and thyroid data.) We used a Sun UltraSPARC IIi 333MHz workstation to run the optimized Fortran codes. The time listed in the following tables includes training time of the classifier and recognition of the training and test data by the classifier while training.

7.4.1 Performance Evaluation of the Fuzzy Classifier with Pyramidal Membership Functions

We evaluated performance of the fuzzy classifier with pyramidal membership functions. We used piecewise linear and quadratic membership functions, and we set the initial values of the tuning parameters α_{ij} at 1, the margin to control α_{ij} setting δ at 0.1, and the margin to control center setting δ^c at 0.1. Except for the thyroid data we set the minimum edge length ε at 0.01. For the thyroid data, since a smaller value of ε showed better recognition rates, we set ε at 1.0×10^{-5}. For the iris data and the numeral data, we set the maximum allowable number of misclassifications in tuning slopes plus 1, l_M, at 10 and the maximum allowable number of misclassifications in tuning centers plus 1, l_{M_c}, at 10.

The performance measures we used were the recognition rate of the test data, the number of generated rules and training time.

Iris Data. Table 7.1 shows the recognition rate of the test (training) data for
the initial and final epochs of tuning, when preclustering and postclustering
were not performed. In the first column, C and S denote that only the centers
and the slopes were tuned, respectively and SC denotes that both the slopes
and centers were tuned. Symbols L and Q denote that the piecewise linear
and quadratic membership functions were used, respectively.

With piecewise linear membership functions, when only the centers were
tuned, initially the recognition rate of the training data was 93.33% and it
was increased to 98.67%, while that of the test data remained the same. But
when the slopes and the centers were tuned, although the final recognition
rate of the training data was increased to 100%, the recognition rate of the
test data was decreased to 94.67%. Thus overfitting occurred. It also occurred
for tuning slopes.

With quadratic membership functions, when slopes and centers or centers
alone were tuned, the maximum recognition rates were obtained both for the
training and test data. When slopes alone were tuned, overfitting occurred.

The number of epochs of training was two to three and the training finished instantly.

Table 7.1. Performance for the iris data

Method	Initial (%)	Final (%)	Time (s)
SC & L	97.33 (93.33)	94.67 (100)	0.01
C & L	97.33 (93.33)	_97.33_ (98.67)	0.03
S & L	97.33 (93.33)	96.00 (96.00)	0.02
SC, C & Q	97.33 (93.33)	_97.33_ (100)	0.01
S & Q	97.33 (93.33)	96.00 (96.00)	0.01

Numeral Data. Table 7.2 shows the performance when preclustering and
postclustering were not performed. The classifier with quadratic membership
functions performed better than that with piecewise linear membership functions and tuning both slopes and centers performed better than or comparable
to tuning slopes or centers. The training was finished within a second.

For the iris and numeral data, when quadratic membership functions were
used, tuning slopes and centers performed better than tuning slopes or centers. Thus in the following study we tuned both centers and slopes. In addition, since quadratic membership functions performed better than piecewise
linear membership functions we only used quadratic membership functions.

Thyroid Data. Table 7.3 lists the initial and the final recognition rates
when l_M was set equal to l_{M_c} and were changed without clustering the data.

Table 7.2. Performance for the numeral data

Method	Initial (%)	Final (%)	Time (s)
L	99.15 (99.63)	99.27 (99.88)	0.3
SC, S & Q	99.51 (99.75)	<u>99.63</u> (100)	0.2
C & Q	99.51 (99.75)	99.51 (99.88)	0.3

The maximum recognition rate of 95.74% was obtained for the test data. The change of $l_M (= l_{M_c})$ did not have much effect on the recognition rates, but when $l_M = l_{M_c} = 60$, the best recognition rate of the training data was obtained. Thus, in the following study with quadratic membership functions, we set $l_M = l_{M_c} = 60$.

Table 7.3. Performance for the thyroid data changing l_M and l_{M_c} (with quadratic membership functions)

l_M, l_{M_c}	Initial (%)	Final (%)	Time (s)
10	21.73 (19.35)	94.60 (95.68)	7
20	21.73 (19.35)	94.08 (95.78)	4
30	21.73 (19.35)	93.90 (95.10)	5
40	21.73 (19.35)	95.19 (96.53)	8
50	21.73 (19.35)	95.27 (96.37)	6
60	21.73 (19.35)	<u>95.74</u> (<u>96.77</u>)	8
70	21.73 (19.35)	95.36 (96.24)	7

Table 7.4 shows the results when the training data were preclustered by the fuzzy min-max clustering algorithm discussed in Section 6.4. When the maximum average edge length θ was 0.4, the maximum recognition rate of 98.05% was obtained for the test data.

Table 7.5 lists the recognition rates when postclustered. When $N_c = 30$ and 40, 5 fuzzy rules were generated and the maximum recognition rate of 97.20% was obtained for the test data. For the thyroid data, the classification performance by preclustering was better than that by postclustering.

Blood Cell Data. Table 7.6 lists the recognition rates of the classifier when the training data were not divided and l_M and l_{M_c} were changed from 10 to 70. When l_M and l_{M_c} were 40, the maximum recognition rates were obtained both for the test and training data.

Table 7.4. Performance for the thyroid data by the fuzzy min-max clustering algorithm ($l_M = l_{M_c} = 60$)

θ	Initial (%)	Final (%)	Rules	Time (s)
—	21.73 (19.35)	95.74 (96.77)	3	8
0.7	32.32 (29.32)	97.84 (98.62)	4	7
0.6	44.87 (37.91)	97.52 (98.75)	6	14
0.5	82.88 (80.83)	94.08 (94.72)	8	8
0.4	90.26 (89.10)	98.05 (99.50)	14	17
0.3	92.53 (93.03)	97.52 (99.26)	27	40
0.2	93.96 (95.41)	97.11 (99.55)	49	51
0.1	95.27 (96.31)	96.09 (99.55)	105	111

Table 7.5. Performance for the thyroid data by postclustering ($L_M = L_{M_c} = 60$)

N_c	Rate (%)	Rules	Time (s)
—	95.74 (96.77)	3	8
40, 30	97.20 (98.78)	5	15
20	97.20 (98.81)	6	16
10	97.20 (98.97)	8	19

Table 7.6. Performance for the blood cell data changing l_M and l_{M_c} (with quadratic membership functions)

l_M, l_{M_c}	Initial (%)	Final (%)	Time (s)
10	78.87 (81.53)	86.00 (91.57)	10
20	78.87 (81.53)	86.81 (93.19)	11
30	78.87 (81.53)	88.13 (93.48)	12
40	78.87 (81.53)	88.52 (93.87)	12
50, 60, 70	78.87 (81.53)	87.61 (93.74)	12

Since the maximum recognition rate of the training data was obtained when $l_M = l_{M_c} = 40$, we set $l_M = l_{M_c} = 40$ and we preclustered the training data by the fuzzy min-max clustering algorithm and evaluated the performance changing θ. Table 7.7 shows the result. When $\theta = 0.21$, the maximum recognition rate of 90.03% was obtained for the test data. This was better than that without clustering by 1.5%.

Table 7.7. Performance for the blood cell data by the fuzzy min-max clustering algorithm ($l_M = l_{M_c} = 40$)

θ	Initial (%)	Final (%)	Rules	Time (s)
—	78.87 (81.53)	88.52 (93.87)	12	12
0.4	78.06 (80.79)	87.87 (93.57)	14	11
0.3	82.87 (87.12)	87.19 (94.51)	20	19
0.21	82.97 (87.34)	90.03 (95.87)	51	42
0.2	83.23 (87.89)	88.74 (95.83)	60	55

Table 7.8 lists the performance when postclustered. The maximum recognition rate of the test data was obtained when $N_c = 5$, which was slightly better than that without clustering and was lower than that with preclustering.

Table 7.8. Performance for the blood cell data by postclustering ($l_M = l_{M_c} = 40$)

θ	Rate (%)	Rules	Time (s)
—	88.52 (93.87)	12	12
20	88.55 (93.93)	14	16
10	88.81 (94.12)	17	18
5	88.97 (94.83)	55	36

Hiragana Data. We evaluated the performance of the classifier using the hiragana-50, hiragana-105, and hiragana-13 data. Since for the thyroid and blood cell data classification performance by preclustering was better than that by postclustering, we only evaluated performance by preclustering.

Table 7.9 lists the recognition rates of the hiragana-50 data when the training data were not divided and l_M and l_{M_c} were changed. When l_M and

Table 7.9. Performance for the hiragana-50 data changing l_M and l_{M_c} (with quadratic membership functions)

l_M, l_{M_c}	Initial (%)	Final (%)	Time (s)
5	79.35 (82.15)	<u>87.18</u> (<u>92.75</u>)	525
10	79.35 (82.15)	86.18 (92.41)	476
15	79.35 (82.15)	87.05 (92.54)	428
20	79.35 (82.15)	85.12 (91.80)	409

l_{M_c} were 5, the maximum recognition rates were obtained both for the test and training data.

Since the maximum recognition rate of the training data was obtained when $l_M = l_{M_c} = 5$, we set $l_M = l_{M_c} = 5$, preclustered the training data by the fuzzy min-max clustering algorithm, and evaluated the performance changing θ. Table 7.10 shows the result. When $\theta = 0.3$, the maximum recognition rate of 92.02% was obtained for the test data.

Table 7.10. Performance for the hiragana-50 data by the fuzzy min-max clustering algorithm ($l_M = l_{M_c} = 5$)

θ	Initial (%)	Final (%)	Rules	Time (s)
0.4	88.07 (92.08)	91.17 (97.94)	157	1076
0.3	91.41 (97.57)	<u>92.02</u> (99.24)	258	616
0.2	91.28 (99.33)	91.71 (<u>99.96</u>)	493	384

Table 7.11 lists the recognition rates of the hiragana-105 data when the training data were not divided and l_M and l_{M_c} were changed. When l_M and l_{M_c} were 30, the maximum recognition rates were obtained both for the test and training data.

Since the maximum recognition rate of the training data was obtained when $l_M = l_{M_c} = 30$, we set $l_M = l_{M_c} = 30$, preclustered the training data by the fuzzy min-max clustering algorithm, and evaluated the performance changing θ. Table 7.12 shows the result. When $\theta = 0.3$, the maximum recognition rate of 99.65% was obtained for the test data.

Table 7.13 lists the recognition rates of the hiragana-13 data when the training data were not divided and l_M and l_{M_c} were changed. When l_M and l_{M_c} were 10, the maximum recognition rates were obtained both for the test and training data.

Table 7.11. Performance for the hiragana-105 data changing l_M and l_{M_c} (with quadratic membership functions)

l_M, l_{M_c}	Initial (%)	Final (%)	Time (s)
10	86.61 (86.58)	94.89 (96.31)	3613
20	86.61 (86.58)	95.15 (96.69)	3284
30	86.61 (86.58)	95.44 (96.91)	3702
40	86.61 (86.58)	94.94 (96.42)	2840

Table 7.12. Performance for the hiragana-105 data by the fuzzy min-max clustering algorithm ($l_M = l_{M_c} = 30$)

θ	Initial (%)	Final (%)	Rules	Time (s)
0.5	97.42 (97.64)	98.18 (99.45)	185	4584
0.4	98.91 (99.33)	99.15 (99.87)	341	3633
0.3	99.63 (99.93)	99.65 (100)	667	665

Table 7.13. Performance for the hiragana-13 data changing l_M and l_{M_c} (with quadratic membership functions)

l_M, l_{M_c}	Initial (%)	Final (%)	Time (s)
5	75.14 (75.65)	89.09 (91.65)	237
10	75.14 (75.65)	91.48 (93.93)	260
15	75.14 (75.65)	91.10 (93.72)	276
20	75.14 (75.65)	90.71 (92.90)	232

Since the maximum recognition rate of the training data was obtained when $l_M = l_{M_c} = 10$, we set $l_M = l_{M_c} = 10$, preclustered the training data by the fuzzy min-max clustering algorithm, and evaluated the performance changing θ. Table 7.14 shows the result. When $\theta = 0.1$, the maximum recognition rate of 97.42% was obtained for the test data. A large number of fuzzy rules were generated.

Discussions. Although all the results were not included in the book, classification performance using quadratic membership functions was better than that using piecewise linear membership functions. When quadratic membership functions are used, the fuzzy classifier with pyramidal membership functions are equivalent to the fuzzy classifier with ellipsoidal regions when the

Table 7.14. Performance for the hiragana-13 data by the fuzzy min-max clustering algorithm ($l_M = l_{M_c} = 10$)

θ	Initial (%)	Final (%)	Rules	Time (s)
0.4	82.28 (82.64)	92.20 (94.56)	68	457
0.3	88.36 (89.10)	93.61 (96.50)	142	744
0.2	93.96 (95.18)	95.85 (98.78)	403	735
0.1	97.70 (99.47)	<u>97.42</u> (<u>99.98</u>)	1442	1153

diagonal covariance matrices are used as discussed in Section 4.2.7. The difference is the weight matrices that calculate the tuned distances. But except for the iris and numeral data, the former classifier required clustering of the training data, while the latter classifier required no clustering as will be shown in Section 7.4.3. The major reason is as follows: since the input variables are usually correlated, the principal axes of the training data distributions are not parallel to the input variables; thus while the covariance matrix well expresses the data distributions, the hyperboxes that are parallel to the input variables are not.

7.4.2 Performance Evaluation of the Fuzzy Classifier with Polyhedral Regions

We evaluated the performance of the fuzzy classifier with polyhedral regions for the iris and thyroid data sets listed in Table 1.1 on page 19. We did not tune the slopes of the membership functions. Thus, we set the tuning parameters α_{ij} at 1. In addition, we did not cluster the training data.

Since the thyroid data set consists of more than 3000 training data with 21 input variables, it took much time in generating and modifying facets. Thus we reduced the number of input variables by the backward ellipsoidal selection method discussed in Section 11.3, and we used the five input variables: the 3rd, 17th, 18th, 19th, and 21st variables (see Table 11.9 on page 232).

In addition, to reduce the number of facet modifications, we processed data that were far away from the class center with the high priority, and we imposed the upper bound on the number of facets generated:

$$N_f = m_{c,i}\,\beta \tag{7.74}$$

where $m_{c,i}$ is the dimension of the class i convex hull and β is a positive parameter.

Iris Data. Fig. 7.14 shows the recognition rate for the number of facets generated. The best recognition rates were achieved for the test and training data when all the training data were used for generating convex hulls. As listed in Table 7.15, the recognition rate of the test data was 96.00%.

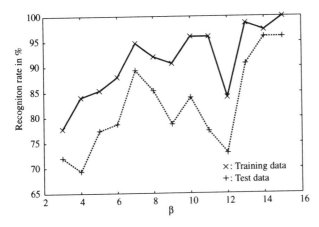

Fig. 7.14. Recognition rate of the iris data for the number of facets

Table 7.15. Recognition rates for the benchmark data sets

Data	All data used (%)	Maximum recognition rate (%)
Iris	96.00 (100)	96.00 (100)
Thyroid	93.67 (94.01)	98.07 (98.41)

Thyroid Data. Fig. 7.15 shows the number of Class 3 facets generated against the number of training data processed. As mentioned before, the data that were far away from the Class 3 center were processed with the high priority and Fig. 7.16 shows the recognition rate for the number of facets. When all the training data were used for generating the facets, the recognition rates both for the test and training data decreased. As listed in Table 7.15, the recognition rate of the test data was 93.67% when all the data were used and the maximum recognition rate was 98.07%.

Discussions. The improvement of the recognition rate of the training data was not monotonic for the increase of the training data processed. This might be caused by processing the data that were far away from the class center with the high priority. By processing the data that were far away from the center, the convex hull might be skewed and thus it deteriorated the classification performance.

For the thyroid data about 1000 facets were generated when all the Class 3 data were used for approximating the class region. But since the maximum recognition rate was achieved with a smaller number of training data, it may be important to select appropriate data for convex hull generation.

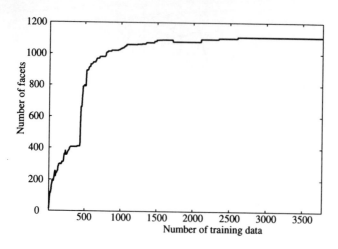

Fig. 7.15. The number of facets generated for the thyroid data

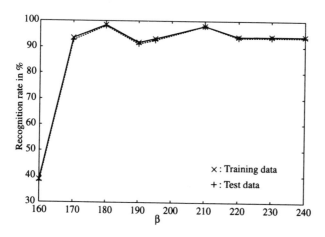

Fig. 7.16. Recognition rate of the thyroid data for the number of facets

In this study we did not tune the membership functions. Thus, there is a possibility for improving classification performance if the membership functions are tuned.

7.4.3 Performance Evaluation of the Fuzzy Classifier with Ellipsoidal Regions

We evaluated the performance of the fuzzy classifier with ellipsoidal regions for the data listed in Table 1.1 on page 19. We tuned the slopes and locations of the membership functions by the direct methods. Unless otherwise stated, we set the tuning parameters α_{ij} at 1, the margin control parameters δ and δ^c at 0.1, the maximum numbers of allowable misclassifications plus 1, l_M and l_{M_c}, at 10. Except for the thyroid and blood cell data, we did not cluster the data belonging to a class and the number of fuzzy rules generated was the number of classes. For the thyroid and blood cell data we postclustered the training data as discussed in Section 5.4.2.

Iris Data. Table 7.16 shows the classification performance when slopes (S) and centers (C) of the membership functions were tuned. The columns "Initial" and "Final" show the recognition rates of the test (training) data when the tuning parameters α_{ij} or the centers were not tuned and tuned, respectively. For the iris data, without tuning, the recognition rate of the test data was 98.67% (one datum misclassified). With one iteration of tuning slopes, all the training data were correctly classified but the recognition rate of the test data decreased to 97.33% (two data misclassified). Namely, overfitting occurred. But when the centers of the membership functions were tuned, the recognition rate of the training data reached 100% and that of the test data remained the same.

Table 7.16. Performance for the iris data

Method	Initial (%)	Final (%)	Time (s)
S	98.67 (98.67)	97.33 (100)	0.01
C	98.67 (98.67)	<u>98.67</u> (100)	0.02

Numeral Data. Table 7.17 lists the results for the numeral data. When the slopes of the membership function was tuned, the recognition rate of the test data was decreased from 99.63% to 99.39%. Thus overfitting occurred. But when the centers of the membership functions were tuned, the recognition rate of the training data reached 100% and that of the test data remained the same.

Table 7.17. Performance for the numeral data

Method	Initial (%)	Final (%)	Time (s)
S	99.63 (99.63)	99.39 (99.88)	0.8
C	99.63 (99.63)	<u>99.63</u> (100)	0.7

Thyroid Data. Table 7.18 shows the effect of l_M on the recognition rate without clustering. In the table, the "Initial" column lists the recognition rates of the test (training) data without tuning slopes of the membership functions, and the "Final" column lists the recognition rates of the test (training) data after tuning. For $l_M = 1$, the final recognition rate of the test data was 87.08% and there was little improvement by tuning, but when we increased l_M, the recognition rate was drastically improved. When l_M was equal to or larger than 10, the recognition rates were almost the same. Thus it was a good choice to set $l_M = 10$.

Table 7.18. Performance for the thyroid data when the slopes were tuned

l_M	Initial (%)	Final (%)	Epochs	Time (s)
1	86.41 (86.77)	87.08 (87.67)	2	3
5	86.41 (86.77)	94.37 (94.99)	8	9
10	86.41 (86.77)	95.60 (96.02)	7	8
15	86.41 (86.77)	95.62 (96.05)	5	6
20	86.41 (86.77)	95.60 (96.05)	5	6

Without clustering, the recognition rates for the test and training data were not so good when slopes were tuned. Thus we postclustered the training data. Table 7.19 lists the results when N_c was changed. Without postclustering, after tuning, 118 data belonging to Class 2 were misclassified into Class 3. Thus if N_c was set to be larger than 118, clusters were not generated.

For $N_c = 40$, two additional clusters were generated, and the recognition rates of both the test data and the training data were improved greatly and the improvement for N_c smaller than 40 was small.

Table 7.19. Performance for the thyroid data by postclustering

N_c	Rates (%)	Rules	Time (s)
—	95.60 (96.02)	3	8
40	97.00 (98.12)	5	13
30	97.26 (98.44)	8	28
20	97.17 (98.67)	10	42
10	97.29 (99.02)	13	60

Table 7.20 lists the performance for the thyroid data when the centers were tuned without clustering. When $l_{M_c} = 80$, the maximum recognition rate of 97.37% was achieved with 3 rules. This was better than that (97.29% with 13 rules) when the slopes were tuned and the training data were postclustered.

Table 7.20. Performance for the thyroid data when the centers were tuned (without clustering)

l_{M_c}	Initial (%)	Final (%)	Time (s)
20	86.41 (86.77)	96.62 (97.85)	40
40	86.41 (86.77)	97.20 (98.25)	48
60	86.41 (86.77)	96.85 (97.83)	49
80	86.41 (86.77)	97.37 (98.44)	61

Blood Cell Data. Table 7.21 shows the performance of the fuzzy classifier with ellipsoidal regions when the slopes of the membership functions were tuned for different l_M. The training data were not clustered; hence the number of fuzzy rules was 12. When l_M was larger than 5, the recognition rates of the test data were almost the same. Thus in this case also, it was reasonable to set $l_M = 10$.

Table 7.21. Performance for the blood cell data when the slopes were tuned

l_M	Initial (%)	Final (%)	Epochs	Time (s)
1	87.45 (92.64)	87.77 (93.15)	2	5
5	87.45 (92.64)	90.32 (95.00)	4	13
10	87.45 (92.64)	91.65 (95.41)	4	12
15	87.45 (92.64)	91.39 (95.41)	3	11
20	87.45 (92.64)	91.55 (95.38)	5	14

Table 7.22 shows the performance of the fuzzy classifier with ellipsoidal regions when the slopes of the membership functions were tuned and the training data were postclustered. When postclustered, the recognition rates of the test data were better than that without postclustering and the best recognition rate was achieved for $N_c = 25$, in which one additional cluster was generated.

Table 7.23 lists the performance for the blood cell data when the centers were tuned without clustering. The maximum recognition of the test data

Table 7.22. Performance for the blood cell data by postclustering

N_c	Rates (%)	Rules	Time (s)
—	91.65 (95.41)	12	13
25	<u>92.13</u> (95.96)	13	14
20	92.10 (96.32)	14	19
15	92.03 (96.29)	14	16
10	92.03 (96.29)	15	19
5	91.81 (<u>97.03</u>)	29	34

was 89.23%, which was lower than that (91.65%) when the slopes were tuned, although the recognition rates of the training data were almost the same.

Table 7.23. Performance for the blood cell data when the centers were tuned (without clustering)

l_{M_c}	Initial (%)	Final (%)	Time (s)
10	87.45 (92.64)	89.13 (95.38)	42
20	87.45 (92.64)	89.16 (<u>95.41</u>)	34
40	87.45 (92.64)	<u>89.23</u> (95.25)	33
60	87.45 (92.64)	<u>89.23</u> (95.25)	33

Hiragana Data. In the hiragana-50 and hiragana-105 data, there were classes that did not satisfy (5.36). Namely, some covariance matrices became singular. So we evaluated the improvement in the recognition rate when singular values were controlled in calculating the pseudo-inverse.

Table 7.24 shows the initial and final recognition rates of the test (training) data and the training time when the singular values smaller than η in (5.39) were set to zero and the slopes of the membership functions were tuned. When $\eta = 10^{-4}$, the recognition rate of the test data was the highest. Thus by setting the small nonzero singular values to zero, the generalization ability was improved.

In the hiragana-50 data, the number of classes that satisfied (5.36) was 9 and the average number of data was 28.8. Thus the average number of nonzero singular values was about 28. When $\eta = 10^{-7}$, the number of singular covariance matrices was 14 and the average number of nonzero singular values was 27.6. Thus when $\eta = 10^{-7}$, the nonzero singular values were assumed to be treated as nonzero in the calculation. When η was increased, the non-singular covariance matrices become singular and when $\eta = 10^{-4}$, all the covariance matrices became singular and the average number of nonzero singular values was 27.9.

Table 7.24. Performance for the hiragana-50 data when the slopes were tuned

η	Initial (%)	Final (%)	Time (s)
10^{-7}	69.39 (95.84)	80.20 (100)	338
10^{-6}	70.33 (95.88)	82.43 (100)	346
10^{-5}	71.82 (94.53)	87.96 (100)	608
10^{-4}	80.13 (91.65)	94.51 (99.89)	946
10^{-3}	84.64 (90.15)	93.97 (97.81)	1274

Table 7.25 shows the initial and final recognition rates of the test (training) data and the training time when the singular values smaller than η in (5.39) were set to zero and the centers of the membership functions were tuned. The maximum allowable number of misclassifications plus 1, l_{M_c}, was set to 20. When $\eta = 10^{-4}$ and 10^{-3}, the recognition rates were higher than the corresponding recognition rates when the slopes were tuned and the maximum recognition rate was 95.49%. Since the number of parameters for tuning centers was 50 times as large as that for tuning slopes, tuning was slower.

Table 7.25. Performance for the hiragana-50 data when the centers were tuned

η	Initial (%)	Final (%)	Time (s)
10^{-7}	69.39 (95.84)	76.49 (100)	480
10^{-6}	70.33 (95.88)	78.59 (100)	576
10^{-5}	71.82 (94.53)	83.47 (100)	925
10^{-4}	80.13 (91.65)	95.49 (99.91)	1710
10^{-3}	84.64 (90.15)	94.64 (99.31)	3783

Table 7.26 lists the initial and final recognition rates of the hiragana-105 test (training) data and the training time when the singular values smaller than η in (5.39) were set to zero and the slopes or the centers of the membership functions were tuned. For $\eta = 10^{-6}$, 10^{-5}, and 10^{-4}, the recognition rates of the training data reached 100% without tuning the slopes or centers. When $\eta = 10^{-3}$, the recognition rate of the test data was 99.99% by tuning

the slopes; thus only one datum was misclassified and a 100% recognition rate was obtained for the training data. The recognition rate of the test data when the centers were tuned was slightly lower than when the slopes were tuned.

Table 7.26. Performance for the hiragana-105 data

Method	η	Initial (%)	Final (%)	Time (s)
	10^{-6}	97.61 (100)	–	1113
S, C	10^{-5}	99.86 (100)	–	1119
	10^{-4}	99.98 (100)	–	1087
S	10^{-3}	99.94 (99.92)	99.99 (100)	2467
C	10^{-3}	99.94 (99.92)	99.96 (100)	3441

By calculating the central moments of hiragana-105 data, hiragana-13 data were generated. Table 7.27 lists the results when slopes or centers of the membership functions were tuned. When the slopes were tuned, without postclustering, the final recognition rate of the training data was 99.99% for $l_M = 1$ and larger. Namely, one datum failed to be correctly classified. Thus although the number of data was large, classification was not difficult. When we tuned the centers, we restricted the singular values. From Table 7.27, the maximum recognition rate of the test data of 99.19% was obtained when $\eta = 10^{-5}$. The computation time was 79 seconds. As seen from Tables 7.26 and 7.27 the recognition rate of the hiragana-13 test data was by 0.8% lower than that of the hiragana-105 test data. But the training time was exceedingly faster; only 79 seconds compared to 2467 seconds or 41 minutes.

Table 7.27. Performance for the hiragana-13 data

Method	η	Initial (%)	Final (%)	Rules	Time (s)
S	—	98.36 (99.84)	98.79 (99.99)	38	36
	10^{-7}	98.36 (99.84)	98.67 (100)	38	49
	10^{-6}	98.67 (99.84)	98.98 (100)	38	49
C	10^{-5}	99.16 (99.79)	99.19 (99.94)	38	79
	10^{-4}	98.56 (99.15)	98.66 (99.67)	38	176

Discussions. For blood cell data, performance improvement by postclustering was small. In addition, for the thyroid data, the recognition rate of the test data with tuning centers and without postclustering was better than that with tuning slopes and with postclustering. Thus, one cluster for one class was enough to realize high generalization ability. In addition, training was very fast when the number of input variables was not so large; but since the training time for the hiragana-105 data with 105 input variables was 41 minutes, the training time for a large-sized problem was still acceptable.

When the number of training data per class was smaller than that of the input features, the generalization ability could be improved by controlling the number of singular values for the covariance matrices. Controlling the number of singular values is equivalent to the principal component analysis (cf. Section 11.2.1) applied separately to the training data belonging to a class. PCA is a well-known dimensionality reduction method. But when it is applied to feature extraction, the principal components of the training data for all the classes are used instead of the principal components for each class. Thus the principal components do not necessarily reflect the separability of classes. But when PCA is applied separately to each class, the principal components become the principal features of a class as seen from the simulation results.

There are two cases when the covariance matrices become singular: 1) the training data are scarce to realize high generalization ability; and 2) the training data are sufficient to realize high generalization ability but are not sufficient to make the covariance matrices regular. The hiragana-50 data correspond to case 1). Namely, although the recognition rate of the training data could be 100%, that of the test data was much lower than that. The hiragana-105 data correspond to case 2). In this case by properly controlling the number of singular values, the recognition rate of the test data reached nearly 100%.

8. Robust Pattern Classification

In developing a classification system, we assume that the training data do not include outliers. But outliers may occur in many occasions and the detection of outliers is difficult especially for multi-dimensional data. If outliers are included, they affect classifiers' performance. For example, in generating the fuzzy min-max classifier (cf. Section 9.1), additional hyperboxes may be generated by outliers. Or in generating the fuzzy classifier with ellipsoidal regions by postclustering, the center and covariance matrices may be affected by outliers and thus this may further cause mal-tuning of the membership functions. Outliers may be excluded by preprocessing, but in this chapter we consider excluding outliers while generating fuzzy rules. We focus our discussions on a robust training method for a fuzzy classifier with ellipsoidal regions. First, we define a fuzzy rule for each class. Next, we determine the weight for each training datum by the two-stage method in order to suppress the effect of outliers. Then, using these weights, we calculate the center and covariance matrix of the ellipsoidal region for each class and tune the fuzzy rules. After tuning, to further improve generalization ability, we introduce interclass tuning parameters to tune fuzzy rules between two classes using the training data in the class boundary. We demonstrate the effectiveness of the above method using four benchmark data sets.

8.1 Why Robust Classification Is Necessary?

In the previous chapters, we assume that the training data do not include outliers. But in practice there are several occasions that outliers are included, for example, by malfunction of measuring devices, human errors, transmission errors, etc. To prevent inclusion of outliers in the training data, we check whether the data are within the physical range. This may be enough for single-input training data. But for the multi-dimensional data, this is not enough [74, pp. 1–9]. Consider the two-dimensional data distributed as shown in Fig. 8.1. Assume that these data belong to the same class. Then, the values of x_1 and x_2 for Data 1 and 2 are within the respective ranges of the remaining data. But according to the two-dimensional distribution of the data, Data 1 and 2 are considered to be outliers. Thus only checking the range of each input is not enough to prevent outliers from inclusion.

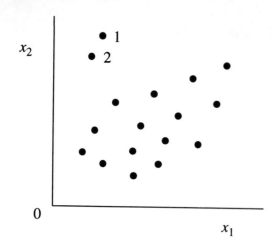

Fig. 8.1. An example of outliers in a two-dimensional input space

To overcome this problem several statistical methods have been developed [74, 75] and applied to neural networks and fuzzy systems [76, 77, 78, 79]. In [78], to realize a robust radial basis function neural network, the center of a hidden neuron is calculated by the median of selected training samples assigned to the hidden neuron and the ith diagonal element of the covariance matrix is calculated by the median absolute deviation given by

$$\sigma_i = \operatorname*{med}_j |x_{ij} - \operatorname*{med}_k x_{ik}|, \tag{8.1}$$

where x_{ij} is the ith input for the jth sample.

In this chapter we discuss robust pattern classification using the fuzzy classifier with ellipsoidal regions [79]. First we summarize the fuzzy classifier with ellipsoidal regions and discuss how outliers affect classification performance.

In Section 5.2.3 plural fuzzy rules are defined for one class, but since usually one rule is sufficient for one class, here we only define a rule for each class.

In classifying an input vector $\mathbf{x}$ into one of n classes, we define the following fuzzy rule for class i:

$$R_i : \text{If } \mathbf{x} \text{ is } \mathbf{c}_i \text{ then } \mathbf{x} \text{ belongs to class } i, \tag{8.2}$$

where $\mathbf{c}_i$ is the center of class i and is calculated by

$$\mathbf{c}_i = \frac{1}{|X_i|} \sum_{\mathbf{x} \in X_i} \mathbf{x}. \tag{8.3}$$

Here X_i is the set of training data belonging to class i and $|X_i|$ is the number of data in X_i.

The center c_i given by (8.3) is not robust in that a single outlier may move the center indefinitely. Rewriting (8.3) we obtain

$$c_i = \frac{x_1}{|X_i|} + \frac{1}{|X_i|} \sum_{x \in X_i - \{x_1\}} x. \tag{8.4}$$

Thus, if all the elements of x_1 approach infinity, the elements of c_i become larger without bounds. Since the estimation of the mean breaks down with one outlier among $|X_i|$ samples, we say the breakdown point is 0 assuming the infinite number of samples. If we take the median instead of the mean, the breakdown point is 0.5, meaning that the median does not break down even if 50% of samples are outliers.

The degree of membership of x for class i, $m_i(x)$, is given by

$$m_i(x) = \exp(-h_i^2(x)), \tag{8.5}$$

$$h_i^2(x) = \frac{d_i^2(x)}{\alpha_i}, \tag{8.6}$$

$$d_i^2(x) = (x - c_i)^t Q_i^{-1} (x - c_i), \tag{8.7}$$

where $h_i(x)$ is a tuned distance between x and c_i, $d_i(x)$ is a weighted distance between x and c_i, α_i is a tuning parameter for class i, and Q_i is a covariance matrix for class i and is given by

$$Q_i = \frac{1}{|X_i|} \sum_{x \in X_i} (x - c_i)(x - c_i)^t. \tag{8.8}$$

The covariance matrices Q_i given by (8.8) is also not robust against a single outlier.

To improve the recognition rate, we tune the tuning parameters α_i as discussed in Section 7.2.1. (In this section we do not consider tuning of the centers.) For an unknown input x, if the degree of membership $m_k(x)$ is the largest among m degrees of membership, x is classified into class k. Thus outliers, if included in the training data, may affect the values of the centers and the covariance matrices erroneously and classification performance may be deteriorated.

In statistical theory, robust techniques for estimating centers and covariance matrices are developed [74, pp. 248–272]. In Stahel and Donoho's method [74, pp. 256–265], using the deviations of the training data from the median, weights are calculated for the training data. Then using the weights, the center and the covariance matrix are calculated. This method is shown to have a breakdown point of 0.5, and cancels the effect of outliers. But since each training datum is assigned a weight, classification performance may be deteriorated if outliers are not included.

Since we do not know in advance whether outliers are included in the training data, we need to develop a robust method in which the generalization ability is not deteriorated even when there are no outliers. In the following we discuss the robust fuzzy classifier with ellipsoidal regions extending Stahel and Donoho's method and introducing the interclass parameters between two classes. Namely, first we determine the weights for the training data belonging to a class based on Stahel and Donoho's method and calculate the center and the covariance matrix. Then we classify the training data and select the training data around the class boundary. Next, using the data around the class boundary, we determine the threshold for each class, and determine the weights for the data that exceed the threshold. The center and the covariance matrix for each class are recalculated by the renewed weights. (In the tuning method discussed in Chapter 7, if outliers are included, classification performance is affected by them.) Then to suppress the effect of outliers, we tune the interclass parameters between two classes considering outliers.

In the following we discuss the robust training method and demonstrate the validity of the method by computer simulations.

8.2 Robust Classification

In the pattern classification method discussed in the previous section, the centers and the covariance matrices are calculated by (8.3) and (8.8), respectively, assuming that all the training data are correct. Since centers and covariance matrices are affected by outliers considerably, classification performance may be degraded by the inclusion of outliers. But for multi-dimensional inputs it is difficult to determine whether the training data include outliers or not in advance.

In this section we discuss a robust two-stage training method; in the first stage of training, a weight is calculated for each training datum using Stahel and Donoho's method and with these weights the centers and the covariance matrices are temporarily calculated. In the second stage of training, using the centers and covariance matrices calculated in the first stage, the training data are classified and the data that are near the class boundaries are determined. Using the data near the boundaries for each class a threshold for the weights are determined and the centers and covariance matrices are recalculated. In addition to the two-stage training, to further improve the recognition rate, we introduce interclass tuning parameters and tune these values between two classes using the data near the class boundary.

8.2.1 The First Stage

For each m-dimensional training datum $\mathbf{x}_k$ ($\in X_i, i = 1, \cdots, n$) belonging to class i, we calculate the deviation ratio u_k as follows:

$$
u_k = \max_{l=1,\cdots,m} \frac{\left| x_{kl} - \operatorname*{med}_{\mathbf{x}_b \in X_i} x_{bl} \right|}{\operatorname*{med}_{\mathbf{x}_a \in X_i} \left| x_{al} - \operatorname*{med}_{\mathbf{x}_b \in X_i} x_{bl} \right|},
\tag{8.9}
$$

where x_{kl} is the lth element of $\mathbf{x}_k$, $\underset{\mathbf{x}_b \in X_i}{\mathrm{med}}\ x_{bl}$ is the median of the lth elements of data belonging to class i. The numerator of (8.9) shows the absolute deviation of x_{lk} from the median of the lth elements of data belonging to class i. The denominator is the median of the absolute deviations for all the class i training data. Thus, by dividing the numerator by the denominator, we normalize the absolute deviation of x_{lk} from the median. Further by taking the maximum among the m normalized deviations, we pick up the salient normalized absolute deviation as the deviation ratio of $\mathbf{x}_k$.

Next, we determine the weight to each datum using the following monotonic function:

$$w_k^{(1)}(u_k) = \frac{1}{1 + u_k}. \tag{8.10}$$

Fig. 8.2 shows the weight function given by (8.10).

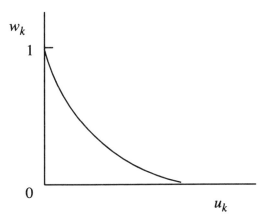

Fig. 8.2. The weight function by Stahel and Donoho's method

Then using the weights we calculate the temporary centers and covariance matrices, respectively, by

$$\mathbf{c}_i^{(1)} = \frac{\displaystyle\sum_{\mathbf{x}_j \in X_i} w_j^{(1)}(u_j)\,\mathbf{x}_j}{\displaystyle\sum_{\mathbf{x}_j \in X_i} w_j^{(1)}(u_j)}, \tag{8.11}$$

$$Q_i^{(1)} = \frac{\displaystyle\sum_{\mathbf{x}_j \in X_i} w_j^{(1)}(u_j)\,(\mathbf{x}_j - \mathbf{c}_i^{(1)})\,(\mathbf{x}_j - \mathbf{c}_i^{(1)})^t}{\displaystyle\sum_{\mathbf{x}_j \in X_i} w_j^{(1)}(u_j)}. \tag{8.12}$$

8.2.2 The Second Stage

Using the temporary centers given by (8.11) and the covariance matrices given by (8.12), we classify the training data with $\alpha_i = 1.0$ (namely, without tuning the tuning parameters), and select the training data that are near class boundaries. Let $\mathbf{x}_k$ belong to class i. Then for $\mathbf{x}_k$ we calculate the following value:

$$\gamma_k = \frac{D(2)_k - D(1)_k}{D(1)_k}, \tag{8.13}$$

where $D(1)_k$ is the minimum weighted distance among the weighted distances from $\mathbf{x}_k$ to the m centers and $D(2)_k$ is the second minimum weighted distance.

Let B_i be the set of class i data that are near the class boundary. Then we define $\mathbf{x}_k$ as belonging to B_i when $\gamma_k \leq 0.5$ and either of the associated classes of $D(1)_k$ and $D(2)_k$ is i.

In pattern classification, the data near class boundaries play an important role in determining the class boundaries. Thus in calculating the centers and the covariance matrices, we set the weights of important data to be 1. To do this, we set the threshold of deviation ratios as the maximum u_k in B_i:

$$\varepsilon_i = \max_{B_i \in \mathbf{x}_k} u_k. \tag{8.14}$$

Using the threshold ε_i, we recalculate the weights as follows:

$$w_k^{(2)}(u_k) = \begin{cases} 1 & \text{for } u_k \leq \varepsilon_i, \\ \dfrac{1}{1 + u_k - \varepsilon_i} & \text{otherwise.} \end{cases} \tag{8.15}$$

Fig. 8.3 shows the weight function given by (8.15). The difference from Fig. 8.2 is whether the threshold is introduced or not.

Then using the weights, we recalculate the centers and the covariance matrices:

$$c_i^{(2)} = \frac{\displaystyle\sum_{\mathbf{x}_j \in X_i} w_j^{(2)}(u_j)\,\mathbf{x}_j}{\displaystyle\sum_{\mathbf{x}_j \in X_i} w_j^{(2)}(u_j)}, \tag{8.16}$$

$$Q_i^{(2)} = \frac{\displaystyle\sum_{\mathbf{x}_j \in X_i} w_j^{(2)}(u_j)\,(\mathbf{x}_j - \mathbf{c}_i^{(2)})\,(\mathbf{x}_j - \mathbf{c}_i^{(2)})^t}{\displaystyle\sum_{\mathbf{x}_j \in X_i} w_j^{(2)}(u_j)}. \tag{8.17}$$

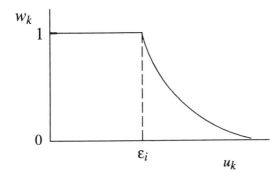

Fig. 8.3. The weight function with the threshold

When an input variable takes on 0 or 1, and the center is largely biased toward 0 or 1, we determine the center by

$$c_{ki}^{(2)} = \operatorname*{med}_{\mathbf{x} \in X_i} x_k. \tag{8.18}$$

Using the centers $\mathbf{c}_i^{(2)}$ and the covariance matrices $Q_i^{(2)}$, we tune α_i as discussed in Section 7.2.1.

8.2.3 Tuning Slopes near Class Boundaries

In tuning α_i, we do not consider the effect of outliers, thus outliers may deteriorate classification performance. In addition, if more than one class overlap, classification performance may be improved if class boundaries can be tuned between any two classes, separately.

Thus, to enhance classification performance and robustness, we introduce the interclass tuning parameters β_{ij} between classes i and j ($i < j$) in addition to the tuning parameters α_i. We define the tuned distance as follows:

$$h_i^2(\mathbf{x}) = \frac{d_i^2(\mathbf{x})}{\alpha_i + \beta_{ij}}. \tag{8.19}$$

In tuning α_i we set $\beta_{ij} = 0$. After tuning α_i, we tune β_{ij} using the data that are near the boundary of classes i and j.

Classification of unknown data is performed as follows. First we classify a datum $\mathbf{x}_k$ setting $\beta_{ij} = 0$ and calculate (8.13). If the datum satisfies $\gamma_k \leq 0.5$ and the associated classes for $D(1)_k$ and $D(2)_k$ are i and j ($i < j$), we recalculate the tuned distance for class i using (8.19) and classify the datum using the tuned distance.

Fig. 8.4 shows the concept of interclass tuning. Data 1 and 5 belong to Class 1 and Data 2, 3, and 4 belong to Class 2. Datum 1 is correctly classified into Class 1 but Data 2, 3, and 4 are misclassified into Class 1. Then if we

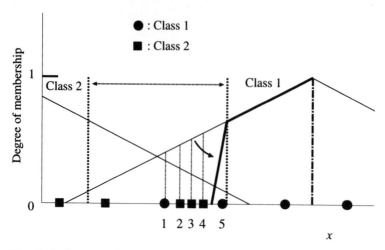

Fig. 8.4. Concept of interclass tuning

decrease the value of β_{12} so that the Class 1 membership function lies at the thick line, Data 2, 3, and 4 are correctly classified although Datum 1 is now misclassified. Then the number of correctly classified data is increased by two. In this way we tune the interclass tuning parameter β_{ij} allowing new misclassification.

In the following we discuss tuning of the interclass tuning parameter β_{ij} for classes i, j $(i < j)$ more in detail. After tuning α_i, we select the data $\mathbf{x}_k$ that belong to class i or j and that satisfy $\gamma_k \leq 0.5$ with one of the classes associated with $D(1)_k$ and $D(2)_k$ being i or j.

Let $U_{ij}^{(2)}(l)$ and $L_{ij}^{(2)}(l)$ be the upper and lower bounds that $l-1$ correctly classified data are misclassified, respectively. Then if the value of β_{ij} is set to the value in $(L_{ij}^{(2)}(2), L_{ij}^{(2)}(1)]$, one correctly classified datum is misclassified. By increasing or decreasing the value of β_{ij}, the correctly classified data may by misclassified and misclassified data may be correctly classified. Then let $\delta_{ij}(l)$ be smaller than $U_{ij}^{(2)}(l)$ and be the upper bound that makes misclassified data be correctly classified, and let $\theta_{ij}(l)$ be larger than $L_{ij}^{(2)}(l)$ and be the lower bound that makes misclassified data be correctly classified.

Next, we find out the interval that realizes the maximum recognition rate when the value of β_{ij} is set to the value in the interval among the intervals $(L_{ij}^{(2)}(l), \theta_{ij}(l))$ and $(\delta_{ij}(l), U_{ij}^{(2)}(l))$ $(l = 1, \cdots)$ and we set the center value of the interval to β_{ij}. In the following, we discuss how to calculate the upper bounds $U_{ij}^{(2)}(l)$, lower bounds $L_{ij}^{(2)}(l)$, and the range of β_{ij} that resolves misclassification.

8.2.4 Upper and Lower Bounds Determined by Correctly Classified Data

Here we calculate the upper bounds of β_{ij}, $U_{ij}^{(2)}(l)$, and the lower bounds of β_{ij}, $L_{ij}^{(2)}(l)$, allowing $l - 1$ new misclassifications. After tuning the tuning parameters α_i, we chose the training data that satisfy

$$\gamma_k = \frac{D(2)_k - D(1)_k}{D(1)_k} \leq 0.5, \tag{8.20}$$

where $D(1)_k$ and $D(2)_k$ are, respectively, the minimum and second minimum tuned distances. We generate the following sets using the data that satisfy (8.20).

1. Put the class i training data into the set:
   a) T_i if the associated class of $D(1)_k$ is class i and that of $D(2)_k$ is class j,
   b) F_i if the associated class of $D(1)_k$ is class j and that of $D(2)_k$ is class i.
2. Put the class j training data into the set:
   a) T_j if the associated class of $D(1)_k$ is class j and that of $D(2)_k$ is class i,
   b) F_j if the associated class of $D(1)_k$ is class i and that of $D(2)_k$ is class j.

Then we calculate the lower bound to keep $\mathbf{x} \in T_i$ correctly classified as follows:

$$L_{ij}^{(2)}(\mathbf{x}) = \frac{\alpha_i h_i^2(\mathbf{x})}{h_j^2(\mathbf{x})} - \alpha_i. \tag{8.21}$$

Then the lower bound $L_{ij}^{(2)}(1)$ that does not make new misclassification is given by

$$L_{ij}^{(2)}(1) = \max_{\mathbf{x} \in T_i} L_{ij}^{(2)}(\mathbf{x}). \tag{8.22}$$

To make discussions clear, we assume that $L_{ij}^{(2)}(\mathbf{x})$ are all different. Then the lower bound that allows one misclassification, $L_{ij}^{(2)}(2)$, is given by the second minimum of $L_{ij}^{(2)}(\mathbf{x})$:

$$L_{ij}^{(2)}(2) = \max_{\mathbf{x} \in T_i,\, L_{ij}^{(2)}(\mathbf{x}) \neq L_{ij}^{(2)}(1)} L_{ij}^{(2)}(\mathbf{x}). \tag{8.23}$$

In general,

$$L_{ij}^{(2)}(l) = \max_{\mathbf{x} \in T_i,\, L_{ij}^{(2)}(\mathbf{x}) \neq L_{ij}^{(2)}(1),\cdots,L_{ij}^{(2)}(l-1)} L_{ij}^{(2)}(\mathbf{x}). \tag{8.24}$$

Similarly, we can calculate the upper bounds $U_{ij}^{(2)}(l)$ using T_j. The upper bound $U_{ij}^{(2)}(\mathbf{x})$ that keeps $\mathbf{x} \in T_j$ correctly classified is given by

$$U_{ij}^{(2)}(\mathbf{x}) = \frac{\alpha_i h_i^2(\mathbf{x})}{h_j^2(\mathbf{x})} - \alpha_i. \qquad (8.25)$$

Thus the upper bound $U_{ij}^{(2)}(1)$ that does not make any new misclassification is given by

$$U_{ij}^{(2)}(1) = \min_{\mathbf{x} \in T_j} U_{ij}^{(2)}(\mathbf{x}). \qquad (8.26)$$

Similarly, the upper bound $U_{ij}^{(2)}(l)$ of β_{ij} that allows $l-1$ new misclassifications is given by

$$U_{ij}^{(2)}(l) = \min_{\mathbf{x} \in T_j,\, U_{ij}^{(2)}(\mathbf{x}) \neq U_{ij}^{(2)}(1),\cdots,U_{ij}^{(2)}(l-1)} U_{ij}^{(2)}(\mathbf{x}). \qquad (8.27)$$

8.2.5 Range of the Interclass Tuning Parameter that Resolves Misclassification

The class i data $\mathbf{x}$ ($\in F_i$) that are misclassified into class j are correctly classified if

$$\beta_{ij} > V_{ij}(\mathbf{x}) = \frac{\alpha_i h_i^2(\mathbf{x})}{h_j^2(\mathbf{x})} - \alpha_i. \qquad (8.28)$$

Let ucor(l) be the number of misclassified data that are correctly classified if β_{ij} is set in the interval $(U_{ij}^{(2)}(l-1), U_{ij}^{(2)}(l))$, and if $V_{ij}(\mathbf{x})$ is in $(U_{ij}^{(2)}(l-1), U_{ij}^{(2)}(l))$, add 1 to ucor($l$). In general $\delta_{ij}(l)$, which is smaller than $U_{ij}^{(2)}(l)$ and is the upper bound that makes misclassified data correctly classified, is given by

$$\delta_{ij}(l) = \max_{\mathbf{x} \in F_i,\, U_{ij}^{(2)}(l-1) < V_{ij}(\mathbf{x}) < U_{ij}^{(2)}(l)} V_{ij}(\mathbf{x}). \qquad (8.29)$$

Similarly, we can determine $\theta_{ij}(l)$ that is larger than $L_{ij}^{(2)}(l)$ and is the lower bound that makes misclassified data correctly classified. First the class j data ($\in F_j$) that are misclassified into class i are correctly classified if

$$\beta_{ij} < K_{ij}(\mathbf{x}) = \frac{\alpha_i h_i^2(\mathbf{x})}{h_j^2(\mathbf{x})} - \alpha_i. \qquad (8.30)$$

Let lcor(l) be the number of misclassified data that are correctly classified if β_{ij} is set in the interval $(L_{ij}^{(2)}(l), L_{ij}^{(2)}(l-1))$. If $K_{ij}(\mathbf{x})$ is in the interval $(L_{ij}^{(2)}(l), L_{ij}^{(2)}(l-1))$, we add 1 to lcor($l$). Then $\theta_{ij}(l)$ is given by

$$\theta_{ij}(l) = \min_{\mathbf{x} \in F_j,\, L_{ij}^{(2)}(l) < K_{ij}(\mathbf{x}) < L_{ij}^{(2)}(l-1)} K_{ij}(\mathbf{x}). \qquad (8.31)$$

Next, to determine at which interval the recognition rate is maximized, for ucor(l) and lcor(l) we search l's that satisfy

$$\max_l \left(\sum_{o=1}^{l} \text{ucor}(o) - o + 1 \right), \tag{8.32}$$

$$\max_l \left(\sum_{o=1}^{l} \text{lcor}(o) - o + 1 \right). \tag{8.33}$$

If there are plural l's that satisfy (8.32) or (8.33), we select the smallest l. When the value given by (8.32) is larger than that by (8.33), we tune β_{ij} as follows:

$$\beta_{ij} = \frac{\delta_{ij}(l) + U_{ij}^{(2)}(l)}{2}. \tag{8.34}$$

If the value given by (8.33) is larger than that by (8.32), we tune β_{ij} as follows:

$$\beta_{ij} = \frac{\theta_{ij}(l) + L_{ij}^{(2)}(l)}{2}. \tag{8.35}$$

If the values given by (8.32) and (8.33) are the same, we break the tie comparing the number of elements in F_i and F_j. Namely, if $|F_i| > |F_j|$, we use (8.34) and if $|F_i| < |F_j|$, we use (8.35).

Otherwise, for $|F_i| = |F_j|$ we set

$$\beta_{ij} = 0. \tag{8.36}$$

8.3 Performance Evaluation

We evaluated performance of the outlier detection methods using the blood cell data, the hiragana-13 data, the thyroid data, and the numeral data listed in Table 1.1 on page 19. First we compared performance of the methods for the four data sets when outliers were not included. Then we evaluated performance of the methods for the blood cell data and the hiragana-13 data when outliers were included. To generate outliers, we randomly selected some training data and then modified the randomly selected input features so that the values became 2 to 3 times larger than the average values of the features.

8.3.1 Classification Performance without Outliers

We compared performance of four methods when outliers were not included. Method 1 was the conventional fuzzy classifier with ellipsoidal regions in which the centers and the covariance matrices were calculated by (8.3) and (8.8), respectively. Method 2 was based on Stahel and Donoho's method in which the centers and the covariance matrices were calculated by (8.11) and (8.12), respectively, using the weights given by (8.10).

Method 3 was based on the two-stage learning in which the centers and the covariance matrices were calculated by (8.16) and (8.17), respectively, using the weights given by (8.15).

Method 4 was based on Method 3 plus interclass tuning.

Table 8.1 lists the recognition rates for the test and training data without outliers. The numerals in the brackets show the recognition rates for the training data. From the table, for the blood cell data classification performance of Method 3, which is based on the two-stage learning, was better than that of Method 2 based on Stahel and Donoho's method. Classification performance of Method 4 was comparable with that of Method 1 for the hiragana and numeral data and better than that of Method 1 for the blood cell and thyroid data. Thus, interclass tuning was effective for the blood cell and thyroid data.

Table 8.1. Comparison of recognition rates for the benchmark data without outliers [79, p. 437]

Data	Method 1 (%)	Method 2 (%)	Method 3 (%)	Method 4 (%)
Blood cell	91.65 (95.41)	89.32 (94.38)	91.29 (95.38)	<u>91.84</u> (<u>95.61</u>)
Hiragana-13	98.79 (99.99)	98.76 (99.99)	<u>98.80</u> (99.99)	<u>98.80</u> (99.99)
Thyroid	95.60 (96.02)	96.62 (<u>97.38</u>)	96.56 (97.11)	<u>96.73</u> (97.22)
Numeral	99.39 (99.75)	<u>99.51</u> (99.75)	<u>99.51</u> (<u>99.88</u>)	<u>99.51</u> (<u>99.88</u>)

To improve the recognition rate of the thyroid data, we applied the post-clustering technique discussed in Section 5.4.2 to Methods 1 and 4. Table 8.2 lists the recognition rates when the minimum number of training data to generate fuzzy rules, N_c, was changed. By Method 4 the maximum recognition rate of 97.72% was achieved for the test data compared with 97.29% by Method 1. The training time was 23 seconds compared with 60 seconds by Method 1 using a Sun UltraSPARC IIi 333MHz workstation. The shorter training time was due to the smaller number of fuzzy rules.

8.3.2 Classification Performance with Outliers

To evaluate the methods we added outliers to the blood cell data and the hiragana-13 data. First we randomly selected 0% to 10% data of the total training data and then from the selected data we randomly selected one feature and changed the value so that it became two to three times larger than the average value of the feature. Using the training data including outliers,

Table 8.2. Performance for the thyroid data changing N_c [79, p. 438]

N_c	Method 1 (%)	Rules	Method 4 (%)	Rules
—	95.60 (96.02)	3	96.73 (97.22)	3
40	97.00 (98.12)	5	96.91 (97.51)	4
30	97.26 (98.44)	8	97.49 (98.67)	6
20	97.17 (98.67)	10	<u>97.72</u> (98.75)	6
10	<u>97.29</u> (<u>99.02</u>)	13	97.58 (<u>99.20</u>)	8

we trained the classifier by the four methods and evaluated the recognition rates of the test data. The recognition rates were average recognition rates of 10 trials changing the random numbers.

Figs. 8.5 and 8.6 show the recognition rates of the blood cell test data and the hiragana-13 test data, respectively, when the classifier was trained using three methods. The recognition rates of Methods 2 and 3 were better than that of Method 1, and the recognition rate of Method 3 was better than that of Method 2. From Fig. 8.5, when the ratio of outliers was less than or around 2%, the recognition rates by Methods 1 and 3 were comparable. This might be due to the tuning method that allows new misclassification if the recognition rate is improved.

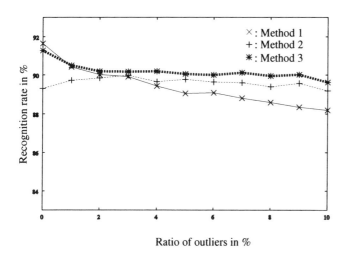

Fig. 8.5. Relation between the ratio of outliers and the recognition rate of the blood cell data (Outliers were included in one of the training inputs) [79, p. 438]

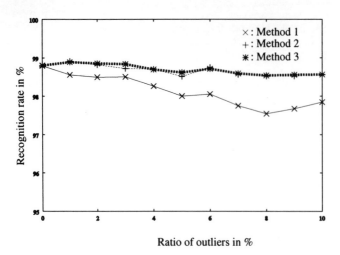

Fig. 8.6. Relation between the ratio of outliers and the recognition rate of the hiragana-13 data (Outliers were included in one of the training inputs) [79, p. 438]

Figs. 8.7 and 8.8 show the recognition rates for the blood cell test data and the hiragana-13 data, respectively, by Methods 3 and 4. Especially for the blood cell data, improvement in the recognition rates by interclass tuning was apparent. The reason that the improvement for the hiragana-13 data was small was that by Method 3 the recognition rates were sufficiently high and data for interclass tuning were scarce.

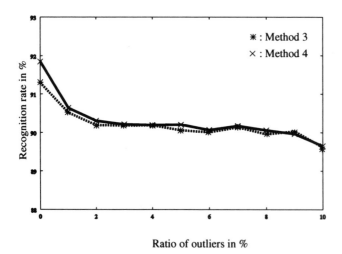

Fig. 8.7. Relation between the ratio of outliers and the recognition rate of the blood cell data (Outliers were included in one of the training inputs) [79, p. 439]

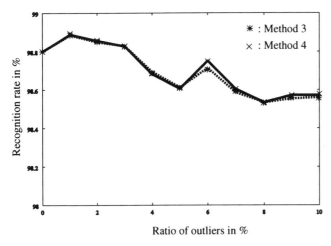

Fig. 8.8. Relation between the ratio of outliers and the recognition rate of the hiragana-13 data (Outliers were included in one of the training inputs) [79, p. 439]

In our next experiment, we randomly selected 0% to 10% data of the training data and modified all the training inputs. We tried 10 times changing the random numbers and compared the average recognition rates of the test data. Figs. 8.9 and 8.10 show, respectively, the recognition rates of the blood cell test data and the hiragana-13 test data for the classifier trained by Methods 1, 2, and 3. The effectiveness of Method 3 was evident; the recognition rates by Method 3, which was based on two-stage learning, were better than those by Methods 1 and 2.

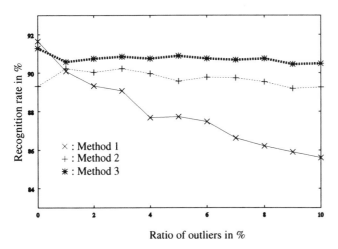

Fig. 8.9. Relation between the ratio of outliers and the recognition rate of the blood cell data (Outliers were included in all of the training inputs) [79, p. 439]

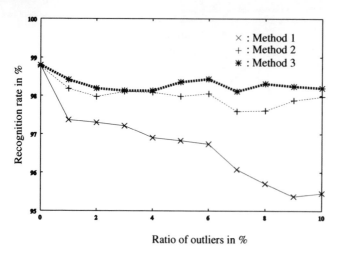

Fig. 8.10. Relation between the ratio of outliers and the recognition rate of the hiragana-13 data (Outliers were included in all of the training inputs) [79, p. 439]

Figs. 8.11 and 8.12 show, respectively, the recognition rates of the blood cell test data and the hiragana-13 test data for the classifier trained by Methods 3 and 4. The recognition rate of Method 4, which included interclass tuning in addition to Method 3, was better than that of Method 3.

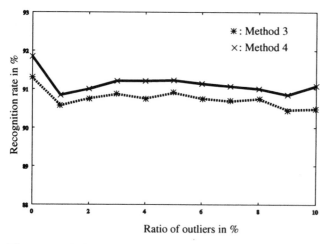

Fig. 8.11. Relation between the ratio of outliers and the recognition rate of the blood cell data (Outliers were included in all of the training inputs) [79, p. 439]

From Tables 8.1 and 8.2, and from Figs. 8.5 to 8.12, the recognition rates by Method 4 were comparable to or better than those by the conventional methods. Thus, the two-stage learning and interclass tuning were shown to

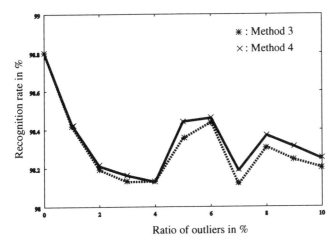

Fig. 8.12. Relation between the ratio of outliers and the recognition rate of the hiragana-13 data (Outliers were included in all of the training inputs) [79, p. 440]

be the robust training method irrespective of the inclusion of outliers in the training data.

9. Dynamic Fuzzy Rule Generation

In this chapter, we discuss two types of fuzzy classifiers with hyperbox regions: a fuzzy min-max classifier [80] and a fuzzy min-max classifier with inhibition [81].

These fuzzy classifiers have truncated rectangular pyramidal membership functions and are based on dynamic clustering: the fuzzy min-max classifier is based on incremental training and the fuzzy min-max classifier with inhibition is based on batch training.

In the fuzzy min-max classifier, hyperboxes for the same class are allowed to overlap but the hyperboxes for different classes are not allowed. Thus during incremental training, hyperboxes are generated, expanded, and contracted. In the fuzzy min-max classifier with inhibition, two types of hyperboxes are used: activation hyperboxes that allow the existence of data and the inhibition hyperboxes that do not. The classifiers' performance can be improved by tuning the membership functions associated with hyperboxes as discussed in Section 7.2.1 [23].

9.1 Fuzzy Min-max Classifiers

9.1.1 Concept

In the fuzzy min-max classifier, hyperboxes are generated, according to the existence of the training data in the input space, by a single scan of the training data. One type of hyperbox is used and hyperboxes belonging to the same class are allowed to overlap but hyperboxes belonging to different classes are not.

Using Fig. 9.1 we explain the training procedure. When a datum is inputted, a new hyperbox including only this datum is generated if there is no hyperbox of the same class within a specified distance (see Fig. 9.1 (a)). Otherwise, the hyperbox is expanded to include that datum (see Fig. 9.1 (b)). Then if the hyperbox is expanded, the overlap is checked between the expanded hyperbox and hyperboxes of different classes. If there exists an overlap, the hyperbox is contracted to resolve the overlap (see Fig. 9.1 (c)).

In the following we discuss the training procedure more in detail.

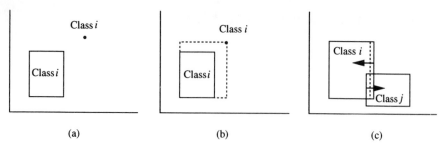

(a)   (b)   (c)

Fig. 9.1. Training process. **(a)** Generation of a new hyperbox which includes only a datum. **(b)** Expansion of an existing hyperbox. **(c)** Contraction of hyperboxes to resolve overlap

9.1.2 Approximation of Input Regions

Let class i $(i = 1, \ldots, n)$ regions with the m-dimensional input vector $\mathbf{x}$ be approximated by hyperbox A_{ij}:

$$A_{ij} = \{\mathbf{x} \mid v_{ijk} \leq x_k \leq V_{ijk}, k = 1, \ldots, m\} \qquad \text{for} \quad j = 1, \ldots, \tag{9.1}$$

where x_k is the kth element of $\mathbf{x}$ and v_{ijk} and V_{ijk} are, respectively, the minimum and maximum values of hyperbox A_{ij} with respect to x_k. Here we allow the overlap of A_{ij} and $A_{ij'}$ but do not allow the overlap of A_{ij} and $A_{i'j'}$ where $i \neq i'$. Namely,

$$A_{ij} \cap A_{i'j'} = \phi \qquad \text{for} \quad i \neq i'. \tag{9.2}$$

On the hyperbox A_{ij}, we can define either of the rectangular pyramidal membership functions and the truncated rectangular pyramidal membership functions discussed in Sections 4.2.2 and 4.2.3, respectively. Simpson used the truncated rectangular pyramidal membership functions with the average operator.

Namely, for the kth input, the one-dimensional membership function is given by

$$
m_{A_{ij}}(x_k) =
\begin{cases}
0 & \text{for} \quad x_k > V_{ij} + \frac{1}{\gamma}, \\
1 - \gamma(x_k - V_{ij})^s & \text{for} \quad V_{ij} + \frac{1}{\gamma} \geq x_k \geq V_{ij}, \\
1 & \text{for} \quad V_{ij} > x_k \geq v_{ij}, \\
1 - \gamma(v_{ij} - x_k)^s & \text{for} \quad v_{ij} > x_k \geq v_{ij} - \frac{1}{\gamma}, \\
0 & \text{for} \quad x_k < v_{ij} - \frac{1}{\gamma},
\end{cases}
\tag{9.3}
$$

where $s = 1$ for piecewise linear membership functions and $s = 2$ for quadratic membership functions. In [80], piecewise linear membership functions were used. Instead of α in (4.12) we use $\gamma = 1/\alpha$ where $\gamma \, (> 0)$ is a sensitivity parameter which controls the slope of the membership function; if we increase γ, the slope increases and if we decrease γ the slope decreases. Then the m-dimensional membership function with the average operator is given by

$$m_{A_{ij}}(\mathbf{x}) = \frac{1}{m} \sum_{k=1}^{m} m_{A_{ij}}(x_k). \tag{9.4}$$

We can define a fuzzy rule for the hyperbox A_{ij}:

If $\mathbf{x}$ is A_{ij} then $\mathbf{x}$ belongs to class i. $\tag{9.5}$

For the input vector $\mathbf{x}$ we calculate the degrees of membership of the if-parts of all the fuzzy rules using (9.4), and classify $\mathbf{x}$ into the class which has the maximum degree of membership.

9.1.3 Fuzzy Rule Extraction

Let X_i be the training data set for class i. Through one scan of the training data we extract fuzzy rules as follows.

- Expansion of hyperboxes: detect the nearest hyperbox A_{ij} from $\mathbf{x}\,(\in X_i)$ that is expandable. If there is no such hyperbox, define a new hyperbox.
- Overlap checking: check if there is an overlap between hyperboxes of different classes.
- Contraction of hyperboxes: contract hyperboxes which overlap each other.

In the following we discuss the procedure in more detail.

Expansion of Hyperboxes. For the training datum $\mathbf{x}$ belonging to class i, calculate the degrees of membership for hyperboxes A_{ik} $(k = 1, \ldots)$ using (9.4). Let A_{ij} have the maximum degree of membership. Now assume that the maximum size of a hyperbox, i.e., the maximum sum length of m edges emanating from a vertex, is bounded by $m\theta$ where θ is a positive parameter. Then if

$$m\theta \geq \sum_{k=1}^{m} (\max(V_{ijk}, x_k) - \min(v_{ijk}, x_k)), \tag{9.6}$$

hyperbox A_{ij} can be expanded. Now we change the minimum values of hyperbox A_{ij} by

$$v_{ijk} \leftarrow \min(v_{ijk}, x_k) \qquad \text{for} \quad k = 1, \ldots, m \tag{9.7}$$

and the maximum values by

$$V_{ijk} \leftarrow \max(V_{ijk}, x_k) \qquad \text{for} \quad k = 1, \ldots, m. \tag{9.8}$$

If there is no expandable hyperbox, we define a new hyperbox for class i:

$$v_{ijk} = V_{ijk} = x_k \qquad \text{for} \quad k = 1, \ldots, m. \tag{9.9}$$

Overlap Checking. If there is no overlap between the expanded or newly defined hyperbox A_{ij} and the hyperbox of a different class A_{kl}, for at least one input variable, e.g. for the oth input variable, there is no overlap between intervals $[v_{ijo}, V_{ijo}]$ and $[v_{klo}, V_{klo}]$. If two hyperboxes overlap we need to resolve their overlap for the input variable with the smallest overlapping interval. Here, we assume that each input variable is normalized in $[0, 1]$ and the initial maximum overlapping interval is $\delta^{old} = 1$. We check the following conditions for the oth ($o = 1, \ldots, m$) input variable and determine the overlapping interval δ^{new}.

1. For $v_{ijo} < v_{klo} < V_{ijo} < V_{klo}$

$$\delta^{new} = V_{ijo} - v_{klo}. \tag{9.10}$$

2. For $v_{klo} < v_{ijo} < V_{klo} < V_{ijo}$

$$\delta^{new} = V_{klo} - v_{ijo}. \tag{9.11}$$

3. For $v_{ijo} < v_{klo} < V_{klo} < V_{ijo}$, the edge of A_{ij} in the oth input variable includes that of A_{kl}. Thus the overlap can be resolved either by decreasing V_{ijo} as far as v_{klo} or increasing v_{ijo} as far as V_{klo}. Therefore the overlapping interval is given by

$$\delta^{new} = \min(V_{klo} - v_{ijo}, V_{ijo} - v_{klo}). \tag{9.12}$$

4. For $v_{klo} < v_{ijo} < V_{ijo} < V_{klo}$, similar to 3 the overlapping interval is given by

$$\delta^{new} = \min(V_{klo} - v_{ijo}, V_{ijo} - v_{klo}). \tag{9.13}$$

If none of these conditions is satisfied for the oth input variable, there is no overlap between two hyperboxes. Thus we stop calculations. If at least one of them is satisfied, and δ^{old} is smaller than or equal to δ^{new}, we iterate the above procedure for the $(o+1)$st input variable. If at least one of them is satisfied, and δ^{old} is larger than δ^{new}, we set $\delta^{old} = \delta^{new}$ and $\Delta = o$, where Δ denotes the input variable number with the minimum overlap, and we iterate the above procedure for the $(o + 1)$st input variable.

Contraction of Hyperboxes. The overlap of the hyperboxes A_{ij} and A_{kl} is resolved by resolving the overlap of intervals in the Δth input variable as follows.

1. For $v_{ij\Delta} < v_{kl\Delta} < V_{ij\Delta} < V_{kl\Delta}$

$$V_{ij\Delta} = v_{kl\Delta} \leftarrow \frac{V_{ij\Delta} + v_{kl\Delta}}{2}. \tag{9.14}$$

2. For $v_{kl\Delta} < v_{ij\Delta} < V_{kl\Delta} < V_{ij\Delta}$

$$V_{kl\Delta} = v_{ij\Delta} \leftarrow \frac{V_{kl\Delta} + v_{ij\Delta}}{2}. \tag{9.15}$$

3. For $v_{ij\Delta} < v_{kl\Delta} < V_{kl\Delta} < V_{ij\Delta}$

$$v_{ij\Delta} \leftarrow V_{kl\Delta} \qquad \text{for} \quad V_{kl\Delta} - v_{ij\Delta} \leq V_{ij\Delta} - v_{kl\Delta}, \tag{9.16}$$

$$V_{ij\Delta} \leftarrow v_{kl\Delta} \qquad \text{for} \quad V_{kl\Delta} - v_{ij\Delta} > V_{ij\Delta} - v_{kl\Delta}. \tag{9.17}$$

4. For $v_{kl\Delta} < v_{ij\Delta} < V_{ij\Delta} < V_{kl\Delta}$

$$V_{kl\Delta} \leftarrow v_{ij\Delta} \qquad \text{for} \quad V_{kl\Delta} - v_{ij\Delta} \leq V_{ij\Delta} - v_{kl\Delta}, \tag{9.18}$$

$$v_{kl\Delta} \leftarrow V_{ij\Delta} \qquad \text{for} \quad V_{kl\Delta} - v_{ij\Delta} > V_{ij\Delta} - v_{kl\Delta}. \tag{9.19}$$

The above method can create fuzzy rules by one scan of the training data. If the same data are not included in different classes, as we set a smaller value to θ, the degree of the overlap becomes smaller. Thus for a sufficiently small θ the recognition rate of the training data becomes 100%. But as we set a smaller value to θ, the number of rules increases and for $\theta = 0$, for each training datum a fuzzy rule is defined.

9.1.4 Performance Evaluation

We evaluated the performance of the fuzzy min-max classifier using the data sets listed in Table 1.1 on page 19. As the value of the sensitivity parameter γ becomes larger, the fuzzy rules become crisper. Therefore, for a large value of γ, there may be cases where a datum is not classified to any class because the degrees of membership for all classes are zero. Thus to avoid an unclassified situation, we set the sensitivity parameter $\gamma = 1$ because all the input ranges of the data sets used in our study were normalized to $[0, 1]$. We evaluated the performance changing θ. The optimized c code was run on a Sun UltraSPARC IIi 333MHz workstation. The time listed in the following tables includes the time for extracting fuzzy rules and evaluating the recognition rates of the training and test data.

Iris Data. Table 9.1 shows the recognition rates of the test (training) data, the numbers of rules generated for various θ, and the execution time. For $\theta = 0.17$, the maximum recognition rates were achieved both for the test and training data; for the test data two data among 75 data were misclassified. The rule extraction was finished in less than a second.

Numeral Data. Table 9.2 shows the results for the numeral data. For $\theta = 0.3$, 23 rules were generated and the maximum recognition rates were achieved for the test and training data; only 2 data among the 820 test data were misclassified.

Thyroid Data. Table 9.3 lists the performance for the thyroid data. For $\theta = 0.9$, three fuzzy rules were generated (one rule per one class) and the recognition rate of the test data was 99.15%. For $\theta = 0.7$ the maximum recognition rate of 99.42% was obtained for the test data with 20 fuzzy rules (about 7 rules per class). This was the best recognition rate among the classifiers evaluated in this book for the thyroid data as will be shown in Table 10.3 on page 200.

Table 9.1. Performance for the iris data

θ	Rates (%)	Rules	Time (s)
0.20	93.33 (98.67)	6	0.02
0.18	97.33 (98.67)	6	0.03
0.17	97.33 (100)	8	0.01
0.10	93.33 (100)	23	0.04

Table 9.2. Performance for the numeral data

θ	Rates (%)	Rules	Time (s)
0.5	98.90 (98.89)	10	0.4
0.4	99.63 (99.62)	12	0.5
0.3	99.76 (100)	23	0.7
0.2	99.76 (99.88)	68	1.7

Table 9.3. Performance for the thyroid data

θ	Rates (%)	Rules	Time (s)
0.9	99.15 (99.89)	3	2
0.8	99.15 (99.89)	4	2
0.7	99.42 (99.65)	20	5
0.6	99.18 (99.89)	43	10

Blood Cell Data. Table 9.4 lists the performance for the blood cell data. Some of the blood cell classes heavily overlapped and classification was difficult. In addition, the input variables such as a perimeter and an area size of a kernel were highly correlated. This was reflected to the classification results; the maximum recognition rate of 90.45% was achieved for the test data for $\theta = 0.08$ with an extremely large number of fuzzy rules (887 fuzzy rules; 74 rules per class or one rule for 3.5 data on average).

Hiragana Data. Table 9.5 lists the results for the hiragana-50 data. The maximum recognition rate of 94.64% was obtained for the test data when the recognition rate of the training data reached 100%. The number of fuzzy rules was 470; 12 fuzzy rules per class on average.

Table 9.4. Performance for the blood cell data

θ	Rates (%)	Rules	Time (s)
0.15	74.10 (74.68)	259	25
0.10	87.68 (96.64)	606	56
0.09	89.55 (98.39)	753	69
0.08	90.45 (99.52)	887	81
0.07	90.16 (99.97)	1077	98

Table 9.5. Performance for the hiragana-50 data

θ	Rates (%)	Rules	Time (s)
0.6	81.73 (90.30)	71	41
0.5	87.14 (96.29)	127	68
0.4	91.93 (99.02)	236	121
0.3	94.64 (100)	470	236

Table 9.6 lists the performance for the hiragana-105 data. The best recognition rate of 99.44% was obtained for $\theta = 0.4$ with 538 rules; 14 rules per class.

Table 9.6. Performance for the hiragana-105 data

θ	Rates (%)	Rules	Time (s)
0.6	98.43 (99.74)	175	339
0.5	99.08 (99.99)	297	565
0.4	99.44 (100)	538	1014

Table 9.7 lists the performance for the hiragana-13 data. When $\theta = 0.06$ the recognition rate of the training data reached 100% and the recognition rate of the test data was 98.22%. Quite a large number of fuzzy rules were generated; 5162 fuzzy rules (136 rules per class or one rule per 1.6 data on average).

The hiragana-13 data were generated by calculating the central moments of the hiragna-105 data; the feature reduction deteriorated generalization ability. This is explained as follows: by calculating the central moments of the original gray-level images, correlation of input variables increased and the hyperbox representation of class regions became inadequate. Thus a large number of fuzzy rules were necessary to realize a high recognition rate of the test data.

Table 9.7. Performance for the hiragana-13 data

θ	Rates (%)	Rules	Time (s)
0.3	67.09 (68.57)	502	122
0.2	80.68 (81.67)	981	231
0.1	96.76 (99.21)	3058	702
0.08	97.97 (99.95)	3956	910
0.06	<u>98.22</u> (100)	5162	1200

Discussions. The fuzzy min-max classifier has a simple architecture and can be trained by one scan of the training data.

The fuzzy min-max classifier showed good classification performance for the data sets used in the study, but for complicated or large sized problems a large number of fuzzy rules were necessary; e.g. 5162 fuzzy rules for the hiragana-13 data. The reason is that the hyperboxes are not suitable for representing the class regions with correlated input variables.

A rule explosion is prohibitive for practical use. But since training of the classifier is very fast, we can use the classifier for prototyping of the classification system.

9.2 Fuzzy Min-max Classifiers with Inhibition

9.2.1 Concept

Here we generate fuzzy rules using two types of hyperboxes: activation hyperboxes that allow existence of data and inhibition hyperboxes that do not allow existence of these data.

Now illustrate the procedure of fuzzy rule extraction using the two-dimensional data belonging to on of two classes as shown in Fig. 9.2. In the figure, an activation hyperbox, which allows the existence of data, is defined by calculating the minimum and maximum values of the data belonging to the same class in each input variable. If two hyperboxes overlap, we define the overlapping regions as inhibition hyperboxes. If there are data in the inhibition hyperbox, we recursively define activation hyperboxes and inhibition hyperboxes until there is no overlap between two classes.

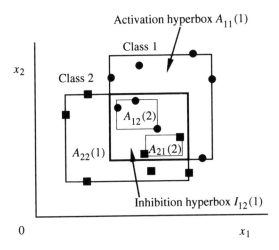

Fig. 9.2. Recursive fuzzy rule extraction

9.2.2 Fuzzy Rule Extraction

We generate fuzzy rules for classifying data with an m-dimensional input vector $\mathbf{x}$ into one of n classes. First assume that we have a training data set of input data X_i for class i, where $i = 1, \ldots, n$. Using X_i, we define an activation hyperbox of level 1, denoted as $A_{ii}(1)$, which is the maximum region of class i data:

$$A_{ii}(1) = \{\mathbf{x} \mid v_{iik}(1) \leq x_k \leq V_{iik}(1), k = 1, \ldots, m\}, \tag{9.20}$$

where x_k is the kth element of input vector $\mathbf{x} \in X_i$, $v_{iik}(1)$ is the minimum value of x_k, and $V_{iik}(1)$ is the maximum value of x_k.

If the activation hyperboxes $A_{ii}(1)$ and $A_{jj}(1)$ $(j \neq i, j = 1, \ldots, n)$ do not overlap, we obtain a fuzzy rule of level 1 for class i as follows:

$$\text{If } \mathbf{x} \text{ is } A_{ii}(1) \text{ then } \mathbf{x} \text{ belongs to class } i. \tag{9.21}$$

If the activation hyperboxes $A_{ii}(1)$ and $A_{jj}(1)$ overlap, we resolve the overlap recursively in which we define the overlapping region as the inhibition hyperbox of level 1 denoted as $I_{ij}(1)$:

$$I_{ij}(1) = \{\mathbf{x} \mid w_{ijk}(1) \leq x_k \leq W_{ijk}(1), k = 1, \ldots, m\}, \tag{9.22}$$

where $v_{iik}(1) \leq w_{ijk}(1) \leq W_{ijk}(1) \leq V_{iik}(1)$. The minimum and maximum values of inhibition hyperbox $I_{ij}(1)$ are given as follows.

1. For $v_{jjk}(1) \leq v_{iik}(1) \leq V_{jjk}(1) < V_{iik}(1)$

$$w_{ijk}(1) = v_{iik}(1), \quad W_{ijk}(1) = V_{jjk}(1). \tag{9.23}$$

2. For $v_{iik}(1) < v_{jjk}(1) \leq V_{iik}(1) \leq V_{jjk}(1)$

$$w_{ijk}(1) = v_{jjk}(1), \quad W_{ijk}(1) = V_{iik}(1). \tag{9.24}$$

3. For $v_{jjk}(1) \le v_{iik}(1) \le V_{iik}(1) \le V_{jjk}(1)$

$$w_{ijk}(1) = v_{iik}(1), \quad W_{ijk}(1) = V_{iik}(1). \tag{9.25}$$

4. For $v_{iik}(1) < v_{jjk}(1) \le V_{jjk}(1) < V_{iik}(1)$

$$W_{ijk}(1) = v_{jjk}(1), \quad W_{ijk}(1) = V_{jjk}(1). \tag{9.26}$$

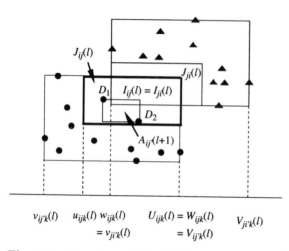

Fig. 9.3. Expansion of the inhibition hyperbox ($j' = i, i' = j$ for $l = 1$, and $j' = j, i' = i$ for $l \ne 2$) [81, p. 24 ©IEEE 1995]

However, the inhibition hyperbox defined in this way has a drawback, that is, data existing on the surface of the inhibition hyperbox may not be classified into either of the two classes. To overcome this problem, we expand the originally defined inhibition hyperbox $I_{ij}(1)$, associated with $A_{ii}(l)$ and $A_{jj}(l)$, in the way shown in Fig. 9.3. We denote the expanded inhibition hyperbox as $J_{ij}(1) = \{\mathbf{x} \mid u_{ijk} \le x_k \le U_{ijk}, k = 1, \ldots, m\}$. The expanded inhibition hyperboxes for $A_{ij}(1)$ and $A_{ji}(1)$ are $J_{ij}(1)$ and $J_{ji}(1)$, respectively, which are different. The expanded inhibition hyperbox $J_{ij}(1)$ is defined as follows.

1. For $v_{jjk}(1) \le v_{iik}(1) \le V_{jjk}(1) < V_{iik}(1)$

$$u_{ijk}(1) = v_{iik}(1),$$
$$U_{ijk}(1) = V_{jjk}(1) + \alpha(V_{iik}(1) - V_{jjk}(1)), \tag{9.27}$$

where $\alpha \, (> 0)$ is an expansion parameter.

2. For $v_{iik}(1) < v_{jjk}(1) \le V_{iik}(1) \le V_{jjk}(1)$

$$u_{ijk}(1) = v_{jjk}(1) - \alpha(v_{jjk}(1) - v_{iik}(1)),$$
$$U_{ijk}(1) = V_{iik}(1). \tag{9.28}$$

3. For $v_{jjk}(1) \leq v_{iik}(1) \leq V_{iik}(1) \leq V_{jjk}(1)$ In this case we do not expand the inhibition hyperbox for class i since we need not calculate the degree of membership for the x_k axis. Namely,

$$u_{ijk}(1) = v_{iik}(1),$$
$$U_{ijk}(1) = V_{iik}(1). \tag{9.29}$$

4. For $v_{iik}(1) < v_{jjk}(1) \leq V_{jjk}(1) < V_{iik}(1)$

$$u_{ijk}(1) = v_{jjk}(1) - \alpha(v_{jjk}(1) - v_{iik}(1)),$$
$$U_{ijk}(1) = V_{jjk}(1) + \alpha(V_{iik}(1) - V_{jjk}(1)). \tag{9.30}$$

Then we define a fuzzy rule of level 1 with inhibition by

If $\mathbf{x}$ is $A_{ii}(1)$ and $\mathbf{x}$ is not $J_{ij}(1)$ then $\mathbf{x}$ belongs to class i. $\qquad$ (9.31)

If $A_{ii}(1)$ is included in $A_{jj}(1)$, i.e., (9.25) holds for all $k, k = 1, \ldots, m$, $A_{ii}(1)$ coincides with $I_{ij}(1)$. In this case (9.31) is a void rule (i.e., it is not generated), since no $\mathbf{x}$ can satisfy (9.31). If some data belonging to X_i exist in $J_{ij}(1)$, we define the activation hyperbox of level 2 denoted as $A_{ij}(2)$ within the expanded inhibition hyperbox $J_{ij}(1)$ by calculating the minimum and maximum values of x_k based on the data in $J_{ij}(1)$:

$$A_{ij}(2) = \{\mathbf{x} \,|\, v_{ijk}(2) \leq x_k \leq V_{ijk}(2), k = 1, \ldots, m\}, \tag{9.32}$$

where $\mathbf{x} \in X_i$ and $\mathbf{x}$ is in $J_{ij}(1)$, $v_{ijk}(2)$ is the minimum value of x_k, $V_{ijk}(2)$ is the maximum value of x_k, and

$$u_{ijk}(1) \leq v_{ijk}(2) \leq x_k \leq V_{ijk}(2) \leq U_{ijk}(1). \tag{9.33}$$

If there is only one activation hyperbox of level 2 or there are two activation hyperboxes but they do not overlap, we define a fuzzy rule of level 2 for class i by

If $\mathbf{x}$ is $A_{ij}(2)$ then $\mathbf{x}$ belongs to class i. $\qquad$ (9.34)

If $A_{ij}(2)$ and $A_{ji}(2)$ overlap, the overlapping region of level 2 is denoted as $I_{ij}(2)$:

$$I_{ij}(2) = \{\mathbf{x} \,|\, w_{ijk}(2) \leq x_k \leq W_{ijk}(2), k = 1, \ldots, m\}, \tag{9.35}$$

where $v_{ijk}(2) \leq w_{ijk}(2) \leq W_{ijk}(2) \leq V_{ijk}(2)$.

Similar to what has been described for level 1, we define the expanded inhibition hyperbox $J_{ij}(2)$:

$$J_{ij}(2) = \{\mathbf{x} \,|\, u_{ijk}(2) \leq x_k \leq U_{ijk}(2), k = 1, \ldots, m\}, \tag{9.36}$$

where $u_{ijk}(2) \leq w_{ijk}(2) \leq W_{ijk}(2) \leq U_{ijk}(2)$.

Then we define a fuzzy rule of level 2 with inhibition:

If $\mathbf{x}$ is $A_{ij}(2)$ and $\mathbf{x}$ is not $J_{ij}(2)$ then $\mathbf{x}$ belongs to class i. $\qquad$ (9.37)

Fuzzy rules of levels higher than 2 can be defined in a similar manner if an overlap can be defined. In a general form, the fuzzy rule $r_{ij'}(l)$ of level $l (\geq 1)$ without inhibition can be expressed as follows:

If $\mathbf{x}$ is $A_{ij'}(l)$ then $\mathbf{x}$ belongs to class i, $\qquad$ (9.38)

where $j' = i$ for $l = 1$ and $j' = j$ for $l \geq 2$. Likewise, the fuzzy rule $r_{ij'}(l)$ of level l with inhibition can be expressed as follows:

If $\mathbf{x}$ is $A_{ij'}(l)$ and $\mathbf{x}$ is not $J_{ij}(l)$ then $\mathbf{x}$ belongs to class i. $\qquad$ (9.39)

The recursive process for defining fuzzy rules terminates when $A_{ij'}(l)$ and $A_{ji'}(l)$ do not overlap or $A_{ij'}(l) = A_{ji'}(l) = I_{ij}(l-1)$ holds. In the latter case, since the overlap cannot be resolved by the recursive process, instead of defining $A_{ij'}(l)$ and $A_{ji'}(l)$, for each datum of class i and/or j in $I_{ij}(l-1)$ we define an activation hyperbox which includes only that datum. We do not further define inhibition and activation hyperboxes of levels higher than l, because as long as no identical data exist in both classes i and j, no overlap exists between the activation hyperboxes of level l.

9.2.3 Fuzzy Rule Inference

Membership Function for Activation Hyperboxes. As the membership function for the activation hyperbox, we use the truncated rectangular pyramidal membership function with the minimum operator. Namely,

$$m_{A_{ij'}(l)}(\mathbf{x}) = \min_{k=1,\ldots,m} m_{A_{ij'}(l)}(x_k), \qquad (9.40)$$

$$m_{A_{ij'}(l)}(x_k) = \begin{cases} 0 & \text{for} \quad x > V_{ijk}(l), \\ 1 - \gamma(x_k - V_{ijk}(l)) & \text{for} \quad V_{ijk}(l) + \frac{1}{\gamma} \geq x \geq V_{ijk}(l), \\ 1 & \text{for} \quad V_{ijk}(l) > x_k \geq v_{ijk}(l), \\ 1 - \gamma(v_{ijk}(l) - x_k) & \text{for} \quad v_{ijk}(l) > x \geq v_{ijk}(l) - \frac{1}{\gamma}, \\ 0 & \text{for} \quad x_k < v_{ijk}(l) - \frac{1}{\gamma}, \end{cases} \qquad (9.41)$$

where γ is the sensitivity parameter. (If different variables are assigned to the sensitivity parameters of different membership functions, these parameters can be tuned as discussed in Section 7.2.1.)

The degree of membership of $\mathbf{x}$ for a fuzzy rule $r_{ij'}(l)$ given by (9.38) is

$$m_{r_{ij'}(l)}(\mathbf{x}) = m_{A_{ij'}(l)}(\mathbf{x}). \qquad (9.42)$$

Membership Function for Inhibition Hyperboxes. The degree of membership of $\mathbf{x}$ for a fuzzy rule given by (9.39) is 1 when $\mathbf{x}$ is in the activation hyperbox but not within the expanded inhibition hyperbox, i.e., $\mathbf{x}$ is in $\overline{A_{ij'}(l) - J_{ij}(l)}$, where $\bar{S}$ denotes the closed region of region S and $j' = i$ for $l = 1$ and $j' = j$ for $l > 1$. If $\mathbf{x}$ moves away from this region the degree of membership decreases. Namely, in this case we also define the membership function so that the contour surface is parallel to, and lies at an equal distance from the surface of $\overline{A_{ij'}(l) - J_{ij}(l)}$ as shown in Fig. 9.4. (If $A_{ij'}(l) = I_{ij}(l)$, i.e., if the rule is void, we do not calculate the degree of membership for this rule.) To realize this membership function we first define a region $H_{ij}(l)$ associated with $A_{ij'}(l)$ and $I_{ij}(l)$ as follows:

$$
\begin{aligned}
H_{ij}(l) = \{\mathbf{x}\,|\, x_k \leq U_{ijk}(l) \quad &\text{for} \quad v_{ji'k}(l) \leq v_{ij'k}(l) \leq V_{ji'k}(l) < V_{ij'k}(l), \\
x_k \geq u_{ijk}(l) \quad &\text{for} \quad v_{ij'k}(l) < v_{ji'k}(l) \leq V_{ij'k}(l) \leq V_{ji'k}(l), \\
-\infty < x_k < \infty \quad &\text{for} \quad v_{ji'k}(l) \leq v_{ij'k}(l) \leq V_{ij'k}(l) \leq V_{ji'k}(l), \\
u_{ijk}(l) \leq x_k \leq U_{ijk}(l) \quad &\text{for} \quad v_{ij'k}(l) < v_{ji'k}(l) \leq V_{ji'k}(l) < V_{ij'k}(l), \\
&\qquad\qquad k = 1, \ldots, m\}, \quad (9.43)
\end{aligned}
$$

where $j' = i$ and $i' = j$ for $l = 1$, $j' = j$ and $i' = i$ for $l \geq 2$, and $H_{ij}(l)$ and $H_{ji}(l)$ are in general different. According to the definition

$$H_{ij}(l) \supset J_{ij}(l). \qquad (9.44)$$

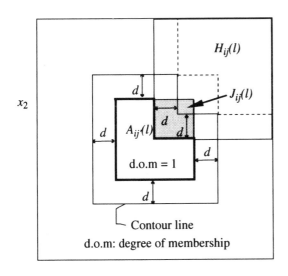

Fig. 9.4. The contour line of the membership function for the activation and inhibition hyperboxes [81, p. 21 ©IEEE 1995]

The region $H_{ij}(l)$ defines an input region where the expanded inhibition hyperbox affects the degree of membership of the rule given by (9.39). If $\mathbf{x} \notin H_{ij}(l)$, the degree of membership for a fuzzy rule $r_{ij'}(l)$ given by (9.39) is the same as (9.42). For $\mathbf{x} \in J_{ij}(l)$ the degree of membership $m_{J_{ij}(l)}(\mathbf{x})$ is given by [81]

$$m_{J_{ij}(l)}(\mathbf{x}) = \max_{k=1,\ldots,m} m_{J_{ij}(l)}(x_k), \tag{9.45}$$

where $m_{J_{ij}(l)}(x_k)$ is the degree of membership of x_k and is calculated as follows.

1. For $v_{ji'k}(l) \le v_{ij'k}(l) \le V_{ji'k}(l) < V_{ij'k}(l)$

$$m_{J_{ij}(l)}(x_k) = 1 - \max(0, \min(1, \gamma(U_{ijk}(l) - x_k))). \tag{9.46}$$

2. For $v_{ij'k}(l) < v_{ji'k}(l) \le V_{ij'k}(l) \le V_{ji'k}(l)$

$$m_{J_{ij}(l)}(x_k) = 1 - \max(0, \min(1, \gamma(x_k - u_{ijk}(l)))). \tag{9.47}$$

3. For $v_{ji'k}(l) \le v_{ij'k}(l) \le V_{ij'k}(l) \le V_{ji'k}(l)$

Since $x_k = v_{ij'k}(l)$ and $x_k = V_{ij'k}(l)$ do not constitute the surface of $\overline{A_{ij'}(l) - J_{ij}(l)}$, it is not necessary to define a membership function in the x_k axis. Thus we set

$$m_{J_{ij}(l)}(x_k) = 0. \tag{9.48}$$

Equation (9.48) holds for all k, where $k = 1,\ldots,m$, only when $A_{ji'} \supset A_{ij'} = I_{ij}(l)$, in other words, when the rule is a void rule. Thus, the x_k axis is ignored when calculating the degree of membership using (9.48) and (9.45).

4. For $v_{ij'k}(l) < v_{ji'k}(l) \le V_{ji'k}(l) < V_{ij'k}(l)$

$$m_{J_{ij}(l)}(x_k) = \begin{cases} 1 - \max(0, \min(1, \gamma(U_{ijk}(l) - x_k))) \\ \quad \text{for} \quad (u_{ijk}(l) + U_{ijk}(l))/2 \le x_k \le U_{ijk}(l), \\ 1 - \max(0, \min(1, \gamma(x_k(l) - u_{ijk}(l)))) \\ \quad \text{for} \quad u_{ijk}(l) \le x_k \le (u_{ijk}(l) + U_{ijk}(l))/2. \end{cases} \tag{9.49}$$

Then the degree of membership for $\mathbf{x} \in H_{ij}(l)$ and $\mathbf{x} \notin J_{ij}(l)$ is obtained by calculating both $m_{A_{ij'}(l)}(\mathbf{x})$ and $m_{J_{ij}(l)}(\mathbf{x})$, and taking the minimum, i.e., $\min(m_{A_{ij'}(l)}(\mathbf{x}), m_{J_{ij}(l)}(\mathbf{x}))$. Thus $m_{r_{ij'}(l)}(\mathbf{x})$ for (9.39) is given by

$$m_{r_{ij'}(l)}(\mathbf{x}) = \begin{cases} m_{A_{ij'}(l)}(\mathbf{x}) & \text{for} \quad \mathbf{x} \notin H_{ij}(l), \\ m_{J_{ij}(l)}(\mathbf{x}) & \text{for} \quad \mathbf{x} \notin J_{ij}(l), \\ \min(m_{A_{ij'}(l)}(\mathbf{x}), m_{J_{ij}(l)}(\mathbf{x})) & \text{for} \quad \mathbf{x} \in H_{ij}(l) \text{ and } \mathbf{x} \notin J_{ij}(l). \end{cases} \tag{9.50}$$

Since $m_{A_{ij'}(l)}(\mathbf{x}) = 1$ for $\mathbf{x} \in J_{ij}(l)$, (9.50) can be rewritten as follows:

$$m_{r_{ij'}(l)}(\mathbf{x}) = \begin{cases} m_{A_{ij'}(l)}(\mathbf{x}) & \text{for} \quad \mathbf{x} \notin H_{ij}(l), \\ \min(m_{A_{ij'}(l)}(\mathbf{x}), m_{J_{ij}(l)}(\mathbf{x})) & \text{for} \quad \mathbf{x} \in H_{ij}(l). \end{cases} \tag{9.51}$$

Classification. The degree of membership of $\mathbf{x}$ for a set of fuzzy rules $\{r_{ij}(l) \mid l = 1, \ldots\}$ denoted as $m_{r_{ij}}(\mathbf{x})$ is given by

$$m_{r_{ij}}(\mathbf{x}) = \max_{l=1,\ldots} m_{r_{ij}(l)}(\mathbf{x}) \qquad \text{for} \quad j = 1, \ldots, n, j \neq i. \tag{9.52}$$

We take the maximum because the activation hyperbox $A_{ij}(l+1)$, if it exists, is included in the expanded inhibition hyperbox $J_{ij}(l)$, and thus each fuzzy rule in $\{r_{ij}(l) \mid l = 1, \ldots\}$ is exclusive of the others.

Now the degree of membership of $\mathbf{x}$ for class i denoted as $m_i(\mathbf{x})$ is given by

$$m_i(\mathbf{x}) = \min_{\substack{j \neq i, j=1,\ldots,n \\ A_{ii}(1) \cap A_{jj}(1) \neq \phi}} m_{r_{ij}}(\mathbf{x}). \tag{9.53}$$

When the activation hyperbox of class i overlaps with those of classes j and k, we resolve the conflict, independently, first between classes i and j, then between classes i and k. This process is reflected by taking the minimum in (9.53). For example, if $m_{r_{ij}}(\mathbf{x}) = 1$ and $m_{r_{ik}}(\mathbf{x}) = 0$, this means that $\mathbf{x}$ is in the region inhibited by the inhibition hyperbox between classes i and k and thus $\mathbf{x}$ should not be classified into class i. Since the inference is performed by first taking the maximum and then the minimum of the membership functions, we call this inference min-max inference.

The input vector $\mathbf{x}$ is finally classified into class i if $m_i(\mathbf{x})$ is the maximum among $m_j(\mathbf{x})$, where $j = 1, \ldots, n$.

The fuzzy inference procedure described above can be represented by a neural network-like architecture as shown in Fig. 9.5 in which only the portion for class i is shown for simplicity. Different classes have different numbers of units for the second to fourth layers of the network and there is no connection among units of different classes. The second layer units consist of fuzzy rules and they calculate the degrees of membership based on the input vector $\mathbf{x}$. The third layer units take the maximum values of inputs from the second layer (cf. (9.52)), which are the degrees of membership generated by resolving overlaps between two classes. The number of third layer units for class i is determined by the number of classes that overlap with class i. Therefore, if there is no overlap between class i and any other classes, the network for class i reduces to two layers. The fourth layer unit for class i takes the minimum value among the maximum values (cf. (9.53)); each of them is associated with a two-class overlap. Therefore, if class i overlaps with only one class, the network for class i reduces to three layers; in other words, the "Min" node in the fourth layer is not required. Calculation of a minimum in the fourth layer resolves overlaps among more than two classes.

9.2.4 Performance Evaluation

We evaluated the performance of the fuzzy min-max classifier with inhibition using the data sets listed in Table 1.1 on page 19. As the value of the sensitivity parameter γ becomes larger, the fuzzy rules become crisper. Therefore,

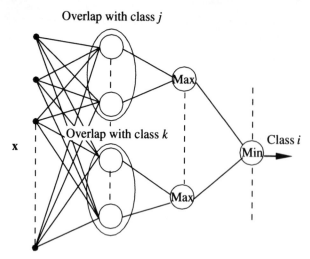

Fig. 9.5. Architecture of a fuzzy classifier (only the network for class i is shown) [81, p. 19 ©IEEE 1995]

for a large value of γ, there may be cases where a datum is not classified to any class because the degrees of membership for all classes are zero. Thus to avoid an unclassified situation, we set the sensitivity parameter $\gamma = 1$ because all the input ranges of the data sets used in our study were normalized to [0, 1]. The rule extraction time was measured by a Sun UltraSPARC IIi 333MHz.

Iris Data. Table 9.8 shows the numbers of rules generated and the recognition rates for the test (training) data for various values of the expansion parameter α. Since rule generation is continued until the overlaps between classes are resolved, a 100% recognition rate is obtained for the training data if there are no identical data in different classes. Thus there were no misclassifications of the training data and also for Class 1 of the test data. As α increased, the recognition rate of the test data increased, and when α was equal to 0.9, the recognition rate was 97.33%. Since the number of rules for Class 1 was always 1, that clearly indicated that no overlap existed between Classes 1 and 2, and between Classes 1 and 3. The rule extraction was finished in less than a second.

Numeral Data. Table 9.9 shows the results for the numeral data. For $\alpha = 0.001$ and 0.01, 11 rules were generated; two rules were generated for Class 7 and one rule was generated for the remaining classes. Thus each class was well separated and only 3 data among the 820 test data were misclassified.

Thyroid Data. Table 9.10 lists the performance for the thyroid data. Rule extraction was finished in a second for 3772 training data and the maximum recognition rate of 99.19% was achieved. Generalization ability of the fuzzy

Table 9.8. Performance for the iris data

α	Rates (%)	Rules	Time (s)
0.001	92.00 (100)	5	0.03
0.1	93.33 (100)	7	0.03
0.2	93.33 (100)	7	0.03
0.3	93.33 (100)	9	0.03
0.4	93.33 (100)	9	0.03
0.5	94.67 (100)	9	0.03
0.6	94.67 (100)	11	0.03
0.7	96.00 (100)	11	0.03
0.8	96.00 (100)	11	0.03
0.9	<u>97.33</u> (100)	17	0.03

Table 9.9. Performance for the numeral data

α	Rates (%)	Rules	Time (s)
0.001	99.63 (100)	11	0.1
0.01	99.63 (100)	11	0.1
0.1	99.63 (100)	13	0.1

min-max classifier with inhibition is high when correlation of the input variables is small. The thyroid data consisted of 21 input features; 6 continuous features and 15 discrete features. Thus the correlation among input variables was small and it led to the high generalization ability.

Table 9.10. Performance for the thyroid data

α	Rates (%)	Rules	Time (s)
0.001	<u>99.19</u> (100)	10	0.44
0.01	<u>99.19</u> (100)	10	0.42
0.1	98.84 (100)	13	0.46

Blood Cell Data. Table 9.11 lists the performance for the blood cell data. Some of the blood cell classes heavily overlapped and classification was difficult. In addition, the input variables such as a perimeter and an area size of a kernel were highly correlated. This was reflected to the classification results; although the recognition rate of the training data was 100%, the maximum recognition rate of the test data was 86.52% and 217 rules were generated. But training was finished in a second for 3097 training data.

Table 9.11. Performance for the blood cell data

α	Rates (%)	Rules	Time (s)
0.001	85.16 (100)	160	0.58
0.01	85.35 (100)	160	0.56
0.1	85.51 (100)	183	0.63
0.2	<u>86.52</u> (100)	217	0.66
0.3	86.26 (100)	252	0.76

Hiragana Data. Table 9.12 lists the results for the hiragana-50 data. The maximum recognition rate of 88.20% was obtained for the test data. The number of fuzzy rules was 620; 16 fuzzy rules per class on average. But training was finished in about 5 seconds.

Table 9.12. Performance for the hiragana-50 data

α	Rates (%)	Rules	Time (s)
0.001	85.47 (100)	230	4.5
0.01	85.49 (100)	232	4.5
0.1	86.33 (100)	283	4.6
0.2	86.79 (100)	343	4.7
0.3	87.66 (100)	424	4.9
0.4	87.87 (100)	520	5.2
0.5	<u>88.20</u> (100)	620	5.3
0.6	87.46 (100)	693	5.6

Tables 9.13 and 9.14 list the performance for the hiragana-105 and hiragana-13 data, respectively. For the hiragana-105 data, the best recognition rate of 96.33% was obtained for $\alpha = 0.9$ with 1049 rules; almost 30 rules per class. While for the hiragana-13 data the best recognition rate of

95.34% was obtained for $\alpha = 0.4$ and 0.5 with 1493 rules; almost 40 rules per class.

The recognition rates of hiragna-105 and hiragana-13 test data with $\alpha = 0.001$ were 94.52% with 190 rules and 91.52% with 601 rules, respectively. The hiragana-13 data were generated by calculating the central moments of the hiragna-105 data; the feature reduction deteriorated generalization ability. This is explained as follows: By calculating the central moments of the original gray-level images, correlation of input variables increased and the hyperbox representation of class regions became inadequate.

Table 9.13. Performance for the hiragana-105 data

α	Rates (%)	Rules	Time (s)
0.001	94.05 (100)	190	13
0.01	94.08 (100)	193	13
0.1	94.47 (100)	222	13
0.2	94.97 (100)	301	13
0.3	95.15 (100)	365	14
0.4	95.44 (100)	453	14
0.5	95.57 (100)	553	14
0.6	95.81 (100)	666	15
0.7	96.00 (100)	787	16
0.8	96.21 (100)	929	17
0.9	<u>96.33</u> (100)	1049	18

Table 9.14. Performance for the hiragana-13 data

α	Rates (%)	Rules	Time (s)
0.001	91.52 (100)	601	2.3
0.01	91.69 (100)	610	2.3
0.1	93.88 (100)	775	2.5
0.2	94.65 (100)	977	2.8
0.3	95.10 (100)	1216	3.2
0.4	<u>95.34</u> (100)	1493	3.6
0.5	<u>95.34</u> (100)	1493	3.6

Discussions. Since fuzzy rules are generated until overlaps between classes are resolved, a 100% recognition rate is obtained for the training data. But generalization ability depends on the characteristics of the data; if the correlation of input variables is low, the high generalization ability is achieved as seen from performance for the thyroid data. But if the correlation is high, generalization ability is not so good as seen from performance for the blood cell data. When the correlation of input variables is high, by increasing the expansion parameter α, generalization ability is improved. But as α is increased the number of fuzzy rules increases and for the hiragana-105 data, the number of generated rules of 190 for $\alpha = 0.001$ was increased to 1049 for $\alpha = 0.9$; a 5.5 times increase in the number of rules for a 2.3% increase in the recognition rate.

For the fuzzy min-max classifier with inhibition, feature extraction is not a good strategy since by feature extraction usually correlation is increased and thus it leads to decreasing generalization ability as seen from performance for the hiragana-105 and hiragana-13 data. To improve generalization ability, preclustering of training data [82] and tuning of membership functions [23] are considered. In tuning membership functions we need tuning data in addition to training data since all the training data are within the generated hyperboxes. In [23] the least-squares method is used but the direct method for tuning slopes discussed in Section 7.2.1 are also applicable.

Since fuzzy rules are generated by calculating the minimum and maximum of training inputs, training is extremely fast. For the hiragana-105 with 105 inputs, 38 outputs, and 8375 training data, training was finished in 20 seconds.

10. Comparison of Classifier Performance

In this chapter we compare performance of the classifiers discussed in this book, using the data sets listed in Table 1.1 on page 19. The classifiers compared in this chapter are the k-nearest neighbor classifier (k-NN), the three-layer neural network classifiers trained by BP and SI, the fuzzy classifiers with pyramidal membership functions (PYD) and ellipsoidal regions (ELD), the fuzzy min-max classifier (MM), the fuzzy min-max classifier with inhibition (MMI), and the support vector machine (SVM). The performance data used in this chapter were summarized from the previous chapters and Appendix A.2.

10.1 Evaluation Conditions

We compare the best recognition rates of the test data for the classifiers and the associated numbers of rules and training time. In the following tables, we list the recognition rates of the test data as well as those of the training data in the brackets.

In estimating the classifier's performance, the cross-validation technique such as the leave-one-out method is widely used. The problems in using this technique are as follows.

1. The evaluation time is prohibitively long especially for a real world data set with large number of samples.
2. The classifier's performance depends on the parameter values used. However, there is no systematic way of determining the optimal values. In addition, they change as the classification problems change.

Without setting the optimal values for each classifier, performance comparison is useless. In developing a classification system, high generalization ability is an important factor in selecting one from several candidate classifiers. Thus in this study, we compared the best recognition rates of the test data set that were obtained by changing parameter values of the classifiers.

For classifiers other than fuzzy classifiers, there is no concept of rules. Thus, the number of rules is meaningless for these classifiers. Therefore, we show, as the number of rules, some appropriate value or index that indicates the complexity of the classifier. For k-nearest neighbor classifiers we count the number of templates (training data) as the number of fuzzy rules, and for neural networks the number of hidden neurons. The support vector machine uses n separating hyperplanes in classifying n classes. Since the number of separating hyperplanes is fixed for a given problem, we use the kernel function to indicate the complexity.

In addition, the complexity of fuzzy rules differs as fuzzy classifiers change. For instance the ellipsoidal fuzzy rule is more complex than the fuzzy rule with the pyramidal membership function since the former uses the covariance matrix. Thus the smallest number of fuzzy rules does not necessarily mean that the corresponding classifier has the simplest structure.

The training time was measured by a Sun UltraSPARC IIi 333MHz workstation. The time listed in the following tables has different meaning as classifiers change. For k-nearest neighbor classifiers, the time included the time for classifying the test data, and for the fuzzy classifiers except for the fuzzy min-max classifier with inhibition the time included the time for training the classifier and classifying the training and test data. For the fuzzy min-max classifier with inhibition rule extraction time was measured. Thus a direct comparison of time is meaningless.

10.2 Iris Data

Table 10.1 lists the best recognition rates for the test data and the associated recognition rates for the training data, the numbers of rules, and training time. The support vector machine used the polynomials with the degree of four as the kernel function. The numbers of misclassifications for the 75 test data ranged from one to four. Thus, performance of the classifiers was comparable, but the 1-nearest neighbor classifier showed the worst classification performance and the support vector machines showed the second worst.

10.3 Numeral Data

Table 10.2 lists the best recognition rates of the test data and the associated recognition rates of the training data, the numbers of rules, and training time. The support vector machine used the polynomials with the degree of three as the kernel function. The numbers of misclassifications of the 820 test data were two, three, and five. Thus performance of the classifiers was comparable. Training by BP was the slowest and that of the support vector machine was the second slowest but it was finished within a minute.

10.4 Thyroid Data

Table 10.3 lists the best recognition rates of the test data and the associated recognition rates of the training data, the numbers of rules, and training

Table 10.1. Performance comparison for the iris data

Classifier	Rates (%)	Rules	Time (s)
1-NN	94.67 (100)	75	0.02
BP	<u>98.67</u> (100)	3	0.85
SI	97.33 (100)	2	0.06
PYD	97.33 (100)	3	0.01
MM	97.33 (100)	8	0.01
MMI	97.33 (100)	17	0.03
ELD	<u>98.67</u> (100)	3	0.02
SVM	96.00 (100)	Poly 4	0.11

Table 10.2. Performance comparison for the numeral data

Classifier	Rates (%)	Rules	Time (s)
1-NN	99.51 (100)	810	0.4
BP	99.63 (100)	6	37
SI	99.39 (100)	11	5
PYD	99.63 (100)	10	0.2
MM	<u>99.76</u> (100)	23	0.7
MMI	99.63 (100)	11	0.1
ELD	99.63 (100)	10	0.7
SVM	<u>99.76</u> (100)	Poly 3	24

time. The support vector machine used the polynomials with the degree of one as the kernel function. The fuzzy min-max classifier (MM) showed the best recognition rate of the test data and the fuzzy min-max classifier with inhibition (MMI) showed the second best. The 5-nearest neighbor classifier (5-NN) showed the worst recognition rate. The recognition rates of the neural network classifier trained by SI, the fuzzy classifier with ellipsoidal regions, and the support vector machine were not so good. The reason why the performance of the fuzzy classifier with ellipsoidal regions was not so good was that the thyroid data included 16 discrete input variables among 21 input variables and the classifier is not suited for discrete input variables. This defect was alleviated by using the robust classification technique as discussed in Chapter 8 (97.72%) or transforming the input variables into continuous variables as discusses in Section 11.2.3 (98.07%).

Training by BP and of SVM took more than 10 minutes while the remaining classifiers finished training in less than two minutes.

Table 10.3. Performance comparison for the thyroid data

Classifier	Rates (%)	Rules	Time (s)
5-NN	93.67 (94.96)	3772	10
BP	97.93 (99.15)	3	642
SI	97.43 (98.44)	2	124
PYD	98.05 (99.50)	14	17
MM	99.42 (99.65)	20	5
MMI	99.19 (100)	10	0.4
ELD	97.37 (98.44)	3	61
SVM	97.43 (98.67)	Poly 1	609

10.5 Blood Cell Data

Table 10.4 lists the best recognition rates of the test data and the associated recognition rates of the training data, the numbers of rules, and training time. The support vector machine used the radial basis function network with $\gamma = 0.01$ as the kernel function. The support vector machine (SVM) and the fuzzy classifier with ellipsoidal regions (ELD) showed the best and the second best performance, respectively, and the former classifier took 46 minutes for training but the latter took only 14 seconds. Training by BP was the slowest; 67 minutes.

The fuzzy min-max classifier (MM) needed a large number of fuzzy rules; 3.5 data per class, and the fuzzy min-max classifier with inhibition (MMI) showed the worst recognition rate of the test data. Since blood cell classification was a very difficult classification problem and the input variables were correlated, approximation of the class regions by hyperboxes was not suited.

10.6 Hiragana Data

Table 10.5 lists the best recognition rates of the hiragana-50 test data and the associated recognition rates of the training data, the numbers of rules, and training time. The support vector machine used the radial basis function neural network with $\gamma = 0.1$ as the kernel function. The support vector

Table 10.4. Performance comparison for the blood cell data

Classifier	Rates (%)	Rules	Time (s)
5-NN	90.13 (93.51)	3097	6
BP	91.42 (95.61)	18	4030
SI	91.65 (93.80)	12	135
PYD	90.03 (95.87)	51	42
MM	90.45 (99.52)	887	81
MMI	86.52 (100)	217	1
ELD	92.13 (95.96)	13	14
SVM	<u>92.19</u> (100)	RBF 0.01	2756

machine showed the best recognition rate of the test data and the 1-nearest neighbor classifier showed the second best. The reason why the support vector machine performed well and other classifiers except for the nearest neighbor classifier did not was as follows. For the hiragana-50 data, there were no overlaps between classes because of a small number of data per class. The support vector machine sets the optimal hyperplane so that the margin is maximized, while other classifiers do not have this mechanism. For instance, if there are no overlap between classes, the fuzzy classifier with ellipsoidal regions does not tune the tuning parameters or locations of the centers.

Training by BP took 7 hours and the support vector machine took 2 hours for training, while the fuzzy classifier with ellipsoidal regions took 29 minutes. A large number of fuzzy rules were generated for the fuzzy classifiers with hyperbox regions (PYD, MM, and MMI) due to the correlation of the input variables.

Table 10.6 lists the best recognition rates of the hiragana-105 test data and the associated recognition rates of the training data, the numbers of rules, and training time. The support vector machine used the polynomials with the degree of two as the kernel function. The support vector machine showed the best recognition rate of the test data, and the fuzzy classifier with ellipsoidal regions and the 3-nearest neighbor classifier followed. The support vector machine took 3 hours for training, the fuzzy classifier with ellipsoidal regions took 41 minutes, and the 3-nearest neighbor classifier took 6 minutes for classification. Training by BP was the slowest (32 hours).

A large number of fuzzy rules were generated for the fuzzy classifiers with hyperbox regions (PYD, MM, and MMI) although the recognition rate of the fuzzy classifier with pyramidal membership functions was comparable with the best performance.

Table 10.7 lists the best recognition rates of the hiragana-13 test data and the associated recognition rates of the training data, the numbers of rules,

Table 10.5. Performance comparison for the hiragana-50 data

Classifier	Rates (%)	Rules	Time (s)
1-NN	97.16 (100)	4610	39
BP	95.77 (98.92)	25	24865
SI	95.12 (98.22)	35	2800
PYD	92.02 (99.24)	258	616
MM	94.64 (100)	470	236
MMI	88.20 (100)	620	5
ELD	95.49 (99.91)	39	1710
SVM	<u>99.07</u> (100)	RBF 0.1	7144

Table 10.6. Performance comparison for the hiragana-105 data

Classifier	Rates (%)	Rules	Time (s)
3-NN	99.99 (99.84)	8375	363
BP	98.47 (99.52)	25	113970
PYD	99.65 (100)	667	665
MM	99.44 (100)	538	1014
MMI	96.33 (100)	1049	18
ELD	99.99 (100)	38	2467
SVM	<u>100</u> (100)	Poly 2	10443

and training time. The support vector machine used the radial basis function neural network with $\gamma = 1$ as the kernel function. The support vector machine showed the best recognition rate of the test data, and the 1-nearest neighbor classifier and the fuzzy classifier with ellipsoidal regions followed. The support vector machine took 6 hours for training, the fuzzy classifier with ellipsoidal regions took only 79 seconds, and the 1-nearest neighbor classifier took 42 seconds for classification. Training by BP was the slowest (4 hours).

10.7 Discussions

Except for the thyroid and hiragana-50 data, the fuzzy classifier with ellipsoidal regions showed the best, second or third best recognition rates of the test data. Since the thyroid data included 16 discrete input variables among 21 input variables, the data were not suitable for the classifier. The perfor-

Table 10.7. Performance comparison for the hiragana-13 data

Classifier	Rates (%)	Rules	Time (s)
1-NN	99.55 (100)	8375	42
BP	98.56 (99.64)	25	15380
PYD	97.42 (99.98)	1442	1153
MM	98.22 (100)	5162	1200
MMI	95.34 (100)	1493	4
ELD	99.19 (99.94)	38	79
SVM	<u>99.77</u> (100)	RBF 1	23216

mance could be improved by using the robust classification method discussed in Chapter 8 (97.72%) or by combining the classifier with the neural network feature extractor discussed in Section 11.2.3 (98.07%).

For some classes of the hiragana data, the numbers of the training data were smaller than the numbers of the input variables. When only small singular values were set to zero, training finished without tuning the tuning parameters or locations of the centers. Thus, there were no overlaps between classes. This means that to improve the generalization ability, the tuning parameters should be tuned even when there are no overlaps between classes for the training data.

Training of the classifier with ellipsoidal regions was sufficiently fast except when it was slowed down by the singular value decomposition and the centers were tuned as shown for the hiragana-50 and hiragana-105 data. In addition, one fuzzy rule for a class was sufficient to realize high classification performance. Thus a compact classification system can be realized for difficult classification problems.

Training of the support vector machine was as slow as that by BP. But except for the iris and thyroid data, the support vector machine showed the best recognition rates among the classifiers used for evaluation. For the thyroid data, however, overfitting occurred and the recognition rate was the third worst. Classification performance depended on the kernel functions used. Thus, to realize good generalization performance, trial and error is necessary. Therefore, speedup of training is essential.

The neural network classifier trained by BP showed relatively good classification performance, but the training was very slow. The recognition rates listed in this chapter were the best recognition rates, not the average recognition rates. Thus, in applying the network to large size problems, training time is a problem. The neural network classifier trained by solving inequalities showed comparable performance with that trained by BP and the training was much faster. Thus training by solving inequalities can be an alternative method for BP.

Training of the fuzzy classifier with hyperbox regions was usually faster than that of the remaining classifiers. The fuzzy min-max classifier showed relatively good recognition rates for most of the data sets, but the numbers of fuzzy rules became large for complicated or large size problems. The thyroid data set was the easiest problem for the fuzzy min-max classifier and the fuzzy min-max classifier with inhibition but not for the fuzzy classifier with pyramidal membership functions. The difference may be due to the shapes of the membership functions: whether truncated rectangular pyramidal membership functions or rectangular pyramidal membership functions were used.

The major difference between the fuzzy classifier with ellipsoidal regions and the fuzzy classifier with pyramidal membership functions, when the quadratic functions are used, is that the former uses the covariance matrix while the latter uses the diagonal covariance matrix. For the blood cell and hiragana data, a large number of fuzzy rules were generated for the fuzzy classifier with pyramidal membership functions, while only one rule per class was generated for the fuzzy classifier with ellipsoidal regions. Thus, the correlation of input variables needs to be considered in designing classifiers.

The k-nearest neighbor classifier showed good classification performance for the hiragana data but showed the worst performance for the iris and thyroid data. Thus classification performance was not stable. One reason may be that the classifier performance depends on scaling of input variables.

11. Optimizing Features

In developing a pattern classification system for a given problem, we need to realize a high recognition rate for unknown data, i.e., high generalization ability. The type of classifier used influences the generalization ability, but the most influencing factor is the set of features used. But since there is no systematic way of determining the set of optimal features from scratch, we assume that we have an initial set of features or input variables.

Each input variable has a different physical meaning and hence has a different range of variations. But if the original ranges of variations are used for classification, the classification results may be affected by the variables with large ranges of variations. Thus there may be the optimal ranges. This problem is called the scaling problem. But since we do not know in advance which input variables are important for classification, we usually normalize input variables. Fuzzy classifiers, however, have invariance for a certain kind of transformation of input variables. In Section 11.1, we discuss invariance of fuzzy classifiers.

If we use neural networks, we normalize each input variable into the range of $[0, 1]$. If we change the ranges, the solution obtained by the back-propagation algorithm is also changed even if the initial weight values are the same. Likewise, if input variables are linearly transformed by translation, scaling, and rotation, the solution is changed. Thus neural networks are not invariant under linear transformations of the input variables.

After we normalize the input variables, we determine the optimal set of features. There are two approaches to determine the set of features: feature extraction and feature selection. Feature extraction, linearly or nonlinearly, transforms the original set of features into a reduced one and feature selection selects relevant features from the original features. In Sections 11.2 and 11.3 we discuss feature extraction and feature selection, respectively, and in Section 11.4 we evaluate performance of feature selection methods using the benchmark data.

11.1 Scaling Features

In this section we discuss invariance of fuzzy classifiers under translation, scaling, and rotation of the input variables. In object classification, invariant classification under translation, scaling, and rotation of objects is an important subject [13]. But here we do not address that subject. Now we assume that the m-dimensional input vector $\mathbf{x}$ is transformed into m-dimensional vector $\mathbf{x}'$ by the following linear transformation:

$$\mathbf{x}' = A\mathbf{x} + \mathbf{b}, \tag{11.1}$$

where A is an $m \times m$ matrix and $\mathbf{b}$ is an m-dimensional vector. When A is a non-zero diagonal matrix and $\mathbf{b}$ is a zero vector, the transformation is scaling. If A is a null matrix, the transformation is translation. If A is non-singular matrix with non-zero off-diagonal elements, the transformation is linear transformation including rotation.

In general, classification performance of the classifier depends on translation, scaling, or rotation of the input variables. If a classifier is not affected by these transformations, the classifier is called translation, scaling, and rotation invariant. If the classifier is not affected by any linear transformation, the classifier is called linear invariant. Invariance of the classifier is important; if the classifier is scale invariant, we need not consider scaling of each input. Neural networks are not invariant; any transformation of inputs may make the network converge to a different solution.

11.1.1 Scale and Translation Invariance of the Fuzzy Classifier with Pyramidal Membership Functions

Here we prove that the fuzzy classifier with pyramidal membership functions has scale and translation invariance, assuming that any edge length of the hyperboxes A_{ij} is longer than ε. Let $\mathbf{x}$ be transformed into $\mathbf{x}'$ by scaling and translation:

$$\mathbf{x}' = A\mathbf{x} + \mathbf{b}, \tag{11.2}$$

where A is an $m \times m$ non-singular diagonal matrix and $\mathbf{b}$ is an m-dimensional vector. Thus the center vector $\mathbf{c}_{ij}$ given by (5.5) is transformed into $\mathbf{c}'_{ij}$ as follows:

$$\mathbf{c}'_{ij} = A\mathbf{c}_{ij} + \mathbf{b}. \tag{11.3}$$

Likewise, $\mathbf{V}_{ij} = (V_{ij1}, \ldots, V_{ijm})^t$ and $\mathbf{v}_{ij} = (v_{ij1}, \ldots, v_{ijm})^t$ that define, respectively, the maximum and minimum values of the hyperbox for cluster ij are transformed into $\mathbf{V}'_{ij}$ and $\mathbf{v}'_{ij}$ as follows:

$$\mathbf{V}'_{ij} = A\mathbf{V}_{ij} + \mathbf{b}, \tag{11.4}$$
$$\mathbf{v}'_{ij} = A\mathbf{v}_{ij} + \mathbf{b}. \tag{11.5}$$

Therefore, from (11.2) to (11.5),

$$\mathbf{c}'_{ij} - \mathbf{x}' = A(\mathbf{c}_{ij} - \mathbf{x}), \tag{11.6}$$
$$\mathbf{V}'_{ij} - \mathbf{v}'_{ij} = A(\mathbf{V}'_{ij} - \mathbf{v}'_{ij}). \tag{11.7}$$

From (5.7), (5.8), (11.6), and (11.7), the one-dimensional membership function $m_{ijk}(\mathbf{x})$ given by (5.10) is invariant under scaling and translation:

$$m_{ijk}(\mathbf{x}') = m_{ijk}(\mathbf{x}). \tag{11.8}$$

Thus the membership functions given by (5.11) and (5.12) are invariant under scaling and translation:

$$m_{ij}(\mathbf{x}') = m_{ij}(\mathbf{x}), \tag{11.9}$$

if any edge length of the hyperboxes A_{ij} is longer than ε.

Then, it is not difficult to prove that the fuzzy classifier with pyramidal membership functions which is based on the scale and translation invariant membership functions is scale and translation invariant. The fuzzy classifier with ellipsoidal regions, which is based on the weighted distance called Mahalanobis distance, has linear transformation invariance including rotation, scaling, and translation invariance as will be discussed later. The fuzzy classifier with pyramidal membership functions, which has the similar weighted distance, lacks rotation invariance since it is based on the hyperboxes whose surfaces are parallel to the input axes.

11.1.2 Invariance of the Fuzzy Classifiers with Hyperbox Regions

It is obvious that fuzzy classifiers with hyperbox regions are not rotation invariant. The fuzzy min-max classifier has translation invariance and the fuzzy min-max classifier with inhibition has translation invariance and limited scale invariance [83]. Here, limited scale invariance means that the classifier has invariance for the training data but it does not have invariance for the test data.

In the fuzzy min-max classifier, a hyperbox is expanded when (9.6) is satisfied. Equation (9.6) checks whether the total length of m edges emanating from a node is less than a prescribed value. Since the total length changes according to scaling, the hyperbox that is expanded in some scaling, may not be expanded in another scaling. Thus the classifier is not scale invariant.

But (9.6) is translation invariant since by translation the edge lengths do not change. Thus expansion of hyperboxes is not affected by translation. Likewise, contraction of hyperboxes is not affected by translation. Thus all the hyperboxes generated are shifted in the same amount in each direction by translation. Since the membership function given by (9.4) is based on the one-dimensional distance between the input x_k and the hyperbox surface, and this distance does not change by translation, the membership function is translation invariant. Thus the fuzzy min-max classifier is translation invariant.

In the fuzzy min-max classifier with inhibition, the hyperboxes of level 1 do not change their relative locations by translation and scaling, since they are generated by calculating the minimum and the maximum values of the training data in each input. Then the inhibition hyperboxes of level 1 also do not change their relative location by translation and scaling. Since the inhibition hyperboxes are expanded using (9.27)–(9.30), which are invariant under translation and scaling, the expanded inhibition hyperboxes do not change their relative relation to the activation and inhibition hyperboxes by translation and scaling. Likewise, the hyperboxes of levels higher than 1, if generated, do not change their relative relations to the remaining hyperboxes by translation and scaling.

In the calculation of the membership functions $m_{A_{ij'}(l)}(\mathbf{x})$ in (9.40) and $m_{J_{ij}(l)}(\mathbf{x})$ in (9.45), the minimum and the maximum values of the membership functions $m_{A_{ij'}(l)}(\mathbf{x},k)$ and $m_{J_{ij}(l)}(\mathbf{x},k)$ are calculated, respectively. Thus the membership functions are translation invariant but are not scale invariant; by scaling, the minimum and the maximum values among the membership functions $m_{A_{ij'}(l)}(\mathbf{x},k)$ and $m_{J_{ij}(l)}(\mathbf{x},k)$ may be changed.

Since all the training data are included in the hyperboxes, i.e., the 100% recognition rate is obtained for the training data, the recognition rate of the training data does not change under scaling. But for the test data, some data may be outside of the hyperboxes; thus the recognition rate of the test data may change under scaling. Thus the fuzzy min-max classifier with inhibition has translation invariance and limited scale invariance.

11.1.3 Invariance of the Fuzzy Classifier with Ellipsoidal Regions

The weighted distance given by (5.34) is known as the Mahalanobis distance which is invariant under linear transformation of input variables [46]. The fuzzy classifier with ellipsoidal regions that is based on the Mahalanobis distance has the same transformation invariance as shown below [83].

Let the linearly transformed vector of $\mathbf{x}$, $\mathbf{x}'$, be given by (11.1). The cluster center $\mathbf{c}_{ij}$ of cluster ij is transformed by (11.1) into

$$\mathbf{c}'_{ij} = A\,\mathbf{c}_{ij} + \mathbf{b}. \tag{11.10}$$

Using (11.1), (11.10), and (5.35), the covariance matrix Q'_{ij} for cluster ij in the transformed space is calculated as follows:

$$\begin{aligned}
Q'_{ij} &= \frac{1}{|X_{ij}|} \sum_{\mathbf{x}\in X_{ij}} (\mathbf{x}'-\mathbf{c}'_{ij})(\mathbf{x}'-\mathbf{c}'_{ij})^t \\
&= \frac{1}{|X_{ij}|} \sum_{\mathbf{x}\in X_{ij}} (A\mathbf{x}+\mathbf{b}-A\mathbf{c}_{ij}-\mathbf{b})(A\mathbf{x}+\mathbf{b}-A\mathbf{c}_{ij}-\mathbf{b})^t \\
&= \frac{1}{|X_{ij}|} \sum_{\mathbf{x}\in X_{ij}} \{A(\mathbf{x}-\mathbf{c}_{ij})\}\{A(\mathbf{x}-\mathbf{c}_{ij})\}^t \\
&= \frac{1}{|X_{ij}|} \sum_{\mathbf{x}\in X_{ij}} A(\mathbf{x}-\mathbf{c}_{ij})(\mathbf{x}-\mathbf{c}_{ij})^t A^t \\
&= A\,Q_{ij}\,A^t. \tag{11.11}
\end{aligned}$$

where X_{ij} is the set of training data included in cluster ij and $|X_{ij}|$ is the number of training data in X_{ij}. Thus using (5.34) the weighted distance between $\mathbf{x}'$ and $\mathbf{c}'_{ij}$, $d_{ij}(\mathbf{x}')$, is given by

$$
\begin{aligned}
d_{ij}^2(\mathbf{x}') &= (\mathbf{x}' - \mathbf{c}'_{ij})^t Q_{ij}'^{-1} (\mathbf{x}' - \mathbf{c}'_{ij}) \\
&= \{A(\mathbf{x} - \mathbf{c}_{ij})\}^t (A Q_{ij} A^t)^{-1} \{A(\mathbf{x} - \mathbf{c}_{ij})\} \\
&= (\mathbf{x} - \mathbf{c}_{ij})^t A^t (A^t)^{-1} Q_{ij}^{-1} A^{-1} A(\mathbf{x} - \mathbf{c}_{ij}) \\
&= d_{ij}^2(\mathbf{x}).
\end{aligned}
\tag{11.12}
$$

Therefore the weighted distance does not change by linear transformation. Assuming the same values are set for α_{ij}, from (5.32)–(5.34), the values of the membership functions for $\mathbf{x}$ and $\mathbf{x}'$ are the same.

Now assume that we start training slopes of the two classifiers using the original training data set and the transformed training data set with the same initial values for α_{ij} and δ. Since, at the initial training stage, the degrees of membership for $\mathbf{x}$ and $\mathbf{x}'$ are the same, the recognition rates for the two classifiers are the same. Also since the weighted distances for the original training data and the transformed training data are the same, $U_{ij}, L_{ij}, \gamma_{ij}(l)$, and $\beta_{ij}(l)$ give the same values for the two classifiers. Thus the renewed α_{ij} given by (7.5) and (7.6) have the same values for the two classifiers. Likewise, when the clusters are added, the renewed α_{ij} have the same values for the two classifiers. Thus the two fuzzy classifiers trained by the original training data set and the transformed training data set are the same when we tune the slope of the membership functions. A similar argument holds when we tune the centers of the membership functions. Thus, the fuzzy classifier with ellipsoidal regions is invariant under linear transformation of the input variables.

11.2 Feature Extraction

Feature extraction, linearly or nonlinearly, transforms the original set of features into a reduced one. Principal component analysis (PCA) [84] is a well-known feature extraction method, in which input axes are rotated around the coordinate origin of the original features in the directions of the eigenvectors of the feature covariance matrix, and some of the transformed features are selected from the most significant axes in order.

Based on the fact that the Mahalanobis distances are the optimum features when the distribution of each class is Gaussian, in [85], the calculation of $Q^{-1/2}$, where Q is the covariance matrix, using a two-layer neural network is proposed. In [86, 87, 88], the class boundaries for a given classifier are analyzed and the features that are orthogonal to the class boundaries are extracted.

Many neural networks have been proposed to realize PCA, discriminant analysis, and many other feature extraction methods and their performance has been compared [89, 90, 91, 92].

11.2.1 Principal Component Analysis

Principal component analysis (PCA) is a well-known feature extraction method, in which the principal components of the feature vector are extracted.

Now consider transforming an m-dimensional feature vector $\mathbf{x}$ into an m-dimensional feature vector $\mathbf{y}$ by the orthogonal transformation:

$$\mathbf{y} = S\mathbf{x}, \tag{11.13}$$

where S is the orthogonal matrix. We assume that the first $d\,(< m)$ components include most of the principal components of the feature vector $\mathbf{x}$.

Now to calculate the covariance matrix of $\mathbf{y}$, we first calculate the covariance matrix Q_x of the training data set X as follows:

$$Q_x = \frac{1}{|X|} \sum_{\mathbf{x} \in X} (\mathbf{x} - \mathbf{c}_x)(\mathbf{x} - \mathbf{c}_x)^t, \tag{11.14}$$

where X is the set of training data, $|X|$ is the number of training data and $\mathbf{c}_x$ is the mean vector of the training data and is calculated by:

$$\mathbf{c}_x = \frac{1}{|X|} \sum_{\mathbf{x} \in X} \mathbf{x}. \tag{11.15}$$

From (11.13) and (11.15), the mean vector of $\mathbf{y}, \mathbf{c}_y$, is given by

$$\mathbf{c}_y = S\,\mathbf{c}_x. \tag{11.16}$$

Then the covariance matrix of $\mathbf{y}$, Q_y, is given by

$$Q_y = S\,Q\,S^t. \tag{11.17}$$

Let the ith row vector of S be S_i. Then the variance of y_i is given by $S_i\,Q\,S_i^t$. To maximize $S_i\,Q\,S_i^t$ under the constraint $S_i\,S_i^t = 1$, we use the Lagrange multiplier λ:

$$L = S_i\,Q_x\,S_i^t - \lambda\,(S_i\,S_i^t - 1). \tag{11.18}$$

The condition to maximize (11.18) is given by

$$\frac{1}{2} \frac{\partial L}{\partial S_i} = Q_x S_i^t - \lambda\,S_i^t = 0, \tag{11.19}$$

which means that λ is the eigenvalue of Q_x and S_i^t is the associated eigenvector.

Since Q_x is a positive definite or semi-definite matrix, all the eigenvalues are nonnegative. Assuming that Q_x is positive definite and let $\lambda_1 > \lambda_2 > \cdots > \lambda_m$ be the eigenvalues of Q_x and $S_1^t, S_2^t, \ldots, S_m^t$ be the associated eigenvectors. Then we call $y_i = S_i \mathbf{x}$ the ith principal component of $\mathbf{x}$. The eigenvalue λ_i is the variance of y_i. The trace of Q_x is defined as the sum of the diagonal elements of Q_x:

$$\mathrm{tr}(Q_x) = \sum_{i=1}^{m} Q_{x,ii}. \tag{11.20}$$

Then $\mathrm{tr}(Q_x) = \lambda_1 + \cdots + \lambda_m$ [47, p. 310]. Thus the sum of the variances of $\mathbf{x}$ is the same with the sum of the variances of $\mathbf{y}$. Suppose we select the first d principal components. We define the accumulation of d eigenvalues as follows:

$$A_c(d) = \frac{\displaystyle\sum_{i=1}^{d} \lambda_i}{\displaystyle\sum_{i=1}^{m} \lambda_i} \times 100 \, (\%). \tag{11.21}$$

The accumulation of eigenvalues shows how well the reduced feature vector reflects the characteristics of the original feature vector.

Although PCA is used for feature extraction, there are some drawbacks. Scaling of input features changes the results of PCA. Namely, if we multiply the value of the feature by constant $a \, (> 1)$, the variance of the first feature, σ^2, becomes $a^2\sigma^2$. Thus by scaling the principal components change. Another problem is that since PCA does not distinguish data belonging to one class from the data belonging to another, the principal components are not directly associated with class separability. In Fig. 11.1, the first principal component does not contribute to classifying two classes but the second principal component does.

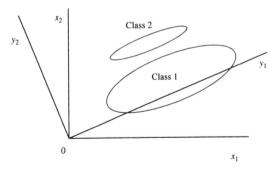

Fig. 11.1. The second principal component well separates classes than the first principal component does

11.2.2 Discriminant Analysis

Linear discriminant analysis maximizes the between-class scatter under the constraint of the constant within-class scatter. The within-class covariance matrix is defined by

$$W = \frac{1}{M} \sum_{i=1}^{n} \sum_{\mathbf{x} \in X_i} (\mathbf{x} - \mathbf{c}_i)(\mathbf{x} - \mathbf{c}_i)^t, \tag{11.22}$$

where $M = \sum_{i=1}^{n} |X_i|$ is the total number of data, $\mathbf{c}_i$ is the center of class i data. The between-class covariance matrix is defined by

$$B = \frac{1}{M} \sum_{i=1}^{n} |X_i| (\mathbf{c}_i - \mathbf{c})(\mathbf{c}_i - \mathbf{c})^t, \tag{11.23}$$

where $\mathbf{c}$ is the center of all the data. Then the total scatter matrix is given by

$$T = \frac{1}{M} \sum_{i=1}^{n} \sum_{\mathbf{x} \in X_i} (\mathbf{x} - \mathbf{c})(\mathbf{x} - \mathbf{c})^t = W + B. \tag{11.24}$$

Liner discriminant analysis transforms m-dimensional vector $\mathbf{x}$ into d-dimensional vector $\mathbf{y}$ by $\mathbf{y} = D\mathbf{x}$ where D is chosen so that

$$J(D) = \frac{\det(D^t B D)}{\det(D^t W D)} \tag{11.25}$$

is maximized, where det denotes the determinant and $d \leq n - 1$. Since the objective function $J(D)$ is interpreted as the ratio of the between class scatter volume to the within class scatter volume [46, pp. 118–121] in the transformed space, its maximization is considered to increase separability of classes. It is shown that such a transform is composed of d eigenvectors associated with the d largest nonzero eigenvalues of $T^{-1}B$.

11.2.3 Neural-network-based Feature Extraction

Three-layer neural networks perform nonlinear transformation and can be used for feature extraction. Namely, for the training inputs we prepare the same training outputs and using these input-output pairs we train the network with the same numbers of input and output neurons and with the smaller number of hidden neurons [93, pp. 255–258]. Then using the two-layer neural network with the input and hidden layers of the trained network, the inputs can be transformed into smaller number of features (see Fig. 11.2). The major problem of this method is that usually the number of inputs is very large and thus we need to train the network with large numbers of input and output neurons. This is especially prohibitive if we train the network by the back-propagation algorithm. This problem is alleviated if we train the network by solving inequalities by the Ho-Kashyap algorithm as discussed in Section 2.5. By this method, we determine the target values of the hidden neuron outputs and solve the inequalities for the weights between the input and hidden layers, and then those for the weights between the hidden and output layers. Thus for feature extraction, we need only to solve the inequalities for the weights between the input and hidden layers.

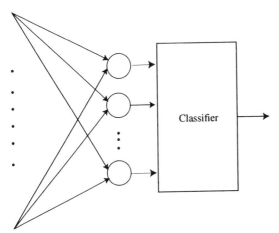

Fig. 11.2. Feature extraction by a two-layer neural network

In the following, we show the performance improvement when features of the thyroid and hiragana data listed in Table 1.1 on page 19 were extracted by two-layer neural networks and then the extracted features were classified by the fuzzy classifier with ellipsoidal regions. We set the maximum allowable number of misclassifications plus 1 to be $l_M = 10$, tuned only the slopes of the membership functions, and did not cluster the training data. We used a Sun UltraSPARC IIi 333MHz workstation for evaluation. The execution time listed in the following tables is feature extraction time plus training time of the classifier and numerals in the brackets show the training time of the classifier.

Table 11.1 shows the results for the thyroid data. In solving the inequalities by the Ho-Kashyap algorithm, we set $\varepsilon(2) = 0.2$. Namely, the network outputs were considered to be converged if they were smaller than 0.2 for the target values of 0, and larger than 0.8 for the target values of 1. The first two rows of the results for the original 21 inputs were from Table 7.19 on page 152. The remaining rows show the results when the margin parameter δ was changed to obtain different numbers of extracted features. By feature extraction, the maximum recognition rate of the test data reached 98.07% without clustering the training data. This was better than that with the 21 inputs, i.e., 97.29%. The reason for the improvement was that by feature extraction, the original 21 features that were the combination of continuous and discrete variables were transformed into continuous features, which were suited for the fuzzy classifier with ellipsoidal regions.

Table 11.1. Performance for the thyroid data

δ	Inputs	Rates (%)	Rules	Time (s)
—	21	95.60 (96.02)	3	8
—	21	97.29 (99.02)	13	60
0.04	2	97.02 (98.25)	3	35 (1)
0.05	3	97.14 (98.22)	3	29 (1)
0.1	3	98.07 (98.73)	3	63 (1)
0.3	3	98.07 (98.73)	3	63 (1)

Table 11.2 shows the results for the hiragana-50 data. In solving the inequalities by the Ho-Kashyap algorithm, we set $\varepsilon(2) = 0.2$. The first row of the results for the original 50 inputs were from Table 7.24 on page 155. The remaining rows show the results when the margin parameter δ was changed to obtain different numbers of extracted features. For the 50 inputs and for the extracted features, the maximum recognition rates of the test data were achieved when the singular values smaller than 10^{-4} were set to zero. For the hiragana-50 data, the best recognition rate of the test data was achieved for the original 50 inputs by controlling the number of singular values.

Table 11.2. Performance for the hiragana-50 data

δ	Inputs	Rates (%)	Time (s)
—	50	94.51 (99.89)	946
0.01	14	92.71 (97.94)	263 (75)
0.05	21	93.55 (99.35)	498 (213)
0.10	27	93.29 (99.85)	483 (147)

Table 11.3 shows the results for the hiragana-105 data. In solving the inequalities by the Ho-Kashyap algorithm, we set $\varepsilon(2) = 0.45$. The first row of the results for the original 105 inputs were from Table 7.26 on page 156 and the second row for the 13 inputs were from Table 7.27 on page 156. The

following rows show the results when the margin parameter δ was changed to obtain different numbers of extracted features. For the 105 inputs the maximum recognition rate of the test data was achieved when the singular values smaller than 10^{-3} were set to zero. For the extracted features the maximum recognition rate of the test data was achieved when the singular values smaller than 10^{-5} were set to zero. When $\delta = 0.1$, with 28 features, the recognition rate of the test data was comparable with that for 105 features.

Table 11.3. Performance for the hiragana-105 data

δ	Inputs	Rates (%)	Time (s)
—	105	99.99 (100)	2467
—	13	98.79 (99.99)	36
0.05	16	99.27 (99.73)	607 (108)
0.10	28	99.81 (99.98)	938 (154)
0.12	33	99.82 (100)	1133 (151)

11.3 Feature Selection

Feature selection selects relevant features from the original features. Suppose a classifier is selected for a given classification problem. Then one way to select the optimum features from given initial features is as follows: generate the classifier for each combination of features; evaluate the recognition rate; and select the features that maximize the recognition rate. But this exhaustive search strategy is time consuming especially when the number of initial features is large and when generation of a classifier takes time. Therefore, usually we use some appropriate search method and some selection criterion that is suitable for the classifier that is to be used.

In the following we discuss several selection criteria and then we explain the search methods based on forward and backward selection searches. In the forward selection search, starting from the empty set of features, we add the most relevant feature for classification to the set one at a time. In the backward selection search, on the other hand, starting from the set of given features, we delete the most irrelevant features from the set one at a time. In the last of this chapter, we compare performance of the forward and backward selection search methods for several benchmark data sets.

11.3.1 Criteria for Feature Selection

In [94] various feature selection criteria, such as the Bhattacharyya proba-
bilistic distance to select the set of features that maximizes class separability,
are discussed. In [95] some fuzzy parameters to measure class separability are
used to select features and in [96] features are selected based on the mutual
information criterion. In [97] and [98], the degree of overlap between classes
is formulated by approximating class regions by hyperboxes and ellipsoids,
respectively.

Neural-network-oriented Feature Selection. The optimal features selected for one classifier may not be optimal for other classifiers. Feature selection techniques that are suited for neural networks are proposed based on
the concept of weight decay that aims to prune unnecessary weights using
neural networks [99, 100, 101].

Karnin [99] proposes to estimate the sensitivity of the error function when
each weight is disconnected from the trained neural network by keeping track
of the changes in each weight during back-propagation training. After training, the weights are deleted from the weight with the smallest sensitivity
value in order.

Setiono and Liu [100] use multilayer neural networks for feature selection.
Instead of the usual squared error function, they use the cross-entropy error function with the penalty term that encourages the weights with small
magnitudes to converge to zero. Features are selected by backward selection.
Initially the network is trained using all the features. Then for each feature
all the weights connecting to that feature are set to zero and the recognition
rate is evaluated using the resultant network. The feature that shows the best
performance, when weights connected to it are set to zero, is deleted from the
set of features. To further delete features, deleting the feature that is found,
retrain the network and iterate the above procedure.

Since training of neural networks is slow, the feature selection methods
that require repetitive training of neural networks are usually impractical for
real applications.

Mutual Information. Battiti [96] uses mutual information as a feature
selection criterion. Let the probability of class i $(i = 1, \ldots, n)$ be $P(c_i)$. Then
the initial uncertainty of classification is measured by the entropy:

$$H(C) = -\sum_{i=1}^{n} P(c_i) \log P(c_i). \tag{11.26}$$

The conditional entropy after the m features are known is given by

$$H(C|F) = -\sum_{j=1}^{m} P(f_j) \left(\sum_{i=1}^{n} P(c_i|f_j) \log P(c_i|f_j) \right), \tag{11.27}$$

where $P(c_i|f_j)$ is the conditional probability of class i given the feature j.
Then the mutual information is given by

$$I(C; F) = H(C) - H(C|F), \tag{11.28}$$

which measures the decrease of uncertainty by providing m features.

Using the mutual information, the feature selection problem is now stated as follows:

For a given set of m features, F, find the subset S ($\subset F$) with k features that maximizes the mutual information $I(C; S)$.

To reduce the computation time for selection, Battiti adopts the forward selection search and uses the following simplified selection criterion:

$$I(C; f_j) - \beta \sum_{s \in S} I(f_j; s), \tag{11.29}$$

where β is the positive parameter (usually between 0.5 and 1) to regulate the relative importance of f_j with respect to the features already selected.

Mutual information is calculated by the training data and he proposes to use Fraser and Swinney's method [102] for a large number of initial features.

Exception Ratio Approximated by Hyperboxes. Difficulties in classification can be measured by the overlaps of class regions. Thus if we can formulate the degree of overlap and can eliminate the features while retaining the degree of overlap for the initial set of features, we can reduce the number of features without deteriorating classification performance.

One way to formulate the degree of overlap is to use the fuzzy regions defined for fuzzy classifiers. In this section we use the fuzzy min-max classifier with inhibition discussed in Section 9.2 in which fuzzy regions are defined by two types of hyperboxes: activation hyperboxes and inhibition hyperboxes. The activation hyperbox of a given class defines the region where data points of the class are allowed to exist. Following the interpretation in [103], the inhibition hyperbox $I_{ij}(l)$ can be regarded as an exception of the activation hyperbox $A_{ij}(l)$. It is noted that, for feature evaluation, we use here the inhibition hyperbox $I_{ij}(l)$, rather than the expanded inhibition hyperbox $J_{ij}(l)$, to exactly represent a region defined.

Suppose two classes approximated by hyperboxes do not overlap as shown in Fig. 11.3. In this case, two classes are linearly separable and we define that the degree of overlap is zero.

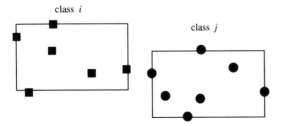

Fig. 11.3. No overlap between classes

If two hyperboxes of different classes overlap, but there are no data in the inhibition hyperbox as shown in Fig. 11.4, we assume that this overlap does not make classification difficult. Thus we define that the degree of overlap is zero if there are no data in the inhibition hyperbox.

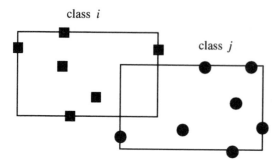

Fig. 11.4. Overlapping hyperboxes with no data in the inhibition hyperbox

If two hyperboxes of different classes overlap and there are some data in the inhibition hyperbox as shown in Fig. 11.5, we consider that the two classes overlap and we define a positive degree of overlap. In the following we discuss how we define hyperboxes to resolve overlaps and overlap functions named exception ratios in detail.

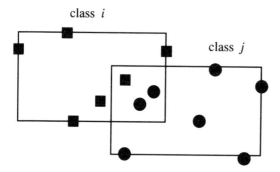

Fig. 11.5. Overlapping hyperboxes with some data in the inhibition hyperbox

Approximation of Class Regions. Here we approximate class regions using the activation and inhibition hyperboxes as discussed in Section 9.2.2. Using the set of class i $(i = 1, \ldots, n)$ training data, X_i, we define an activation hyperbox of level 1, denoted as $A_{ii}(1)$, which is the maximum region of class i data:

$$A_{ii}(1) = \{\mathbf{x} \mid v_{iik}(1) \le x_k \le V_{iik}(1), k = 1, \ldots, m\}, \tag{11.30}$$

where x_k is the kth element of input vector $\mathbf{x} \in X_i$, $v_{iik}(1)$ is the minimum value of x_k, and $V_{iik}(1)$ is the maximum value of x_k.

If the activation hyperboxes $A_{ii}(1)$ and $A_{jj}(1)$ $(j \neq i, j = 1, \ldots, n)$ do not overlap, we finish generating hyperbox between classes i and j.

If the activation hyperboxes $A_{ii}(1)$ and $A_{jj}(1)$ overlap, we resolve the overlap recursively in which we define the overlapping region as the inhibition hyperbox of level 1 denoted as $I_{ij}(1)$:

$$I_{ij}(1) = \{\mathbf{x} \mid w_{ijk}(1) \leq x_k \leq W_{ijk}(1), k = 1, \ldots, m\}, \tag{11.31}$$

where $v_{iik}(1) \leq w_{ijk}(1) \leq W_{ijk}(1) \leq V_{iik}(1)$. The minimum and maximum values of inhibition hyperbox $I_{ij}(1)$ are given as follows.

1. For $v_{jjk}(1) \leq v_{iik}(1) \leq V_{jjk}(1) < V_{iik}(1)$

$$w_{ijk}(1) = v_{iik}(1), \quad W_{ijk}(1) = V_{jjk}(1). \tag{11.32}$$

2. For $v_{iik}(1) < v_{jjk}(1) \leq V_{iik}(1) \leq V_{jjk}(1)$

$$w_{ijk}(1) = v_{jjk}(1), \quad W_{ijk}(1) = V_{iik}(1). \tag{11.33}$$

3. For $v_{jjk}(1) \leq v_{iik}(1) \leq V_{iik}(1) \leq V_{jjk}(1)$

$$w_{ijk}(1) = v_{iik}(1), \quad W_{ijk}(1) = V_{iik}(1). \tag{11.34}$$

4. For $v_{iik}(1) < v_{jjk}(1) \leq V_{jjk}(1) < V_{iik}(1)$

$$W_{ijk}(1) = v_{jjk}(1), \quad W_{ijk}(1) = V_{jjk}(1). \tag{11.35}$$

In Section 9.2.2 we expanded the inhibition hyperbox to control the generalization ability. But in approximating class regions this is not necessary.

Suppose we have defined the inhibition hyperbox of level $l - 1$, $I_{ij}(l - 1)$, and suppose some data belonging to X_i exist in $I_{ij}(l - 1)$, we define the activation hyperbox of level l denoted as $A_{ij}(l)$ within the inhibition hyperbox $I_{ij}(l - 1)$ by calculating the minimum and maximum values of x_k based on the data in $I_{ij}(l - 1)$:

$$A_{ij}(l) = \{\mathbf{x} \mid v_{ijk}(l) \leq x_k \leq V_{ijk}(l), k = 1, \ldots, m\}, \tag{11.36}$$

where $\mathbf{x} \in X_i$ and $\mathbf{x}$ is in $I_{ij}(l)$, $v_{ijk}(l)$ is the minimum value of x_k, $V_{ijk}(l)$ is the maximum value of x_k, and

$$u_{ijk}(l - 1) \leq v_{ijk}(l) \leq x_k \leq V_{ijk}(l) \leq U_{ijk}(l - 1). \tag{11.37}$$

If there is only one activation hyperbox of level l or there are two activation hyperboxes but they do not overlap, we finish generating hyperboxes between classes i and j.

If $A_{ij}(l)$ and $A_{ji}(l)$ overlap, the overlapping region of level l is denoted as $I_{ij}(l)$:

$$I_{ij}(l) = \{\mathbf{x} \mid w_{ijk}(l) \leq x_k \leq W_{ijk}(l), k = 1, \ldots, m\}, \tag{11.38}$$

where $v_{ijk}(l) \leq w_{ijk}(l) \leq W_{ijk}(l) \leq V_{ijk}(l)$.

Hyperboxes of levels higher than l can be defined in a similar manner if an overlap can be defined. The recursion process for defining hyperboxes terminates when $A_{ij'}(l)$ and $A_{ji'}(l)$ do not overlap or $A_{ij'}(l) = A_{ji'}(l) = I_{ij}(l-1)$ holds.

Exception Ratio. Using the activation and inhibition hyperboxes, we define overlap functions called exception ratios.

The larger the size of the inhibition hyperbox, or the exception, in a given activation hyperbox, the less is the contribution of the activation hyperbox to the classification of the corresponding class; thus the higher degree of overlap. However, if the probability that the data exits in the inhibition hyperbox is low, we consider that the degree of overlap is low.

We determine the degree of overlap between two classes based on the exception ratio, the computation of which is described in the following. First, at each overlapping level the ratio of the size of the inhibition hyperbox to the size of the activation hyperbox is computed. Next, since the deeper the level of a rule, the less is the contribution of that rule to the classification, the ratio computed at each level is weighted by the probability to find a datum of the corresponding class inside of the inhibition hyperbox. Finally, the exception ratio is computed by taking the sum for all levels of the weighted ratio.

Let F denote a set of features upon which rules are generated. Then we define the exception ratio $o_{ij}(F)$ between classes i and j as follows:

$$o_{ij}(F) = \sum_{l=1}^{l_{ij}} p_{ij}(l) \frac{B_{I_{ij}}(F,l)}{B_{A_{ij'}}(F,l)}, \tag{11.39}$$

where l_{ij} is the maximum level of recursions between classes i and j, $j' = i$ for $l = 1$ and $j' = j$ for $l > 1$, $B_{I_{ij}}(F,l)$ and $B_{A_{ij'}}(F,l)$ are the sizes of the inhibition hyperbox I_{ij} and the activation hyperbox $A_{ij'}$, respectively, and are given by

$$B_{I_{ij}}(F,l) = \prod_{f \in F} b_{I_{ij}}(f,l),$$

$$B_{A_{ij'}}(F,l) = \prod_{f \in F} b_{A_{ij'}}(f,l),$$

$$b_{I_{ij}}(f,l) = \begin{cases} W_{ijf}(l) - w_{ijf}(l) & \text{for } W_{ijf}(l) - w_{ijf}(l) > \varepsilon, \\ \varepsilon & \text{otherwise,} \end{cases}$$

$$b_{A_{ij'}}(f,l) = \begin{cases} V_{ij'f}(l) - v_{ij'f}(l) & \text{for } V_{ij'f}(l) - v_{ij'f}(l) > \varepsilon, \\ \varepsilon & \text{otherwise.} \end{cases}$$

Here ε is a minimum edge length and

$$p_{ij}(l) = \frac{\text{number of class } i \text{ training data in } I_{ij}(l)}{\text{total number of training data}}.$$

In the above formula, it is necessary to limit the smallest value of $b_{X_{ij}}(f,l)$ to ε. This allows the computation of $o_{ij}(F)$ for the case where there exists a feature f^* in F such that $V_{ijf^*}(l) - v_{ijf^*}(l) = 0$.

Let $O(F)$ denote the exception ratio and be defined as follows:

$$O(F) = \sum_{i,j,i \neq j} o_{ij}(F). \tag{11.40}$$

We use $O(F)$ as the selection criterion.

Exception Ratio Approximated by Ellipsoids. Instead of analyzing the hyperbox class regions, we discuss a feature selection criterion based on analysis of ellipsoidal class regions [98].

We approximate class regions by ellipsoids with the centers and the covariance matrixes calculated by the data belonging to the classes. Then, similar to what was discussed for the exception ratio approximated by hyperboxes, the exception ratio is defined to represent the degree of overlap in the class regions approximated by ellipsoids.

In the following, we first approximate the class regions by ellipsoids and we define the exception ratio.

Approximation of Class Regions. Using the m-dimensional training data belonging to class i we approximate the class region by an ellipsoid with the m-dimensional center vector:

$$\mathbf{c}_i = \frac{1}{|X_i|} \sum_{\mathbf{x} \in X_i} \mathbf{x} \tag{11.41}$$

and the $m \times m$ covariance matrix:

$$Q_i = \frac{1}{|X_i|} \sum_{\mathbf{x} \in X_i} (\mathbf{x} - \mathbf{c}_i)(\mathbf{x} - \mathbf{c}_i)^t, \tag{11.42}$$

where X_i is the set of training data belonging to class i and $|X_i|$ is the number of data in X_i.

We define the degree of membership of input vector $\mathbf{x}$ for class i by

$$m_i(\mathbf{x}) = \exp\left(-\frac{1}{m}d_i^2(\mathbf{x})\right), \tag{11.43}$$

where $d_i(\mathbf{x})$ is the weighted distance of $\mathbf{x}$ from $\mathbf{c}_i$ and is given by

$$d_i^2(\mathbf{x}) = (\mathbf{x} - \mathbf{c}_i)^t Q_i^{-1}(\mathbf{x} - \mathbf{c}_i). \tag{11.44}$$

Since we can show that the mean square weighted distance is m (see Theorem B.3.2 on page 312):

$$\frac{1}{|X_i|} \sum_{\mathbf{x} \in X_i} d_i^2(\mathbf{x}) = m, \tag{11.45}$$

we can normalize the degree of membership given by (11.43) for the different numbers of input variables.

Exception Ratio. Similar to the exception ratio approximated by hyperbox regions, we define the overlap of ellipsoidal class regions as follows. Let $p_i(\mathbf{x})$ be the probability that $\mathbf{x}$ belongs to class i. Then we define the degree of overlap of class i with respect to class j by

$$o_{ij}^2(F) = \frac{\displaystyle\int_{\mathbf{x} \in X_i} dp_j(\mathbf{x})}{\displaystyle\int_{\mathbf{x} \in X_i} dp_i(\mathbf{x})}. \tag{11.46}$$

By integrating $p_j(\mathbf{x})$ over $\mathbf{x}$ belonging to class i, we obtain the accumulated probability of class j for the class i data. The denominator of (11.46) normalizes the numerator and is 1 if the probability $p_i(\mathbf{x})$ is normalized. To approximate $p_i(\mathbf{x})$, we use $m_i(\mathbf{x})$ and we calculate (11.46) using the training data. Namely,

$$o_{ij}(F) = \frac{\displaystyle\sum_{\mathbf{x} \in X_i} m_j(\mathbf{x})}{\displaystyle\sum_{\mathbf{x} \in X_i} m_i(\mathbf{x})}. \tag{11.47}$$

Equation (11.47) gives the measure that the class i region overlaps with the class j region. But if any of the class i data is not misclassified into class j, i.e., $m_i(\mathbf{x}) > m_j(\mathbf{x})$ for $\mathbf{x}$ belonging to class i, the overlap given by (11.47) does not make the classification of class i data difficult. To reflect this, we define the exception ratio by

$$O(F) = \sum_{i,j=1}^{n} p_{ij} o_{ij}(F), \tag{11.48}$$

where $p_{ij} = \dfrac{\text{number of class } i \text{ data misclassified into class } j}{\text{total number of training data}}$.

The exception ratio given by (11.48) has the form similar to that by hyperboxes. The major difference is that in the former, the class region is approximated by one ellipsoid while in the latter, the class region is approximated by nested hyperboxes.

11.3.2 Monotonicity of the Selection Criteria

If we delete (add) some feature from the set of features, the degree of overlap increases (decreases) or remains the same. Thus for the deletion (addition) of features if the value of a selection criterion that represents the degree of overlap increases (decreases) or does not change, the criterion is considered to well represent the complexity of class overlap. This characteristic is called monotonicity of the selection criterion.

We can easily show that the exception ratio approximated by hyperboxes is monotonic for an addition or a deletion of features as follows.

From (11.39) and (11.40), monotonicity of the exception ratio is guaranteed if $p_{ij}(l)$ and $B_{I_{ij}}(F,l)/B_{A_{ij'}}(F,l)$ are monotonic. Assume that we have two overlapping activation hyperboxes $A_{ij'}(l)$ and $A_{ji'}(l)$. If we add a feature, the activation hyperboxes $A_{ij'}(l)$ and $A_{ji'}(l)$ are expanded in the direction of the feature. If the overlap is resolved by this addition, $I_{ij}(l)$ becomes empty. Thus $p_{ij}(l) = 0$. Even if the overlap remains, some of the training data that were in the original inhibition hyperbox may be out of the expanded inhibition hyperbox. But training data that are outside of the original inhibition hyperbox will not go into the expanded inhibition hyperbox. Thus $p_{ij}(l)$ is non-increasing for the addition of features. Similarly, for the deletion of features, $p_{ij}(l)$ is non-decreasing. Thus $p_{ij}(l)$ is monotonic. According to the definition of $B_{I_{ij}}(F,l)/B_{A_{ij'}}(F,l)$, by the addition of a feature f, $b_{I_{ij}}(f,l)/b_{A_{ij'}}(f,l)$ (≤ 1) is multiplied to $B_{I_{ij}}(F,l)/B_{A_{ij'}}(F,l)$. Thus $B_{I_{ij}}(F,l)/B_{A_{ij'}}(F,l)$ is non-increasing (non-decreasing) for the addition (deletion) of a feature. Thus the exception ratio approximated by hyperboxes is monotonic for the addition or deletion of features.

The exception ratio approximated by ellipsoids, on the other hand, does not have monotonicity. Fig. 11.6 shows the exception ratios for the numeral data when features were deleted by the hyperbox selection or added by the ellipsoidal selection one at a time. For the exception ratio approximated by hyperbox, clear monotonicity is visible, but the exception ratio approximated by ellipsoids first decreased and then increased. Thus it is not monotonic.

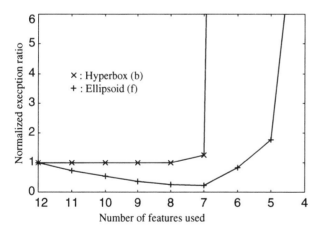

Fig. 11.6. Monotonicity of exception ratios. The exception ratio approximated by hyperboxes is monotonic, while that approximated by ellipsoids is not

11.3.3 Selection Search

The widely used selection search method is based on either the backward selection search or the forward selection search. In the backward selection search, selection starts from the initial set of features and the most irrelevant feature for classification is deleted from the set based on some criterion. In the forward selection search, on the other hand, selection starts from an empty set and adds the most relevant feature for classification to the set.

Siedlecki and Sklansky [104] propose to use the genetic algorithm for feature selection. The entire features are represented by a bit string in which 1 indicates that the associated feature is used and 0 indicates that the associated feature is not. Brill et al. [105] evaluate the fitness by the combination of the recognition rate by the nearest neighbor classifier and the number of features. The latter is multiplied with a constant and the result is subtracted from the former to favor the smaller number of features.

Forward Selection Search. Using the exception ratio we determine the optimal set of features by the forward feature selection. Namely, initially we start with an empty set of features and we add one feature at a time to the set that has the minimum increase in the exception ratio.

Let F_{org} denote the set of the original M features where $M \geq 2$; let F^k denote the set of k selected features; and let F_i^{k+1} be the set of $(k + 1)$ features obtained by temporarily adding f_i^k to F^k, i.e., $F_i^{k+1} = F^k + \{f_i^k\}$, where f_i^k is the ith element of F_{org} not included in F^k. Let F_j^{k+1} satisfy $O(F_j^{k+1}) = \min_i(O(F_i^{k+1}))$. Then we permanently add f_j^k to F^k. Namely, $F^{k+1} = F_j^{k+1}$.

If the exception ratio has monotonicity, we terminate the feature addition when the current exception ratio $O(F_j^{k+1})$ is smaller than or equal to $(1 + \beta) O(F_{org})$:

$$\frac{O(F_j^{k+1}) - O(F_{org})}{O(F_{org})} \leq \beta, \tag{11.49}$$

where β is a small positive parameter.

Since the exception ratio approximated by hyperboxes has monotonicity, we apply (11.49) for the forward hyperbox selection. In Fig. 11.6, let the exception ratio for the forward hyperbox selection be the same with that for the backward hyperbox selection. Then in the figure the exception ratio changes from right to left as features are added. When we add the seventh feature to the selected features, the exception ratio satisfies (11.49) when $\beta = 0.5$. Thus we terminate the selection and seven features are selected.

As shown in Fig. 11.6 the exception ratio approximated by ellipsoids does not have monotonicity. In this case, as the sixth feature is added, the exception ratio satisfies (11.49), and when the eighth feature is added the exception ratio begins to increase. Thus in the forward selection, we allow to add features even if (11.49) is satisfied and when the current exception ratio begins to increase:

$$\frac{O(F_j^{k+1}) - O(F^k)}{O(F_{org})} \geq \delta, \tag{11.50}$$

where δ is a small positive parameter, we terminate the algorithm. For the numeral data, for $\delta = 0.01$ (11.50) is satisfied when the eighth feature is added, and we terminate feature selection. In this case, we do not include the newly added feature. Thus, the number of selected feature is seven.

The exception ratio approximated by ellipsoids is not guaranteed to have monotonicity. But it may have monotonicity for a given selection problem. Since we do not know in advance whether the non-monotonic exception ratio shows monotonicity for the given problem, we apply the termination criterion (11.50). If the exception ratio shows monotonicity for the given problem, the addition of features continues until all the data are added. After that we can select the appropriate features using the criterion (11.49).

The forward feature selection algorithm is as follows.

1. Calculate the exception ratio $O(F_{org})$ using the set of original features.
2. Initialize F^k by setting $F^0 = \phi$, hence $k = 0$.
3. By temporarily adding the feature f_i^k $(i = 1, \ldots, M - 1)$ not included in F^k, compute $O(F_i^{k+1})$.
4. Find the feature f_j^k that satisfies $O(F_j^{k+1}) = \min_i(O(F_i^{k+1}))$.
5. If (11.50) by the ellipsoidal selection is satisfied, terminated the algorithm; otherwise, go to Step 6.
6. Set $F^{k+1} \leftarrow F_j^{k+1}$. ($f_j^k$ is permanently added to F^k.) If (11.49) by the hyperbox selection is satisfied, terminated the algorithm; otherwise, go to Step 7.
7. Set $k \leftarrow k + 1$. If $k = M - 1$, terminate the algorithm; otherwise, go to Step 3.

Backward Selection Search. Backward selection search begins with all the features and eliminates the most irrelevant feature as follows. First, each of the features is temporarily eliminated and the exception ratio after each temporary elimination is computed. Then the feature whose elimination minimizes the exception ratios is deleted permanently. We iterate this procedure and delete features until the stopping criteria discussed below are satisfied.

Let F_{org} be the set of the original M features where $M \geq 2$, F^m denote the set of m remaining features and F_i^{m-1} be the set of $(m - 1)$ features obtained by temporarily eliminating f_i^m from F^m, i.e., $F_i^{m-1} = F^m - \{f_i^m\}$ where f_i^m is the ith element in F^m. Further let F_j^{m-1} be

$$O(F_j^{m-1}) = \min_i(O(F_i^{m-1})). \tag{11.51}$$

Then if the exception ratio has monotonicity, or the exception ratio increases as we delete features, we terminate the feature elimination when the exception ratio for F_j^{m-1} exceeds that of the original exception ratio:

$$\frac{O(F_j^{m-1}) - O(F^{org})}{O(F_{org})} \geq \beta, \tag{11.52}$$

where β is a small positive parameter.

In Fig. 11.6, when we delete the sixth feature by the backward hyperbox selection, the exception ratio satisfies (11.52), and we terminate the selection. In this case, we select seven features.

When the exception ratio does not have monotonicity, namely, the exception ratio decreases as the features are deleted, we terminate the selection when the exception ratio begins to increase:

$$\frac{O(F_j^{m-1}) - O(F^m)}{O(F_{org})} \geq \delta, \tag{11.53}$$

where δ is a small positive parameter.

Assuming that the exception ratio approximated by ellipsoids shown in Fig. (11.6) is generated by the backward selection, the exception ratio begins to increase when the sixth feature is deleted. Thus we terminate the selection and seven features are selected.

The exception ratio based feature elimination algorithm can be described by the following procedure:

1. Initialize F^m by setting $F^m \leftarrow F_{org}$, hence $m = M$.
2. Compute $O(F_i^{m-1})$ for $i = 1, \ldots, m$.
3. Find the feature f_j^m that satisfies $O(F_j^{m-1}) = \min_i (O(F_i^{m-1}))$.
4. If (11.52) by the hyperbox selection or (11.53) by the ellipsoidal selection is satisfied, terminate the algorithm; otherwise go to Step 5.
5. Set $F^{m-1} \leftarrow F_j^{m-1}$. ($f_j^m$ is permanently eliminated from F^m.)
6. Set $m = m - 1$. If $m = 1$, terminate the algorithm; otherwise go to Step 2.

Acceleration of Selection Search. When we use the exception ratio approximated by ellipsoids, we can speed up the calculation of the squares of the weighted distances by the symmetric Cholesky factorization: $Q_i = L_i L_i^t$, where L_i is the lower triangular matrix (cf. Section 5.2.3).

In addition, when we use forward selection, we can further speed up the calculation of the squares of the weighted distances used in the exception ratio. In calculating $O(F_j^{k+1})$, the values calculated using the first to kth features, i.e., $c_i^{(k)}$, $Q_i^{(k)}$, $L_i^{(k)}$, and $d_i^{(k)}(\mathbf{x})$ do not change, where the superscript (k) denotes that the associated variable is calculated using the first to kth features. Thus, if we store these values, we can accelerate the calculation of the exception ratio as follows:

1. When k features are selected, calculate and store $c_i^{(k)}$ and $Q_i^{(k)}$.
2. When f_j^k is temporarily added to F^m, calculate $c_{i,k+1}^{(k+1)}$, $Q_{i,k+1,l}^{(k+1)}$, $L_{i,k+1,l}^{(k+1)}$ ($l = 1, \ldots, k+1$), and $y_{i,k+1}^{(k+1)}$, where

$$Q_i^{(k+1)} = \left(\begin{array}{c|c} Q_i^{(k)} & \begin{array}{c} Q_{i,1,k+1}^{(k+1)} \\ \vdots \\ Q_{i,k,k+1}^{(k+1)} \end{array} \\ \hline Q_{i,k+1,1}^{(k+1)}, \cdots, Q_{i,k+1,k}^{(k+1)} & Q_{i,k+1,k+1}^{(k+1)} \end{array} \right), \qquad (11.54)$$

$$L_i^{(k+1)} = \left(\begin{array}{c|c} \begin{array}{cc} \ddots & 0 \\ L_i^{(k)} & \ddots \end{array} & 0 \\ \hline L_{i,k+1,1}^{(k+1)} \cdots L_{i,k+1,k}^{(k+1)} & L_{i,k+1,k+1}^{(k+1)} \end{array} \right). \qquad (11.55)$$

3. Using the values stored in Step 1 and calculated in Step 2, calculate $(O(F_i^{m+1}))$.

11.4 Performance Evaluation of Feature Selection

We compared performance of the backward and forward feature selection methods discussed in Section 11.3. Hereafter we call the backward/forward selection method using the exception ratio approximated by hyperboxes the backward/forward hyperbox selection and the backward/forward selection method using the exception ratio approximated by ellipsoids the backward/forward ellipsoidal selection.

11.4.1 Evaluation Conditions

We compared the backward/forward hyperbox/ellipsoidal selection methods and PCA using the iris data, numeral data, blood cell data, and thyroid data listed in Table 1.1 on page 19. The training data were used both for selecting features and training classifiers. The test data were used for evaluating the recognition rate of the classifiers. In [97], both the discriminant analysis (DA) and the feature selection method that performs the backward selection search using interclass Euclidean distance as the class separability measure (EDFE) were shown to be inferior to PCA for the above four benchmark data sets. Therefore, we do not include their comparison here.

Three classifiers were used, namely, the fuzzy min-max classifier with inhibition, the fuzzy classifier with ellipsoidal regions, and the three-layer neural network classifier trained by the back-propagation algorithm. Unless explicitly specified, the following sets of parameters were used for the fuzzy classifiers and the neural network classifier.

For the fuzzy min-max classifier with inhibition we set the expansion parameter for controlling the expansion size of the inhibition hyperbox to be

0.001 and the sensitivity parameter for controlling the slope of the membership function to be 1.

For the fuzzy classifier with ellipsoidal regions, we defined one cluster per class and we set the maximum allowable number of misclassifications plus 1, l_M, to be 10. (We tuned only the slopes of the membership functions.)

We trained the three-layer neural network classifier by the back-propagation algorithm with the learning rate of 1 and the momentum coefficient of 0.

The numbers of hidden neurons and epochs are the same with those listed in Table 2.3 on page 41. Since the recognition rate of the neural network classifier varies according to the initial weights, we used the average recognition rate for 10 runs.

The parameters used for the backward hyperbox selection were $\varepsilon = 0.001$ (which specifies the minimum edge length) and $\beta = 0.5$. The parameters for the forward/backward ellipsoidal selection were $\beta = 0.5$ and $\delta = 0.01$.

11.4.2 Selected Features

Table 11.4 lists the numbers of features selected by the backward/forward hyperbox/ellipsoidal selection methods as well as the associated accumulation of eigenvalues by PCA for the number of features determined by the backward hyperbox selection. From the table, the accumulations of eigenvalues for the iris and blood cell data were nearly 100% and the principal components were included even when the features were reduced from 4 to 3 and 13 to 10, respectively. But for the thyroid data the accumulation of eigenvalues was 73% and some principal components were lost when the features were reduced from 21 to 5.

The numbers of features selected by the four methods were similar for each data set except for the forward hyperbox selection for the thyroid data. For the thyroid data, the exception ratio approximated by hyperboxes was 101.6. Thus, by the forward selection, when the exception ratio was below 152.4, the addition of features was terminated. When the first feature (feature 8) was added, the exception ratio was 284. When the second feature (feature 7) was added, it reduced to 151 and the addition of features was terminated. To make comparisons clear, in the following when we compare the recognition rates using the features selected by different methods, we use the numbers of features that were selected by the backward hyperbox selection.

11.4.3 Iris Data

Table 11.5 shows the features of the iris data set listed in the order of importance. Thus by the forward selection, the features were selected from left to right in the sequence and by the backward selection, they were selected from right to left.

Table 11.4. Number of features selected and the associated accumulation of eigenvalues by PCA

Data Set	Number of Features				Acc. Ev.
	Hyperbox (f)	Hyperbox (b)	Ellipsoid (f)	Ellipsoid (b)	(%)
Iris	3	3	2	3	99.65
Numeral	7	7	7	8	93.45
Thyroid	2	5	7	7	73.33
Blood Cell	10	10	7	10	99.53

Table 11.5. Iris features in the order of importance

Method	1	2	3	4
Ellipsoid (f)	4	2	3	1
Ellipsoid (b)	3	1	4	2
Hyperbox (f)	3	4	1	2
Hyperbox (b)	3	4	1	2

The forward and backward hyperbox selection methods gave the same order of sequence and the three methods except for the forward ellipsoidal selection method gave the same three important features.

Table 11.6 lists the recognition rate of the iris test data with the original features, the three features selected by the four methods, and the three features extracted by PCA. We used three classifiers: the fuzzy classifier with ellipsoidal regions (ELD), the fuzzy min-max classifier with inhibition (MMI), and the three-layer neural network classifier trained by the back-propagation algorithm (BP). For a given classifier, if the recognition rate of the data with reduced features was comparable with that of the original features, we considered that the feature reduction was done successfully. From this viewpoint, the features selected by the forward ellipsoidal selection showed recognition rates slightly lower than those by the original features for the fuzzy classifier with ellipsoidal regions and the neural network classifier. The features extracted by PCA showed a lower recognition rate for the fuzzy min-max classifier with inhibition. But since the decreases were small, the five reduction methods were considered to perform well for the iris data.

Fig. 11.7 shows, on the left side ordinate, the recognition rates of the fuzzy classifier with ellipsoidal regions for the iris data, when the features were deleted or added by the ellipsoidal and hyperbox selection methods. The right side ordinate plots the exception ratio of the backward ellipsoidal selection normalized by that with the original features. The normalized exception ratio increased as the features were deleted. Thus the exception ratio approximated

Table 11.6. Recognition rate of the iris data

Features	ELD (%)	MMI (%)	BP (%)
Original	<u>98.67</u>	92.00	97.33
Ellipsoid (f)	97.33	92.00	96.27
Ellipsoid (b)	<u>98.67</u>	<u>93.33</u>	97.33
Hyperbox (f)	<u>98.67</u>	<u>93.33</u>	97.33
Hyperbox (b)	<u>98.67</u>	<u>93.33</u>	97.33
PCA	<u>98.67</u>	90.67	<u>98.13</u>

by ellipsoids showed monotonicity. Since the exception ratio increased from the beginning, the exception ratio satisfies (11.52) when two features were deleted, we chose three features by the backward ellipsoidal selection.

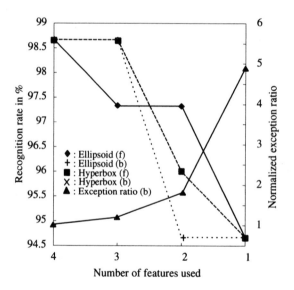

Fig. 11.7. Comparison of feature elimination for the iris data. The recognition rates for the test data were evaluated by the fuzzy classifier with ellipsoidal regions

11.4.4 Numeral Data

Table 11.7 lists the features of the numeral data set listed in the order of importance. Among the seven important features, 5 to 6 features were commonly selected for any two selection methods.

Table 11.7. Numeral features in the order of importance

Method	1	2	3	4	5	6	7	8	9	10	11	12
Ellipsoid (f)	9	10	1	8	6	11	2	3	12	4	7	5
Ellipsoid (b)	9	10	1	8	5	11	2	12	7	4	3	6
Hyperbox (f)	1	5	10	11	8	2	9	12	3	4	6	7
Hyperbox (b)	1	5	11	8	2	12	9	10	7	6	4	3

Table 11.8 lists the recognition rate of the numeral data with the original features, the seven features selected by the four methods, and the seven features extracted by PCA. The recognition rate of the data selected by the forward ellipsoidal selection was lower than 99% when evaluated by the neural network classifier. It also happened for the features extracted by PCA and evaluated by the fuzzy min-max classifier with inhibition. But in general the four selection methods and PCA performed well.

Table 11.8. Recognition rate of the numeral data

Features	ELD (%)	MMI (%)	BP (%)
Original	99.63	99.63	99.43
Ellipsoid (f)	99.51	99.51	98.26
Ellipsoid (b)	99.39	99.51	99.43
Hyperbox (f)	99.39	99.51	99.43
Hyperbox (b)	99.15	99.51	99.26
PCA	99.39	98.90	99.23

Fig. 11.8 shows the recognition rates of the fuzzy classifier with ellipsoidal regions for the numeral data and the normalized exception ratio of the backward ellipsoidal selection. There was no much difference of recognition rates for the features with 7 to 12 features but the recognition rate using the features selected by the forward ellipsoidal selection dropped rapidly when one more feature was deleted. The exception ratio decreased as the features were deleted, and it began to increase when the original features were deleted to seven features. Thus by the backward ellipsoidal selection, eight features were selected.

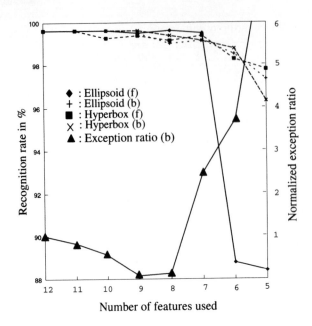

Fig. 11.8. Comparison of feature selection for the numeral data. The recognition rates for the test data were evaluated by the fuzzy classifier with ellipsoidal regions

11.4.5 Thyroid Data

Table 11.9 lists the features of the thyroid data set listed in the order of importance. Among 5 important features, 3 to 4 features were the same for any two selection methods excluding the forward hyperbox selection. Only the third feature was commonly selected for the forward hyperbox selection and other selection methods.

Table 11.9. Thyroid features in the order of importance

Method	1	2	3	4	5	6	7	8	9	10	11	12	13	14	15	16	17	18	19	20	21
Ellipsoid (f)	17	21	3	8	11	15	13	20	7	12	4	5	9	14	16	10	6	19	1	18	2
Ellipsoid (b)	17	21	19	3	18	12	11	5	15	13	16	7	8	9	1	4	2	14	20	10	6
Hyperbox (f)	7	19	6	3	4	8	16	13	5	15	12	2	9	10	14	11	20	1	18	17	21
Hyperbox (b)	21	3	17	18	8	1	19	20	15	13	16	12	14	11	10	9	7	6	5	4	2

Table 11.10 lists the recognition rate of the thyroid data with the original features, the five features selected by the four methods, and the five features extracted by PCA. The recognition rates of the three classifiers with the features extracted by PCA were not good. This was because the 5 principal

components did not give high accumulation of the eigenvalues as listed in Table 11.4. The recognition rates of the three classifiers with the features selected by the forward hyperbox selection were not good, especially for the fuzzy min-max classifier with inhibition. This coincided with the fact that only one feature was commonly selected for the forward hyperbox selection and the remaining selection methods as listed in Table 11.9.

Table 11.10. Recognition rate of the thyroid data

Features	ELD (%)	MMI (%)	BP (%)
Original	95.60	99.19	97.53
Ellipsoid (f)	95.95	98.66	<u>98.06</u>
Ellipsoid (b)	<u>96.65</u>	<u>99.21</u>	97.76
Hyperbox (f)	93.00	41.21	93.41
Hyperbox (b)	<u>96.65</u>	99.01	97.79
PCA	92.42	85.82	92.68

Fig. 11.9 shows the recognition rates of the fuzzy min-max classifier with inhibition for the thyroid data and the normalized exception ratio for the forward ellipsoidal selection. Although the recognition rate by the forward ellipsoidal selection was a little lower than those by the backward ellipsoidal/hyperbox selection methods, the recognition rates by the three were stable between 4 to 21 features.

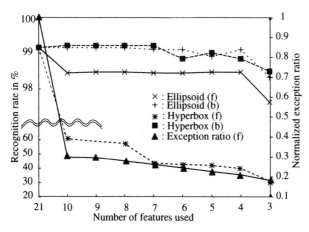

Fig. 11.9. Comparison of feature selection for the thyroid data. The recognition rates for the test data were evaluated by the fuzzy min-max classifier with inhibition

11.4.6 Blood Cell Data

Table 11.11 lists the features of the blood cell data set listed in the order of importance. Among 10 important features, 7 to 9 features were commonly selected by any two of the selection methods.

Table 11.11. Blood cell features in the order of importance

Method	1	2	3	4	5	6	7	8	9	10	11	12	13
Ellipsoid (f)	7	12	2	5	3	11	6	9	13	10	1	4	8
Ellipsoid (b)	7	12	2	5	6	11	1	10	9	3	4	13	8
Hyperbox (f)	7	12	3	5	13	11	6	1	2	4	9	8	10
Hyperbox (b)	7	12	1	2	6	4	3	8	5	11	9	13	10

Table 11.12 lists the recognition rate of the blood cell data with the original features, the ten features selected by the four methods, and the ten features extracted by PCA. The recognition rates for the selected features were comparable with those for the original features when fuzzy classifier with ellipsoidal regions and the fuzzy min-max classifier were used.

By deleting features, training by the back-propagation algorithm became slow and in some trial, training did not converge even after 15000 epochs. Therefore, we deleted the worst recognition rate and calculated the average for 9 trials. Thus when original features were used, the recognition rate of the test data improved from 89.75% to 90.16%. The recognition rates with the selected features were lower than that with the original features. But the recognition rate by PCA showed better than that with the original features.

Table 11.12. Recognition rate of the blood cell data

Features	ELD (%)	MMI (%)	BP (%)
Original	91.65	85.16	90.16
Ellipsoid (f)	91.65	85.45	88.28
Ellipsoid (b)	91.39	84.71	87.58
Hyperbox (f)	91.45	85.61	86.62
Hyperbox (b)	90.61	85.45	88.75
PCA	89.87	83.23	90.85

Fig. 11.10 shows the recognition rates of the fuzzy classifier with ellipsoidal regions for the blood cell data and the normalized exception ratio for

the forward ellipsoidal selection. The normalized exception ratio started to increase when the eighth feature was added. Thus seven features were selected by the forward ellipsoidal selection. The recognition rates dropped for 9 to 7 features using the backward hyperbox selection. The high recognition rates were maintained for 9 to 7 features using the ellipsoidal selection methods.

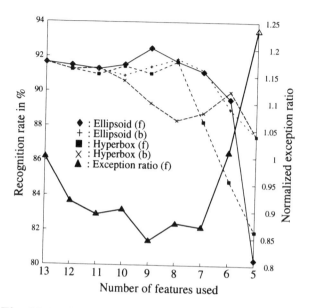

Fig. 11.10. Comparison of feature selection for the blood cell data. The recognition rates for the test data were evaluated by the fuzzy classifier with ellipsoidal regions

Computation Time. We compared the selection time of the forward/backward ellipsoidal/hyperbox selection methods for the thyroid and blood cell data. The time was measured using a Sun ULTRA10 (336MHZ) workstation.

Fig. 11.11 shows the feature selection time of the forward/backward ellipsoidal selection methods for the thyroid data. In the figure, "Forward (I)" denotes the forward ellipsoidal selection without acceleration and "Forward (II)" denotes the forward ellipsoidal selection with symmetric Cholesky factorization and deletion of redundant calculations.

Table 11.13 lists the computation time by the methods. Each of the four selection methods was terminated when 7 features for the thyroid data and 10 features for the blood cell data were selected. For the thyroid data, the forward hyperbox selection was the fastest but for the blood sell data the backward hyperbox selection was the fastest. The reason why the backward hyperbox selection was faster than the forward hyperbox selection was as follows. For the blood cell data three features were deleted by the backward selection but 10 features needed to be added by the forward selection. For the blood cell data, the backward hyperbox selection was the second fastest and for the thyroid data the forward ellipsoidal selection with acceleration was the second fastest. The forward ellipsoidal selection could compute 9 times faster than the backward ellipsoidal selection for the thyroid data.

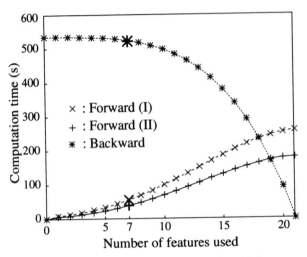

Fig. 11.11. Computation time for the thyroid data

Table 11.13. Computation time for feature selection

Method	Thyroid (s)	Blood Cell (s)
Ellipsoid (f) (I)	57	63
Ellipsoid (f) (II)	39	46
Ellipsoid (b)	524	96
Hyperbox (f)	17	13
Hyperbox (b)	59	10

11.4.7 Discussions

The backward/forward ellipsoidal/hyperbox selection methods are local optimization methods: they delete or add one feature at a time that optimizes the exception ratio. Thus, the global optimality of the features selected by those methods is not guaranteed. But according to the simulations, performance of the four methods showed better than or comparable to that of PCA. The forward/backward ellipsoidal selection methods were robuster than the forward/backward hyperbox selection methods, since the formers could approximates class regions well even when the distribution of class data was not parallel to the feature axes. But the weakness of the ellipsoidal selection is non-monotonicity of the exception ratio. Since the exception ratio approximated by hyperboxes is monotonic, the optimum set of features can be obtained by the branch and bound method [94].

Although the forward hyperbox selection showed inferior performance compared with the backward hyperbox selection, the forward ellipsoidal selection showed comparable performance with the backward ellipsoidal selection, and the selection time was faster than that of the backward ellipsoidal selection.

12. Generation of Training and Test Data Sets

Generation of training and test data sets from gathered data is an important task in developing a classifier with high generalization ability. In some applications, some of the class data, e.g. abnormal class data for diagnosis problems, may be difficult to obtain while the remaining class data (normal class data for diagnosis problems) are easily obtained. Thus, the number of data for each class is unbalanced. In other applications such as recognition of hand-written characters, there may be a huge number of data for each class.

Resampling techniques [106], which are widely used in statistical analysis, are used for model selection by estimating the classification performance of classifiers.

In this chapter, first we discuss the resampling techniques and their possibility in generating training and test data sets [107]. Then we consider partitioning the gathered data into the training and test data sets. When we have sufficient data, which are neither too many or nor too small in number, that represent the true distribution, it is better to divide the data into the training and test data sets with similar distributions [108]. To have the data sets with similar distributions, it is advised to divide the data randomly [109]. Here we discuss partitioning the data so that partitioned classes have similar centers and covariance matrices, and evaluate the usefulness of the method [110].

12.1 Resampling Techniques

Resampling techniques are used for estimating statistics such as the mean and the median by randomly selecting data from the given data set, calculating statistics on these data, and repeating the above procedure many times.

12.1.1 Estimation of Classification Performance

For a given classification problem, it is important to choose the classifier that realizes the best classification performance. This is called model selection. But when the number of gathered data is small, it is difficult to estimate the true classification performance. In such a situation, resampling techniques such as cross-validation and bootstrap are used.

Cross-validation. Cross-validation is used to measure the classification performance of classifiers for a limited number of gathered data. In cross-validation, the M given data are divided into two data sets S_i^{tr} ($i = 1, \ldots, {}_M C_l$) which includes l training data and S_i^{ts} which includes $M - l$ test data. Then for the training data set S_i^{tr} the classifier is trained and tested for the test data set S_i^{ts}. This is repeated for all the combinations $({}_M C_l)$ of the partitioned training and test data sets, and the total recognition rate for all the test data sets is calculated as the estimation of the classification performance. But since this is a time consuming task, a special case of cross-validation, a leave-one-out method (i.e., $l = 1$), is usually used.

When $l > 1$ and the numbers of data for some classes are small, we must be careful not to make the training data set of any class be empty. But for the leave-one-out method we do not need to worry about this.

Bootstrap. The bootstrap estimates the statistic such as the mean and the median by resampling with replacement from the gathered data many times. Using M data $x_1, \ldots, x_M$, we consider estimating the mean by the bootstrap. We resample l data K times from the original M data with replacement, meaning allowing multiple selection of one datum. Let the jth sampled data set be $\{x_1^j, \ldots, x_l^j\}$ and we calculate the mean for the jth sampled data set:

$$\bar{x}^j = \sum_{i=1}^{l} x_i^j. \tag{12.1}$$

Now the estimate of the mean is given by the mean of (12.1):

$$\hat{x} = \sum_{i=K}^{l} \bar{x}^j \tag{12.2}$$

with the standard deviation:

$$\sigma_l = \sqrt{\left(\frac{1}{K-1} \sum_{i=1}^{K} (\bar{x}^i - \hat{x})^2\right)}. \tag{12.3}$$

Instead of the standard deviation we can estimate the bias of the estimate from the true value and the confidence interval [106].

Classification performance can be estimated by bootstrap. In [106, pp. 247–255], the classification performance is estimated by only using the data sampled, i.e., the data used for training the classifier. Here, we estimate the classification performance similar to cross-validation. From the class i data set $S^i = \{\mathbf{x}_1^i, \ldots, \mathbf{x}_{M_i}^i\}$, we resample l data with replacement K times. Let the jth sampled data set be $S^{ij} = \{\mathbf{x}_1^{ij}, \ldots, \mathbf{x}_l^{ij}\}$. Then we train the classifier using S^{ij}, and evaluate the recognition rate using $S^i - S^{ij}$, namely using the data not used for training. Combining the K recognition rates we get the estimate of the classification performance.

12.1.2 Effect of Resampling on Classifier Performance

Cross-validation and bootstrap are useful in estimating the classifier performance for a given problem. In addition to estimating the performance, if we can select the best classifier from among the same type of trained classifiers, the best training and test data sets are also selected. However, there is no systematic way of doing this. Instead of selecting one classifier, we may select plural classifiers and form a committee of classification [30]. Namely, an unknown datum is classified by the classifiers' voting.

By resampling with replacement, we can increase the number of training data if we set $l > M$. In the following, we discuss how this affects performance of fuzzy classifiers. For the fuzzy min-max classifier and the fuzzy min-max classifier with inhibition, the hyperboxes defined with duplicate data in the same class are the same with those without them. For the fuzzy classifier with pyramidal membership functions and the fuzzy classifier with polyhedral regions, the initial fuzzy regions defined with duplicate data in the same class are the same with those without them. However, since the recognition rates change by the inclusion of duplicate data, the tuned membership functions may be different.

For the fuzzy classifier with ellipsoidal regions, the center vectors and thus the covariance matrixes change by the inclusion of duplicate data. Thus, the initial values of the parameters differ when duplicate data are added. But if the covariance matrix of a class is singular, it remains to be singular even if duplicated data are added (see Theorem B.3.1 on page 311). To avoid singularity, it may be better to add a small random noise to duplicate data.

12.2 Division of Data by Pairing

When sufficient data are gathered to represent the true distribution of the data, it is better to divide the data into the training and test data sets with similar distributions. This is because the distribution of the training data set becomes the closest to the true distribution. Similarly, when the gathered data are not sufficient, it is considered to be effective to divide the data into two data sets with similar distributions.

Here we consider that the two data sets belonging to the same class are similar when the associated centers and covariance matrices are similar. To partition the data, we consider a heuristic division algorithm in which we calculate the center and covariance matrix of class data, search a pair of the data which are nearest in some measure, and put one of the pair into the training data set and the other into the test data set [110]. We call this method division of data by pairing (DDP).

Let X_i be a set of data belonging to class i. We calculate the center $\mathbf{c}_i$ and the covariance matrix Q_i for class i data by

$$c_i = \frac{1}{|X_i|} \sum_{x \in X_i} x, \tag{12.4}$$

$$Q_i = \frac{1}{|X_i|} \sum_{x \in X_i} (x - c_i)(x - c_i)^t, \tag{12.5}$$

respectively, where X_i is the set of class i data and $|X_i|$ is the number of data in X_i. We define the Mahalanobis distance of x belonging to class i from the center c_i by $d_i(x)$ given by

$$d_i^2(x) = (x - c_i)^t Q^{-1}(x - c_i). \tag{12.6}$$

Then we assume that the conditions that x and $x' \in X_i$ become a pair are given by

$$|d_i(x) - d_i(x')| < \varepsilon_1, \tag{12.7}$$

$$\cos\theta = \frac{(x - c_i)^t (x' - c_i)}{\|x - c_i\| \|x' - c_i\|} > \varepsilon_2, \tag{12.8}$$

where $\varepsilon_1 > 0$, $1 > \varepsilon_2 > 0$ and

$$\|x - c_i\| = \sqrt{\sum_{j=1}^{m} (x_j - c_{ij})^2}. \tag{12.9}$$

Fig. 12.1 shows the concept of selecting a pair of data. In the figure for a datum x, we search the data that satisfy (12.7) and (12.8). Suppose we find two data x' and x''. We select x' whose Mahalanobis distance is nearer.

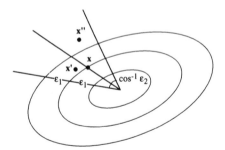

Fig. 12.1. Selection of a pair [110, p. 90]

Instead of using (12.7) and (12.8), we may use the following criterion:

$$(x - x')^t Q_i^{-1}(x - x') < \varepsilon. \tag{12.10}$$

The problem with this criterion is that, as shown in Fig. 12.2, the three pairs of data with different Mahalanobis distances have the same values of (12.10), and thus it is difficult to make the characteristics of the training and test data sets similar.

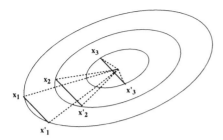

Fig. 12.2. Problem with the criterion given by (12.10) [110, p. 90]

Based on (12.7) and (12.8), the algorithm to divide the set of data belonging to class i $(i = 1, \ldots, n)$, X_i, is as follows:

1. Calculate $\mathbf{c}_i$ and Q_i using (12.4) and (12.5), respectively. Then, calculate the Mahalanobis distance of each datum using (12.6).
2. Sort the elements of X_i in the increasing order of the Mahalanobis distances. We denote the ordered set X_i'.
3. Find a pair of data that satisfies both (12.7) and (12.8).
4. Among the pairs that are found in Step 3, select the pair with the smallest difference of the Mahalanobis distances and delete the pair from the data set. Put one datum in the pair into the training data set and the other into the test data set.
5. Do the above procedure for all the classes. If the number of pairs generated is sufficient, randomly divide the remaining data into the training and test data sets and terminate the algorithm. If not, changing ε_1 and ε_2, iterate the above procedure.

12.3 Similarity Measures

To check whether the centers and covariance matrices of the two data sets are similar, we need to use similarity measures for vectors and matrices. To measure vector similarity, we define the relative difference of the infinite norm of vectors [47] and to measure matrix similarity, we diagonalize two covariance matrices simultaneously [108, pp. 31-34] and use the similarity measure of vectors.

12.3.1 Relative Difference of Center Norms

We define the relative difference of the infinite center norms for class i training and test data as follows:

$$\Delta \mathbf{c}_i = \frac{|\; ||\mathbf{c}_i^{tr}||_\infty \; - \; ||\mathbf{c}_i^{ts}||_\infty \;|}{||\mathbf{c}_i^{tr}||_\infty}, \tag{12.11}$$

where

$$\|\mathbf{c}_i^{tr}\|_\infty = \max_{1 \le i \le m} |c_i^{tr}|, \tag{12.12}$$

$$\|\mathbf{c}_i^{ts}\|_\infty = \max_{1 \le i \le m} |c_i^{ts}|. \tag{12.13}$$

Here, $\mathbf{c}_i^{tr} = (c_{i1}^{tr}, \ldots, c_{im}^{tr})^t$ is the center of class i training data and $\mathbf{c}_i^{ts} = (c_{i1}^{ts}, \ldots, c_{im}^{ts})^t$ is the center of class i test data. Then we define the average relative difference of center norms as follows:

$$\Delta \mathbf{c} = \frac{1}{n} \sum_{i=1}^{n} \Delta \mathbf{c}_i. \tag{12.14}$$

12.3.2 Relative Difference of Covariance Matrix Norms

We denote the covariance matrices of class i training and test data as Q_i^{tr} and Q_i^{ts}, respectively. Let the diagonal matrix with eigenvalues of Q_i^{tr} and the matrix with the associated eigenvectors be Θ and Φ, respectively. Then by the following transformation, we first diagonalize Q_i^{tr}.

$$\Theta^{-1/2} \, \Phi^t \, Q_i^{tr} \, \Phi \, \Theta^{-1/2} = I, \tag{12.15}$$

$$\Theta^{-1/2} \, \Phi^t \, Q_i^{ts} \, \Phi \, \Theta^{-1/2} = K. \tag{12.16}$$

Here let the eigenvalues of Q_i^{tr} be $\lambda_1, \ldots, \lambda_m$. Then $\Theta^{-1/2} = \mathrm{diag}((\lambda_1)^{-1/2}, \ldots, (\lambda_m)^{-1/2})$. If the eigenvalue λ_i is 0, we replace $\lambda_i^{-1/2}$ with 0.

Next, let the matrix consisting of the eigenvectors of K and the diagonal matrix with the eigenvalues be Ψ and I^{ts}, respectively and do the following transformation.

$$\Psi^t \, I \, \Psi = \Psi^t \, \Psi = I, \tag{12.17}$$

$$\Psi^t \, K \, \Psi = I^{ts}. \tag{12.18}$$

By the transformation, Q_i^{tr} remains to be the unit matrix I and Q_i^{ts} becomes the diagonal matrix I^{ts}. Using these two diagonal matrices, we define the relative norm difference ΔQ_i between Q_i^{tr} and Q_i^{ts} as follows:

$$\Delta Q_i = \frac{|\, \|I\|_\infty - \|I^{ts}\|_\infty \,|}{\|I\|_\infty}, \tag{12.19}$$

where

$$\|I^{tr}\|_\infty = \max_{1 \le i \le m} |I_{ii}| = 1, \tag{12.20}$$

$$\|I^{ts}\|_\infty = \max_{1 \le i \le m} |I_{ii}^{ts}|. \tag{12.21}$$

Using (12.19), we define the average relative difference between the covariance matrix norms as follows:

$$\Delta Q = \frac{1}{n} \sum_{i=1}^{n} \Delta Q_i. \tag{12.22}$$

12.4 Performance Evaluation

12.4.1 Evaluation Conditions

Using the blood cell data and the hiragana-50 data listed in Table 1.1 on page 19, we evaluated the recognition improvement for the test data by DDP over the random and original divisions using the fuzzy classifier with ellipsoidal regions.

We tuned only the slopes of membership functions and we set the parameter $\delta = 0.1$ and the maximum allowable number of misclassifications plus one, l_M, to be 10. When the covariance matrices were singular, which happened for the hiragana-50 data, the matrices were diagonalized. We used a Sun UltraSPARC IIi 300MHz workstation.

To evaluate the performance of DDP, we compared the average relative differences of the center and covariance matrix norms for the original data, the data divided randomly, and the data divided by DDP. To divide the data randomly, we assign a uniform random number for each datum and using the random number as the key, we sorted the data and we put the first half of the data into the training data set and the remaining data into the test data set.

In DDP, we divided the data with $\varepsilon_1 = 1.2 \times \text{avr}$ and $\varepsilon_2 = \cos 30°$, where avr is the average distance given by

$$\text{avr} = \frac{1}{n} \sum_{i=1}^{n} \left(\max_{\mathbf{x} \in X_i} d_i(\mathbf{x}) - \min_{\mathbf{x} \in X_i} d_i(\mathbf{x}) \right). \tag{12.23}$$

Then we divided the data not paired with $\varepsilon_1 = 1.5 \times \text{avr}$ and $\varepsilon_2 = \cos 45°$. Further we divided the data with $\varepsilon_1 = 2.0 \times \text{avr}$ and $\varepsilon_2 = \cos 60°$.

12.4.2 Blood Cell Data

Table 12.1 lists the performance for the blood cell data. We divided the data randomly twice. In the table the figures in the brackets show the recognition rates for the training data. The time to divide the data by DDP was 21 seconds and the rate of the paired data was 67.6%. The time to divide the data randomly was 11 seconds. From the table, there were not so much difference among the $\Delta \mathbf{c}$'s for the data obtained by DDP, the randomly divided data, and the original data, but the ΔQ for data obtained by DDP was the smallest and the recognition rate of the test data by DDP was the best. The recognition rate of the randomly divided test data was better than that of the original test data. For the blood cell data, there were some overlapping classes and DDP could divide the data around these boundaries into the training data and the test data. Thus the recognition rate of the test data was improved.

Table 12.1. Effect of the data division methods for the blood cell data [110, p. 92]

Data	ΔQ	Δc	Rates (%)
DDP	<u>1.00</u>	<u>0.00028</u>	<u>94.09</u> (95.39)
Original	2.10	0.00095	91.65 (95.41)
Random 1	2.00	0.00043	92.74 (95.90)
Random 2	2.97	0.00028	92.90 (<u>96.19</u>)

12.4.3 Hiragana-50 Data

Table 12.2 lists the performance for the hiragana-50 data. The execution time of DDP was 162 seconds and the rate of generated pairs was 54.42 %. The execution time of the random division was 120 seconds. For the original data the value of Δc was large but those for DDP and the random division were small and there was not much difference between the two. The value of ΔQ for DDP was the smallest but the recognition rate of the test data by DDP was comparable to those of the random division but better than that of the original data. The reason why the recognition rates of DDP and the random division were similar was that the overlap of the hiragana-50 data was not so heavy compared to the blood cell data and the effect of DDP was small.

Table 12.2. Effect of the data division methods for the hiragana-50 data [110, p. 92]

Data	ΔQ	Δc	Rates (%)
DDP	<u>1.588</u>	0.0026	<u>88.85</u> (<u>99.24</u>)
Original	4.092	0.0182	81.30 (99.11)
Random 1	1.872	<u>0.0020</u>	88.84 (99.22)
Random 2	3.424	0.0026	88.84 (<u>99.24</u>)

12.4.4 Discussions

The recognition rates of the data sets generated by DDP and random division were better than those of the original data sets. Thus, at least random division is necessary in generating the data sets. In addition, for a difficult classification problem such as blood cell classification, DDP was effective in generating the data sets with high generalization ability.

Since some of the covariance matrices were diagonalized for the hiragana-50 data, the recognition rates were not so good. When the singular values of the covariance matrices were restricted, the recognition rates could be improved as discussed in Section 5.2.3. The recognition rate of the original test data was 94.51% with $\eta = 10^{-4}$ (see Table 7.24 on page 155) and that of the test data generated by DDP was 97.06% with $\eta = 10^{-3}$.

Part II

Function Approximation

13. Introduction

Function approximation is to determine the input-output relations using the training input-output pairs. There are many function approximation methods. The linear regression method is used when the input-output relation is linear or quasi-linear. Here our interest is nonlinear function approximation with no a priori knowledge of input-output relations. Multilayer neural networks are well suited for this purpose. But the major drawbacks are slow training by the back-propagation algorithm and the difficulty in analyzing the network behavior.

Conventional fuzzy systems are defined based on the expert's knowledge and their analysis is easily done by the fuzzy rules. But usually it is difficult to define fuzzy rules and in general approximation performance is inferior to that of multilayer neural networks. In addition, since in conventional fuzzy systems the input space is divided into subregions in advance and for each subregion a fuzzy rule is defined, the curse of dimensionality occurs; the number of fuzzy rules explodes as the number of input variables increases. Thus conventional fuzzy systems are limited to the problems with a small number of input variables.

In the second part of the book our discussions focus on the fuzzy systems that define fuzzy rules according to the existence of the training data; thus this solves the curse of dimensionality. In the following we call fuzzy systems that perform function approximation fuzzy function approximators.

13.1 Function Approximators

There are many approaches for function approximation. Here we survey neural networks, conventional fuzzy function approximators, and fuzzy function approximators with learning capability. There are numerous methods that fuse neural networks and fuzzy systems [3, 4], but we do not include them here.

13.1.1 Neural Networks

Originally multilayer neural networks were developed as classifiers, but since they were shown to be universal approximators [28, 29], they have been widely used for function approximation. The use of sigmoid functions (see Fig. 1.4 (c) on page 8), which are global functions, as output functions of neurons makes training of the networks slow. But the generalization ability is relatively good for a wide range of applications. Since the structures of multilayer neural networks for pattern classification and function approximation are the same, and we can use the back-propagation algorithm for training, we do not repeat the details here (cf. Chapter 2).

Radial basis function neural networks consist of three layers and use the Gaussian function (see Fig. 1.4 (d) on page 8) as the output function of hidden neurons and use the liner function as the output function of the output neurons. Radial basis function neural networks are considered to be both neural networks and fuzzy systems since the Gaussian function can be considered to be the membership function of the if-part of a fuzzy rule.

13.1.2 Conventional Fuzzy Function Approximators

Suppose we want to determine the output y using the two inputs x_1 and x_2. In a conventional fuzzy function approximator, we first divide the ranges of input and output variables into several intervals. Suppose we divide each range into two intervals labeled "small" and "large." These intervals are defined as fuzzy sets and their one-dimensional membership functions discussed in Section 4.1 are defined. Let the membership functions for "small" and "large" for the variable z be $m_s(z)$ and $m_l(z)$. In this case, the input space is divided into four subregions and according to the knowledge of the problem we define for each subregion a fuzzy rule. For example, we define the following two fuzzy rules:

$$\text{If } x_1 \text{ is large and } x_2 \text{ is large then } y \text{ is large,} \tag{13.1}$$

$$\text{If } x_1 \text{ is small and } x_2 \text{ is small then } y \text{ is small.} \tag{13.2}$$

Using fuzzy rules given by (13.1) and (13.2), we explain how to determine the output value of y for the crisp input vector (a_1, a_2) by Mamdani's method [2]. The process of determining the output is divided into two stages: determination of the fuzzy set (the membership function) of y for the inputs (a_1, a_2), and defuzzification of the fuzzy set, namely determination of the output value of y using the fuzzy set.

Since the input vector (a_1, a_2) is crisp, we have the proposition:

$$(x_1, x_2) \text{ is } a_1 \cap a_2, \tag{13.3}$$

where a_1 and a_2 are special cases of fuzzy sets called singletons in which for a_i $(i = 1, 2)$ the degree of membership is 1 when $x_i = a_i$ and 0, otherwise, and $\cap$ is the AND operator. Using the proposition given by (13.3) and the if-then rules given by (13.1) and (13.2), we determine the fuzzy set of y.

According to Mamdani's definition, the fuzzy sets in the if-part and the then-part of a fuzzy rule are connected by the AND operator. Thus (13.1) and (13.2) are rewritten as follows:

$$(x_1, x_2, y) \text{ is large} \cap \text{large} \cap \text{large}, \tag{13.4}$$

$$(x_1, x_2, y) \text{ is small} \cap \text{small} \cap \text{small}, \tag{13.5}$$

respectively. Since each fuzzy rule is connected by the OR operator $\cup$, (13.4) and (13.5) reduce to

$$(x_1, x_2, y) \text{ is (large} \cap \text{large} \cap \text{large)} \cup (\text{small} \cap \text{small} \cap \text{small}). \tag{13.6}$$

Since the proposition and the if-then rules are combined by the AND operator, from (13.3) and (13.5), we obtain

$$y \text{ is } (a_1 \cap a_2) \cap \{(\text{large} \cap \text{large} \cap \text{large}) \cup (\text{small} \cap \text{small} \cap \text{small})\}. \tag{13.7}$$

Now we calculate the membership function $m(y)$ associated with the fuzzy set given by (13.7). Assuming that the AND and OR operators correspond to the maximum and minimum operators, respectively, we obtain the following membership function:

$$
\begin{aligned}
m(y) &= (m_{a_1}(a_1) \wedge m_{a_2}(a_2)) \wedge \{(m_l(x_1) \wedge m_l(x_2) \wedge m_l(y)) \\
&\quad \vee (m_s(x_1) \wedge m_s(x_2) \wedge m_s(y))\} \\
&= (m_l(a_1) \wedge m_l(a_2) \wedge m_l(y)) \vee (m_s(a_1) \wedge m_s(a_2) \wedge m_s(y)), \quad (13.8)
\end{aligned}
$$

where $m_{a_1}(x_1)$ and $m_{a_2}(x_2)$ are the membership functions for the singletons and $m_{a_i}(x_i)$ $(i = 1, 2)$ is 1 for $x_i = a_i$ and 0 otherwise, and $\wedge$ and $\vee$ denote the minimum and maximum operators, respectively. In rewriting the second equation into the third equation, we used the associative law. Since $m_l(a_1)$ and $m_l(a_2)$ are constant values, the membership function $m_l(y)$ is bounded by the minimum of $m_l(a_1)$ and $m_l(a_2)$. Likewise, $m_s(y)$ is bounded by the minimum of $m_s(a_1)$ and $m_s(a_2)$.

The second stage is to determine the output value of y using the fuzzy set whose membership function is given by (13.8). We usually determine the output by the center-of-gravity method as follows:

$$\hat{y} = \frac{\int_Y m(y)\, y\, dy}{\int_Y m(y)\, dy}, \tag{13.9}$$

where Y is the range of y.

Fig. 13.1 illustrates the above inference procedure. The minimum value of the degrees of membership of the if-part of each rule is calculated. This in turn restricts the maximum value of the membership function of the then-part. The maximum of the two membership functions is taken and the crisp value is obtained by the center-of-gravity method.

The major advantage of the conventional fuzzy function approximators is that if- and then-parts are written in words such as "x_1 is small," we can easily understand the input-output relations. But since the input space needs to be divided in advance and for each subregion a fuzzy rule is defined, the curse of dimensionality becomes a problem for a large number of input variables. In addition, we need to know the input-output relations in advance. To facilitate fuzzy rule acquisition, training capability similar to that of neural networks

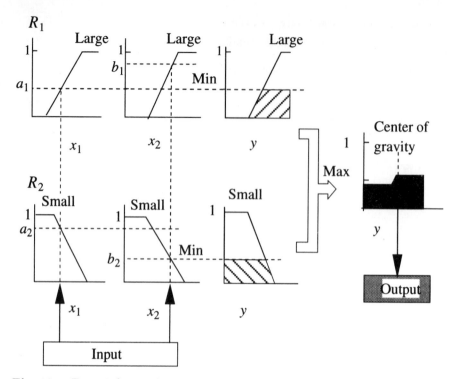

Fig. 13.1. Fuzzy inference by Mamdani's method

has been introduced. To avoid the curse of dimensionality, multi-dimensional membership functions discussed in Section 4.2 have been used and fuzzy rules have been defined according to the distribution of the training data.

13.1.3 Fuzzy Function Approximators with Learning Capability

Conventional fuzzy systems have no training capability but recently many fuzzy systems with training capability have been proposed [2, 7, 60, 111, 112, 113, 114, 115, 116, 117, 118, 119, 120]. In general, these fuzzy systems have faster training capability than, and comparable generalization ability with multilayer neural networks. Similar to the rule generation by fuzzy classifiers, these fuzzy systems generate fuzzy rules either by preclustering or postclustering. But unlike pattern classification, since we do not need to resolve overlaps between fuzzy regions, there is no concept of dynamic clustering for function approximation.

The fuzzy rules extracted by these fuzzy systems are classified into two types according to the input regions: hyperbox regions and ellipsoidal regions. The fuzzy rules generated by preclustering include conventional fuzzy rules that are generated by dividing input and output spaces into subregions in advance [7, 111, 112]. By postclustering, input subregions are defined during fuzzy rule generation [113, 119, 120].

Radial basis function networks are considered to be one type of fuzzy function approximator in which the output is synthesized by the linear combination of the degrees of membership of the fuzzy rules with ellipsoidal regions [60, 116]. Fuzzy rules with ellipsoidal regions are generated by preclustering [115, 117] or postclustering [60, 116]. The center and the covariance matrix of a fuzzy rule need to be determined. The center is determined based on cluster estimation [115, 117] or on the approximation error [60, 116]. The covariance matrix is either fixed [115, 116], estimated by the steepest descent method [60], unsupervised learning [117], or calculated using the training data around the center [119]. In Section 15.3, similar to tuning membership functions for pattern classification, we discuss tuning centers and slopes of the membership functions (covariance matrices) one at a time so that the approximation error is minimized.

14. Fuzzy Rule Representation and Inference

14.1 Fuzzy Rule Representation

We discuss function approximation with the m-dimensional input vector $\mathbf{x}$ and the one-dimensional output y. (Extension to the multi-dimensional output is straightforward.) The fuzzy rules are given by

$$R_i: \quad \text{If } \mathbf{x} \text{ is } A_i \quad \text{then } y = o_i \text{ for } i = 1, \ldots, N, \tag{14.1}$$

where A_i is the ith fuzzy region, o_i is the corresponding output, and N is the number of fuzzy rules. The fuzzy region A_i can be a hyperbox, polyhedral, or ellipsoidal region. In this book we consider only hyperbox and ellipsoidal regions. On A_i, we define one of the membership functions discussed in Section 4.2, but here we use (rectangular) pyramidal membership functions and bell-shaped membership functions. We call the fuzzy rules with (rectangular) pyramidal membership functions the hyperbox fuzzy rules and the fuzzy rules with bell-shaped membership functions the ellipsoidal fuzzy rules. Unlike conventional fuzzy rules, we do not define the membership function for the then-part.

Instead of using constant o_i in (14.1), we can use the linear combination of input variables as follows [121]:

$$o_i = p_{i0} + p_{i1} x_1 + \ldots + p_{im} x_m \quad \text{for } i = 1, \ldots, N, \tag{14.2}$$

where $p_{i0}, p_{i1}, \ldots, p_{im}$ are constants and are determined by the least-squares method. This is called the Takagi-Sugeno type model.

Hyperbox Fuzzy Rules. When hyperboxes are not parallel to input axes, they can express the correlation among input variables [122], but here we do not consider this.

Let the training data be clustered into S_i ($i = 1, \ldots, N$). Then for cluster i we define the hyperbox A_i that includes the subset of the training data, S_i, as follows:

$$A_i = \{\mathbf{x} \mid v_{ik} \leq x_k \leq V_{ik}, k = 1, \ldots, m\}, \tag{14.3}$$

where x_k is the kth element of $\mathbf{x}$, and v_{ik} and V_{ik} are, respectively, the minimum and maximum values of the hyperbox A_i with respect to x_k. Namely,

$$v_{ik} = \min_{\mathbf{x} \in S_i} x_k,$$ (14.4)

$$V_{ik} = \max_{\mathbf{x} \in S_i} x_k.$$ (14.5)

Then the center vector $\mathbf{c}_i$ of the hyperbox A_i is given by

$$\mathbf{c}_i = \left(\frac{V_{i1} + v_{i1}}{2}, \ldots, \frac{V_{im} + v_{im}}{2} \right)^t.$$ (14.6)

Now we define, for A_i, a (truncated) rectangular pyramidal membership function with the minimum, product, or average operator discussed in Sections 4.2.2 and 4.2.3.

Hyperbox Fuzzy Rules with Rectangular Pyramidal Membership Functions. First we define, for the input vector $\mathbf{x}$, a one-dimensional tuned distance $h_{ik}(\mathbf{x})$ $(k = 1, \ldots, m)$ as follows:

$$h_{ik}^s(\mathbf{x}) = \frac{d_{ik}^s(\mathbf{x})}{\alpha_i},$$ (14.7)

where $s = 1$ or 2 and α_i is the tuning parameter that determines the slope of the membership function and the initial value is set to be 1, and $d_{ik}(\mathbf{x})$ is the one-dimensional weighted distance given by

$$d_{ik}(\mathbf{x}) = \frac{|c_{ik} - x_k|}{w_{ik}},$$ (14.8)

where c_{ik} is the kth element of $\mathbf{c}_i$ and

$$w_{ik} = \frac{V_{ik} - v_{ik}}{2}.$$ (14.9)

To avoid zero division, if $w_{ik} < \varepsilon$ where ε is a small positive number, in (14.8) we replace w_{ik} with ε.

Then the one-dimensional membership function for input variable x_k $(k = 1, \ldots, m)$, $m_{ik}(\mathbf{x})$, is defined as follows:

$$m_{ik}(\mathbf{x}) = \begin{cases} 0 & \text{for} \quad x_k > c_{ik} + \alpha_i\, w_{ik}, \\ 1 - h_{ik}^s(\mathbf{x}) & \text{for} \quad c_{ik} + \alpha_i\, w_{ik} \geq x_k \geq c_{ik} - \alpha_i\, w_{ik}, \\ 0 & \text{for} \quad x_k < c_{ik} - \alpha_i\, w_{ik}. \end{cases}$$ (14.10)

Hyperbox Fuzzy Rules with Truncated Rectangular Pyramidal Membership Functions. First, we define a one-dimensional tuned distance for the input variable x_k, $h_{ik}(\mathbf{x})$:

$$h_{ik}(\mathbf{x}) = \frac{d_{ik}(\mathbf{x})}{\alpha} = \begin{cases} \dfrac{x_k - V_{ik}}{\alpha_i} & \text{for} \quad V_{ik} + \alpha_i \geq x_k \geq V_{ik}, \\ 0 & \text{for} \quad V_{ik} > x > v_{ik}, \\ \dfrac{v_{ik} - x_k}{\alpha_i} & \text{for} \quad v_{ik} > x_k \geq v_{ik} - \alpha_i, \end{cases}$$ (14.11)

where $d_{ik}(\mathbf{x})$ is a one-dimensional distance. Using (14.11), the one-dimensional membership function is given by

$$
m_{ik}(\mathbf{x}) = \begin{cases} 0 & \text{for} \quad x_k > V_{ik} + \alpha_i, \\ 1 - h_{ik}^s(\mathbf{x}) & \text{for} \quad V_{ik} + \alpha_i \geq x_k \geq v_{ik} - \alpha_i, \\ 0 & \text{for} \quad x_k < v_{ik} - \alpha_i, \end{cases} \tag{14.12}
$$

where $s = 1$ for linear membership functions and $s = 2$ for quadratic membership functions.

Multi-dimensional Membership Functions. The degree of membership for the input variable $\mathbf{x}$, $m_i(\mathbf{x})$, can be defined in three ways using (14.10) or (14.12). The first one uses the minimum operator:

$$
m_i(\mathbf{x}) = \min_{k=1,\ldots,m} m_{ik}(\mathbf{x}). \tag{14.13}
$$

The second one uses the product operator:

$$
m_i(\mathbf{x}) = \prod_{k=1}^{m} m_{ik}(\mathbf{x}). \tag{14.14}
$$

The third one uses the average operator:

$$
m_i(\mathbf{x}) = \frac{1}{m} \sum_{k=1}^{m} m_{ik}(\mathbf{x}). \tag{14.15}
$$

Instead of (14.10), if we use

$$
m_i = \exp\left(-h_{ik}^2(\mathbf{x})\right) \tag{14.16}
$$

as the one-dimensional membership function and the product operator for generating the m-dimensional membership function, the resulting membership function is equivalent to (14.18) with the diagonal covariance matrix.

Ellipsoidal Fuzzy Rules. The ellipsoidal fuzzy rules are given by

$$
R_i: \quad \text{If } \mathbf{x} \text{ is } \mathbf{c}_i \quad \text{then } y = o_i \quad \text{for} \quad i = 1, \ldots, N, \tag{14.17}
$$

where $\mathbf{c}_i$ is the center of the ith fuzzy rule, o_i is the corresponding output, and N is the number of fuzzy rules. The degree of membership of the fuzzy rule R_i, $m_i(\mathbf{x})$, is given by

$$
m_i(\mathbf{x}) = \exp\left(-h_i^2(\mathbf{x})\right), \tag{14.18}
$$

$$
h_i^2(\mathbf{x}) = \frac{d_i^2(\mathbf{x})}{\alpha_i}, \tag{14.19}
$$

$$
d_i^2(\mathbf{x}) = (\mathbf{x} - \mathbf{c}_i)^t\, Q_i^{-1}\, (\mathbf{x} - \mathbf{c}_i), \tag{14.20}
$$

where $h_i(\mathbf{x})$ is the tuned distance, α_i is the tuning parameter with initial value of 1 and is tuned after fuzzy rules are extracted, $d_i(\mathbf{x})$ is the weighted distance between $\mathbf{x}$ and $\mathbf{c}_i = (c_{i1}, \ldots, c_{im})^t$, Q_i is the $m \times m$ covariance matrix. The covariance matrix is set in three ways:

1. a constant diagonal matrix with the same diagonal element $\sigma^2 (> 0)$;
2. the diagonal matrix calculated using the set of data around the center $\mathbf{c}_i$, S_i, as follows:

$$Q_{i,jj} = \frac{1}{|S_i|} \sum_{\mathbf{x} \in S_i} (x_j - c_{ij})^2, \tag{14.21}$$

where $|S_i|$ is the number of data in S_i. If the diagonal elements are zero, they are replaced by σ^2, and

3. the non-diagonal matrix given by

$$Q_i = \frac{1}{|S_i|} \sum_{\mathbf{x} \in S_i} (\mathbf{x} - \mathbf{c}_{ij}) (\mathbf{x} - \mathbf{c}_{ij})^t. \tag{14.22}$$

If some diagonal elements of Q_i are zero, they are replaced with the positive σ^2.

14.2 Defuzzification Methods

The output of the fuzzy rules R_i $(i = 1, \ldots, N)$, $\hat{y}(\mathbf{x})$, for the input $\mathbf{x}$ can be synthesized by the center-of-gravity method and the linear combination of the degrees of membership. By approximating the center of gravity given by (13.9) on page 253 for the discrete data, $\hat{y}(\mathbf{x})$ is given by

$$\hat{y}(\mathbf{x}) = \frac{\displaystyle\sum_{i=1}^{N} o_i \, m_i(\mathbf{x})}{\displaystyle\sum_{i=1}^{N} m_i(\mathbf{x})}, \tag{14.23}$$

where o_i are constant or calculated by (14.2).

If we use constant o_i, to improve approximation accuracy, we can introduce additional parameters w_i as follows:

$$\hat{y}(\mathbf{x}) = \frac{\displaystyle\sum_{i=1}^{N} o_i \, w_i \, m_i(\mathbf{x})}{\displaystyle\sum_{i=1}^{N} w_i \, m_i(\mathbf{x})}, \tag{14.24}$$

and tune o_i and w_i by the steepest descent method [7], [118]:

$$o_i^{new} = o_i^{old} + \alpha \, [y(\mathbf{x}) - \hat{y}(\mathbf{x})] \frac{w_i \, m_i(\mathbf{x})}{\displaystyle\sum_{i=1}^{N} w_i \, m_i(\mathbf{x})}, \tag{14.25}$$

$$w_i^{new} = w_i^{old} + \alpha \, [y(\mathbf{x}) - \hat{y}(\mathbf{x})] \times$$

$$\frac{o_i \, m_i(\mathbf{x}) \left(\sum_{j=1}^{N} w_j \, m_j(\mathbf{x}) \right) - \left(\sum_{j=1}^{N} o_j \, w_j \, m_j(\mathbf{x}) \right) m_i(\mathbf{x})}{\left(\sum_{j=1}^{N} w_j \, m_j(\mathbf{x}) \right)^2}. \tag{14.26}$$

By the linear combination of the degrees of membership, the output is synthesized by

$$\hat{y} = q_0 + q_1 \, m_1(\mathbf{x}) + \cdots + q_N \, m_N(\mathbf{x}), \tag{14.27}$$

where $q_0, q_1, \ldots, q_N$ are constants and determined by the least-squares method. This architecture is the radial basis function neural network. In this case the outputs o_i are not used for defuzzification.

15. Fuzzy Rule Generation

In fuzzy rule generation, training data are preclustered or postclustered. In preclustering, we cluster the training data in advance and generate a fuzzy rule for each cluster. In postclustering, we start from one fuzzy rule and generate fuzzy rules around the training data with large estimation errors or at the points where the training data gather until the error limit is within the specified limit.

15.1 Fuzzy Rule Generation by Preclustering

In preclustering, there are two types to generate fuzzy rules. The first type is based on clustering of the input space [2]. Namely, first we divide the output space and consider each divided interval as a class. Then we divide the training data into classes according to the output values of the training data, and extract fuzzy rules for each class according to one of the methods discussed in Chapters 5 and 9. The second type clusters the input and output spaces simultaneously [115]. Namely, we combine the input values and output values of the training data and consider them as input data. Then we select the cluster centers, and define fuzzy rules for the cluster centers.

Since approximation of the multi-dimensional output can be easily realized by an extension of that of the one-dimensional output, in the following we discuss function approximation with the m-dimensional input vector $\mathbf{x}$ and the one-dimensional output y.

15.1.1 Clustering of Input Space

First we divide the range of y into n intervals as follows:

$$[y_0, y_1] : y_0 \leq y \leq y_1,$$
$$(y_1, y_2] : y_1 < y \leq y_2,$$
$$\cdots\cdots\cdots\cdots \tag{15.1}$$
$$(y_{n-1}, y_n] : y_{n-1} < y \leq y_n.$$

Let the center of the ith interval be o_i. We consider o_i as the representative value of the ith interval. Then we consider the ith interval as class i, and divide the training data into n classes. We approximate input regions for class i using one of the methods discussed in Chapters 5 and 9 with the training data for class i. The fuzzy rules for class i are given by

$$R_{ij} : \text{ If } \mathbf{x} \text{ is } A_{ij} \text{ then } y = o_i \quad \text{for } j = 1, \ldots, \tag{15.2}$$

where A_{ij} is the jth fuzzy region approximating the input region for class i and associated with it we define one of the membership functions based on a hyperbox or an ellipsoid. If the fuzzy min-max classifier with inhibition is used, an inhibition hyperbox is added to the if-part of (15.2). In the following, we call this function approximator FAMI. In FAMI, the output is synthesized by (14.24) and o_i and w_i are tuned by (14.25) and (14.26), respectively.

15.1.2 Clustering of Input and Output Spaces

Here, we discuss fuzzy rule extraction clustering input and output spaces together. Namely, we first combine the m-dimensional input training data and 1-dimensional output training data into $(m+1)$-dimensional training data. Then, to these data we apply one of the clustering methods discussed in Chapter 6. We may choose cluster centers from the training data or we may determine the centers which differ from the training data. Let $\mathbf{c}_i$ $(i = 1, \ldots, N)$ and o_i be the selected inputs and the associated outputs, respectively. Then the fuzzy rules become

$$R_i : \text{ If } \mathbf{x} \text{ is } \mathbf{c}_i \text{ then } y = o_i, \quad \text{for } i = 1, \ldots, N. \tag{15.3}$$

We set some appropriate membership function, such as the bell-shaped membership function, as the membership function around $\mathbf{c}_i$.

15.2 Fuzzy Rule Generation by Postclustering

In this section we discuss two types of fuzzy function approximators that generate fuzzy rules with ellipsoidal regions: a function approximator based on Takagi-Sugeno type model with the center-of-gravity defuzzification (FACG) and a function approximator based on a radial basis function neural network (FALC).

In FACG, for each training datum the number of the training data that are within the specified distance is calculated and the training datum which has the maximum number of training data is selected as the center of a fuzzy rule and the covariance matrix is calculated using the training data around the center. Then the parameters of the linear equation that defines the output value of the fuzzy rule are determined by the least-squares method using the training data around the center.

In FALC, the training datum with the maximum approximation error is selected as the center of a fuzzy rule. (In the following, for simplicity we say the maximum approximation error instead of the maximum absolute approximation error.) Then using the training data around the center, the covariance matrix is calculated, and the parameters of a linear equation that determines the output value are calculated by the least-squares method.

15.2.1 Concept

In preclustering, the training data are clustered in advance and a fuzzy rule is generated for each cluster. In postclustering, on the other hand, fuzzy rules are generated as needed until the approximation error meets the required error limit. In the former method, since clustering is not directly linked to approximation errors, it is difficult to find suitable clustering to realize the required error limit. Therefore, in this section for FACG and FALC we consider rule generation by postclustering.

The selection method of the center of a fuzzy rule needs to be changed according to which of the fuzzy function approximators are used. In FACG, when $N = 1$, (14.23) reduces to $\hat{y} = o_1$ (see Fig. 15.1). Thus, if o_1 is given by (14.2), the approximation reduces to linear regression. Thus the subsequent rule generation results in compensating the linear regression error. This is not a favorable strategy, since the Takagi-Sugeno type model works to estimate the derivative of the input-output relations and thus can represent the input-output relations with a small number of rules. Thus, for the Takagi-Sugeno type model to work properly, we need to determine the subset of the training data that should be used to determine the parameters p_{ij} (see Fig. 15.2).

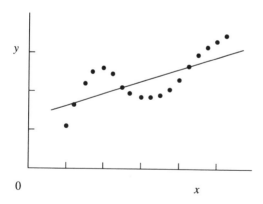

Fig. 15.1. The first rule generation with the center-of-gravity method when all the training data are used for fitting [119, p. 655 ©IEEE 1999]

In FALC, on the other hand, it is effective to add the center of cluster at the point where the approximation error is the maximum [60]. Suppose that N fuzzy rules have been generated and a rule needs to be added to reduce the

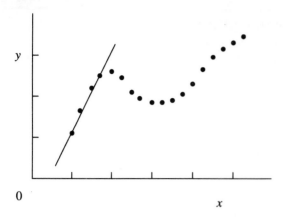

Fig. 15.2. The first rule generation with the center-of-gravity method when the subset of the training data is used for fitting [119, p. 655 ©IEEE 1999]

approximation error. Using N fuzzy rules, the output $\hat{y}$ is given by (14.27). Thus, when the $(N + 1)$st fuzzy rule is generated the output is given by

$$\hat{y} = q_0 + q_1\, m_1(\mathbf{x}) + \cdots + q_N\, m_N(\mathbf{x}) + q_{N+1}\, m_{N+1}(\mathbf{x}). \tag{15.4}$$

Assuming that the values of $q_0, q_1, \ldots, q_N$ are the same as those determined for the N fuzzy rules, the approximation error is reduced best when the center of the $(N + 1)$st fuzzy rule is located at the point where the approximation error is the maximum (see Fig. 15.3).

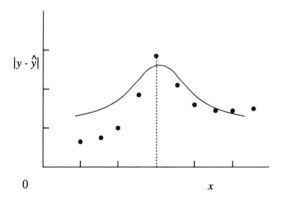

Fig. 15.3. Rule generation when the output is synthesized by a linear combination of the degrees of membership. A new rule is generated at the point where the approximation error is the maximum [119, p. 655 ©IEEE 1999]

15.2.2 Rule Generation for FACG

The center of a fuzzy rule needs to be at the center of a cluster where data gather. There are several clustering techniques as discussed in Chapter 6. Most of them are iterative methods, that is, they iterate the procedure until the clustering is converged. In [115], for each training datum, the potential is calculated and the training datum with the maximum potential is selected as the cluster center. To avoid consuming the calculation time in clustering, here we use the simplified version of this method. Namely, to generate the first fuzzy rule, for each training input, we count the number of the training inputs that are within the specified distance from the training input in consideration. Then we select the training input that has the maximum number of the training inputs within the specified distance as the center of the first fuzzy rule and determine the parameters $p_{10}, p_{11}, \ldots, p_{1m}$ in (14.2) by the least-squares method. If the parameters are not determined because the associated matrix is singular, we used the fixed o_i, where the training output corresponding to the selected training input is set to o_i. (If we use the pseudo-inverse in the least-squares method as discussed in Appendix B.2, we can determine the parameters irrespective of the singularity of the matrix. But in the following study, we do not use the pseudo-inverse.)

The ith ($i > 1$) fuzzy rule is generated as follows: We delete the training inputs that are within the specified distance from the center of the kth ($k = 1, \ldots, i - 1$) fuzzy rule. For each of the remaining training inputs, we count the number of the training inputs that are within the specified distance from the training input in consideration. Then we select the training input that has the maximum number of the training inputs within the specified distance as the center of the ith fuzzy rule and determine the parameters $p_{i0}, p_{i1}, \ldots, p_{im}$ by the least-squares method. If the parameters are not determined because the associated matrix is singular, we used the fixed o_i, where the training output corresponding to the selected training input is set to o_i. If there are no training inputs remained but still the approximation error is not within the specified error limit ε, we select the training input whose approximation error is the maximum as the center of the fuzzy rule. By deleting the training inputs that are within the specified distance from the already selected centers, we can avoid selecting the training inputs that are near some of the centers (see Fig. 15.4).

Selection of Centers. Let S denote the set of the initial training inputs. For each training input $\mathbf{x}_j$, we count the number of the training inputs $\mathbf{x}_k$ that are within the specified, average one-dimensional distance $R (> 0)$:

$$\sqrt{\frac{1}{m} \sum_{l=1}^{m} (x_{jl} - x_{kl})^2} < R, \tag{15.5}$$

where R is an application dependent parameter. But since usually the training time is short, we can easily obtain a suitable value as seen from the examples

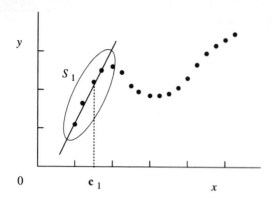

Fig. 15.4. Concept of clustering. After the center $\mathbf{c}_1$ is selected, the data in S_1 are eliminated from selection of the second center [119, p. 656 ©IEEE 1999]

in Section 15.4. Then we select the training input that has the maximum number of the training inputs within R as the center of the fuzzy rule, $\mathbf{c}_1$. Let S_i denote the subset of the training inputs that are within the average one-dimensional distance of R from $\mathbf{c}_i$ and that are included in $S-(S_1\cup\cdots\cup S_{i-1})$.

To select the ith $(i > 1)$ center, $\mathbf{c}_i$, for each of the training inputs in $S - (S_1 \cup \cdots \cup S_{i-1})$, we count the number of the training inputs in $S - (S_1 \cup \cdots \cup S_{i-1})$ whose average one-dimensional distance is within R. Then we select the training input that has the maximum number of the training inputs within R as the center of the fuzzy rule, $\mathbf{c}_i$. If the set $S-(S_1\cup\cdots\cup S_{i-1})$ is empty, and still the approximation error is not within the specified limit, we select the training input where the approximation error is the maximum as the center $\mathbf{c}_i$. For the ith fuzzy rule, we set a constant diagonal matrix to the covariance matrix or calculate the covariance matrix using (14.21) or (14.22).

Determination of Parameters. Let assume that we have already determined parameters $p_{k0},\ldots,p_{km}$ for $k = 1,\ldots,i-1$ and we determine parameters for the ith fuzzy rule, $p_{i0},\ldots,p_{im}$, by the least-squares method using the subset of the training inputs $S_1 \cup \cdots \cup S_i$. The reasons why we use the training inputs included in $S_1 \cup \cdots \cup S_{i-1}$ as well as S_i are that the training inputs in $S_1 \cup \cdots \cup S_{i-1}$ may include data that are within the average one-dimensional distance of R from $\mathbf{c}_i$ and to ensure sufficient data to prevent the associated matrix from being singular.

From (14.23), using the i fuzzy rules, $\hat{y}$ for $\mathbf{x}$ in $S_1 \cup \cdots \cup S_i$ is given by

$$\hat{y}(\mathbf{x}) = a_i(\mathbf{x})\, o_i + b_i(\mathbf{x})$$
$$= a_i(\mathbf{x})\,(1, \mathbf{x}^t)\, \mathbf{p}_i + b_i(\mathbf{x}), \qquad (15.6)$$

where $\mathbf{p}_i = (p_{i0}, p_{i1}, \ldots, p_{im})^t$ and $a_i(\mathbf{x})$ and $b_i(\mathbf{x})$ are given by

$$a_i(\mathbf{x}) = \frac{m_i(\mathbf{x})}{\sum_{k=1}^{i} m_k(\mathbf{x})}, \tag{15.7}$$

$$b_i(\mathbf{x}) = \frac{\sum_{k=1}^{i-1} o_k \, m_k(\mathbf{x})}{\sum_{k=1}^{i} m_k(\mathbf{x})}. \tag{15.8}$$

Thus by the least-squares method, $p_{i0}, \ldots, p_{im}$ can be determined using the training data included in $S_1 \cup \cdots \cup S_i$.

When the maximum approximation error is within the specified limit:

$$\max |y_k - \hat{y}_k| < \varepsilon, \tag{15.9}$$

where ε is a positive value, we terminate generating fuzzy rules.

15.2.3 Rule Generation for FALC

We generate the fuzzy rules successively according to the errors between the training data outputs and the synthesized outputs. In the beginning of rule generation, we consider that the synthesized outputs are zero. We select the training datum that gives the maximum error between the training data outputs and the synthesized outputs. We set the center of the fuzzy rule at the selected datum. We iterate the fuzzy rule generation until the approximation error for the training data is within the specified limit.

Let $(\mathbf{x}_j, y_j)$ $(j = 1, \ldots, M)$ be input-and-output pairs of the training data where M is the number of the training data. We assume that the synthesized outputs $\hat{y}_j$ with no fuzzy rules are zero. In the following we discuss the fuzzy rule generation.

Selection of Centers. Initially, we set $\hat{y}_j = 0$ for $j = 1, \ldots, M$ and calculate the maximum approximation error. Let

$$\arg \max_{j=1,\ldots,M} |y_j - \hat{y}_j| \tag{15.10}$$

be k where arg is the function that returns the subscript of the associated function, in this case max. If

$$|y_k - \hat{y}_k| > \varepsilon, \tag{15.11}$$

where ε is the specified error limit, we generate the first fuzzy rule with the center $\mathbf{c}_1 = \mathbf{x}_k$. For the ith rule $(i > 1)$ we select the training input that is not previously selected as the center of the fuzzy rule, i.e., $\mathbf{c}_i$. Then we calculate the subset S_i of the training inputs that are within the average, one-dimensional distance of R from $\mathbf{c}_i$. We set the constant diagonal matrix to the covariance matrix or calculate the covariance matrix using (14.21) or (14.22).

Determination of Parameters for Output Synthesis. We determine the parameters $q_0, q_1, \ldots, q_i$ by the least-squares method using all the training data. We discard the values for $q_0, q_1, \ldots, q_{i-1}$ previously determined by the least-squares method, since, if we use previously determined values, the solution may go into the local minimum.

15.3 Fuzzy Rule Tuning

15.3.1 Concept

Now we consider tuning of the tuning parameter α_i of the covariance matrix and the center c_{ij} of each membership function successively to improve approximation performance. Here we specify the convergence of tuning by

$$|y_k - \tilde{y}_k| \leq \varepsilon_t \quad \text{for} \quad k = 1, \ldots, M, \tag{15.12}$$

where $\tilde{y}_k$ is the kth tuned output, ε_t is a small positive value and $\varepsilon_t \leq \varepsilon$. To make the maximum approximation error not be worsened after tuning, we impose

$$|y_k - \tilde{y}_k| \leq \max_{k=1,\ldots,M} |y_k - \hat{y}_k| \quad \text{for} \quad k = 1, \ldots, M. \tag{15.13}$$

Similar to the tuning for fuzzy classifiers discussed in Chapter 7, we tune each one of c_{ij} and α_i at a time. For each membership function, we choose the training data that do not satisfy (15.12) but can satisfy (15.12) by moving the center c_{ij} or changing α_i. Among the chosen training data, we determine the training datum with the minimum approximation error under the constraint (15.13) and move c_{ij} or change α_i. Assuming that we use FALC for an approximation problem with a one-dimensional input, we explain the concept of tuning α_i and c_{ij}. By changing α_i, the slope of the ith membership function is changed. Fig. 15.5 shows how to tune α_i. In the figure, for simplicity, the ith membership function $m_i(x)$ is shown as the triangular membership function. For the given membership functions, in which only $m_i(x)$ is shown, the estimate $\hat{y}_k$ does not satisfy (15.12). Assuming $q_i = 1$, (15.12) is satisfied if α_{i1} is set to the value in the interval $[\alpha_i'', \alpha_i']$. To simplify tuning we calculate the approximation errors for α_i' and α_i'' and change α_i to the value with the smaller approximation error.

Now we consider improving the approximation error by moving the center c_i as shown in Fig 15.6. If the center c_i moves to c_i', $\hat{y}_k$ reaches $y_k - \varepsilon$ and if it moves to c_i'', $\hat{y}_k$ reaches $y_k + \varepsilon$. Thus when the center is located at any point in the interval between $[c_i', c_i'']$, (15.12) is satisfied. This also happens when the membership function is located between the locations shown by the dotted lines. To simplify tuning we calculate the approximation errors at the locations where the equality holds and moves the center to the location with the minimum approximation error. To improve tuning performance, initially we set a larger value to ε_t and decrease it during tuning.

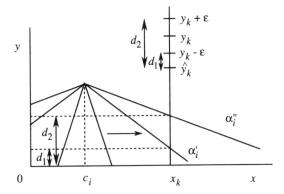

Fig. 15.5. Concept of tuning the slope of the membership function

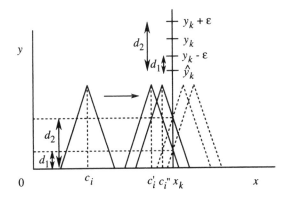

Fig. 15.6. Concept of tuning the location of the membership function

15.3.2 Evaluation Functions

In tuning membership functions we evaluate approximation performance using the following evaluation functions:

$$E(r_{ij}) = \sum_{i=1}^{M} e_k, \tag{15.14}$$

$$e_k = \begin{cases} 0 & \text{for } |y_k - \tilde{y}_k| \leq \varepsilon_t, \\ |y_k - \tilde{y}_k| & \text{for } |y_k - \tilde{y}_k| > \varepsilon_t, \end{cases} \tag{15.15}$$

and

$$E_{\max}(r_{ij}) = \max_{k=1,\dots,M} |y_k - \tilde{y}_k|. \tag{15.16}$$

Here, r_{ij} is either c_{ij} or α_i.

15.3.3 Tuning of Parameters

For the N fuzzy rules, let $\mathbf{x}_k$ do not satisfy (15.12). Then by solving

$$\tilde{y}_k(c_{ij}, \alpha_i) = y_k \pm \varepsilon_t \tag{15.17}$$

for c_{ij} or α_i, we obtain the value of parameter c_{ij} or α_i that satisfies (15.12). First we solve (15.17) for FALC. From (14.27),

$$\tilde{y}_k(c_{ij}, \alpha_i) = \hat{y}_k^i + q_i\, m_i(\mathbf{x}_k), \tag{15.18}$$

where

$$\hat{y}_k^i = q_0 + \sum_{\substack{j=1 \\ j \neq i}}^{N} q_j\, m_j(\mathbf{x}_k). \tag{15.19}$$

Hence from (15.17) and (15.18),

$$m_i(\mathbf{x}_k) = \frac{y_k \pm \varepsilon_t - \hat{y}_k^i}{q_i}. \tag{15.20}$$

When the right hand side of (15.20) is positive, (15.20) is solved for c_{ij} or α_i. Now assume that the right hand side of (15.20) is positive. Then from (14.18) and (14.19),

$$\frac{d_i^2(\mathbf{x}_k)}{\alpha_i} = -\ln\left(\frac{y_k \pm \varepsilon_t - \hat{y}_k^i}{q_i}\right). \tag{15.21}$$

Solving (15.21) for α_i gives

$$\alpha_i = -\frac{d_i^2(\mathbf{x}_k)}{\ln\left(\dfrac{y_k \pm \varepsilon_t - \hat{y}_k^i}{q_i}\right)}. \tag{15.22}$$

Now we solve (15.21) for c_{ij}. We assume the covariance matrix Q_i is diagonal and is given by $Q_i = \operatorname{diag}(\sigma_{i1}^2, \ldots, \sigma_{im}^2)$. Then from (14.20),

$$d_i^2(\mathbf{x}_k) = d_i^{j2}(\mathbf{x}_k) + \frac{(x_{kj} - c_{ij})^2}{\sigma_{ij}^2}, \tag{15.23}$$

where

$$d_i^{j2}(\mathbf{x}_k) = \sum_{\substack{l=1 \\ l \neq j}}^{m} \frac{(x_{kl} - c_{il})^2}{\sigma_{il}^2}. \tag{15.24}$$

Thus from (15.21) and (15.23),

$$\frac{(x_{kj} - c_{ij})^2}{\alpha_i\, \sigma^2} = -\ln\left(\frac{y_k \pm \varepsilon_t - \hat{y}_k^i}{q_i}\right) - \frac{d_i^{j2}(\mathbf{x}_k)}{\alpha_i}. \tag{15.25}$$

When the right hand side of (15.25) is positive, (15.25) is solved for c_{ij} and

$$c_{ij} = x_{kj} \pm \sqrt{-\alpha_i \sigma^2 \left(\ln \left(\frac{y_k \pm \varepsilon_t - \hat{y}_k^i}{q_i} \right) + \frac{d_i^{j2}(\mathbf{x}_k)}{\alpha} \right)}. \tag{15.26}$$

The number of solutions for c_{ij} is 4, 2, or 0 and that for α_i is 2, 1, or 0.

Similarly, we can solve (15.17) for c_{ij} and α_i for FACG. We rewrite (14.23) as follows:

$$\tilde{y}_k(c_{ij}, \alpha_i) = \frac{O_i(\mathbf{x}_k) + o_i\, m_i(\mathbf{x}_k)}{M_i(\mathbf{x}_k) + m_i(\mathbf{x}_k)}, \tag{15.27}$$

where

$$O_i(\mathbf{x}_k) = \sum_{\substack{l=1 \\ l \neq i}}^{N} o_l\, m_l(\mathbf{x}_k), \tag{15.28}$$

$$M_i(\mathbf{x}_k) = \sum_{\substack{l=1 \\ l \neq i}}^{m} m_l(\mathbf{x}_k). \tag{15.29}$$

Thus from (15.17) and (15.27),

$$m_i(\mathbf{x}_k) = \frac{O_i(\mathbf{x}_k) - (y_k \pm \varepsilon_t) M_i(\mathbf{x}_k)}{y_k \pm \varepsilon_t - o_i}, \tag{15.30}$$

where the same + or − operators in ± are taken simultaneously. When the right hand side of (15.30) is positive, (15.30) is solved for α_i as follows:

$$\alpha_i = -\frac{d_i^2(\mathbf{x}_k)}{\ln \left(\dfrac{O_i(\mathbf{x}_k) - (y_k \pm \varepsilon_t)\, M_i(\mathbf{x}_k)}{y_k \pm \varepsilon_t - o_i} \right)}. \tag{15.31}$$

The number of solutions for α_i is 2 or 0.

Similarly, when the right hand side of (15.30) is positive, (15.30) is solved for c_{ij} as follows:

$$\frac{(x_{kj} - c_{ij})^2}{\alpha_i \sigma_{ij}^2} = -\ln \left(\frac{O_i(\mathbf{x}_k) - (y_k \pm \varepsilon_t)\, M_i(\mathbf{x}_k)}{y_k \pm \varepsilon_t - o_i} \right) - \frac{d_i^{j2}(\mathbf{x}_k)}{\alpha_i}, \tag{15.32}$$

where the same + or − operators in ± are taken simultaneously. From (15.32),

$$c_{ij} = x_{kj}$$

$$\pm \sqrt{-\alpha_i \sigma^2 \left(\ln \left(\frac{O_i(\mathbf{x}_k) - (y_k \pm \varepsilon_t) M_i(\mathbf{x}_k)}{y_k \pm \varepsilon_t - o_i} \right) + \frac{d_i^{j2}(\mathbf{x}_k)}{\alpha_i} \right)}, \tag{15.33}$$

where in the root the same + or − operators in ± are taken simultaneously. The number of solutions for c_{ij} is 4, 2, or 0.

Now we discuss tuning of slopes and locations for FALC and FACG. If ε_t is too small, there may be no training data that satisfy (15.17). Thus to make tuning effective, we start with a large value of ε_t and gradually decrease the value until the specified value of ε_t is reached. Let ε_i and $\varepsilon_f\, (\leq \varepsilon_i)$ be the initial and final values of ε_t. The general flow of tuning slopes is as follows.

1. We set ε_i by

$$\varepsilon_i = \frac{1}{M} \sum_{k=1}^{M} |y_k - \hat{y}_k| \qquad (15.34)$$

and set a positive value to ε_t.

2. For the ith fuzzy rule ($i = 1, \ldots, N$), let $\mathbf{x}_k$ ($k = 1, \ldots, M$) be the training input that does not satisfy (15.12) and calculate the values of α_i by (15.22) or (15.31) and the values of c_{ij} by (15.26) or (15.33). Let the set of solutions be X_i. If X_i is empty, we proceed to the next training input. If X_i is not empty, we choose $\delta_{\min}$ that satisfies

$$\delta_{\min} = \arg \min_{\delta \in X_i} E(\delta). \qquad (15.35)$$

To make that the maximum error after tuning does not exceed that before tuning, we impose the following constraint in minimizing (15.35):

$$E_{\max}(\delta) \leq \max_{k=1,\ldots,M} |y_k - \hat{y}_k|. \qquad (15.36)$$

We replace the current α_i or c_{ij} with $\delta_{\min}$. If no α_i nor c_{ij} is modified or the number of iterations exceeds a specified number, we stop tuning. Otherwise, go to Step 3.

3. When Step 2 is finished for all the fuzzy rules, set ε_t with $\beta \varepsilon_t$ where $0 < \beta < 1$. If ε_t is smaller than or equal to ε_f, we stop tuning. Otherwise, return back to Step 2.

We call the execution of Step 2 for all the fuzzy rules one epoch of tuning.

15.4 Performance Evaluation

We evaluated performance of FAMI, FACG, and FALC using two different types of data sets: the noiseless time series data set generated by the Mackey-Glass differential equation [123] and the noisy data set gathered from a water purification plant [124]. We measured the computation time using a Sun UltraSPARC-IIi (335MHz) workstation. When the membership functions for FACG and FALC were tuned, the time included rule generation time as well as the rule tuning time. The maximum number of epochs of tuning was set to be 20.

15.4.1 Mackey-Glass Differential Equation

The Mackey-Glass differential equation generates time series data with a chaotic behavior and is given by

$$\frac{dx(t)}{dt} = \frac{0.2\,x(t-\tau)}{1 + x^{10}(t-\tau)} - 0.1\,x(t), \qquad (15.37)$$

where t and τ denote time and time delay, respectively.

By integrating (15.37), we can obtain the time series data $x(0), x(1), x(2)$, $\ldots, x(t), \ldots$. Using x prior to time t, we predict x after time t. Setting $\tau = 17$, and using four inputs $x(t-18), x(t-12), x(t-6), x(t)$, we estimate $x(t+6)$.

The first 500 data from the time series data $x(118), ..., x(1117)$ were used to generate fuzzy rules and the remaining 500 data were used to test performance. This data set is often used as the benchmark data for function approximators and the normalized root-mean-square error (NRMSE), i.e., the root-mean-square error divided by the standard deviation, of the time series data was used to measure the performance. Therefore, we measured the performance by NRMSE.

Table 15.1 shows the best performance of the several approximators for the Mackey-Glass test data. The performance of FAMI was the worst. The output range was divided into 11 intervals and the sensitivity parameter was set to be 17.5. Both FACG and FALC gave the best performance among them. For FACG a constant covariance matrix with $\sigma = 0.03$ was used, R was set to 0.05, and the membership functions were tuned. For FALC a diagonal covariance matrix was used and R was set to 0.4. For both FACG and FALC performance was measured when 100 rules were generated. In the following we discuss their performance in detail.

Table 15.1. Approximation errors for the Mackey-Glass test data

Approximator	NRMSE
NN [123]	0.02
ANFIS [111]	0.007
Cluster estimation-based [115]	0.014
FAMI [114]	0.092
FACG[1]	0.005
FALC[2]	0.006

1: Constant matrix with $R = 0.05$ and $\sigma = 0.03$ with tuning.
2: Diagonal matrix with $R = 0.4$.

Performance of FACG. First we evaluated FACG with a constant covariance matrix without tuning membership functions. In this case we changed R and σ and calculated the approximation errors when 100 rules were generated. Table 15.2 shows the results. For each run the calculation was completed in about ten seconds. The NRMSE for the training data and that for the test data were almost the same. This meant that the overfitting was not a problem for the Mackey-Glass data; the NRMSE for the test data decreased as that for the training data decreased.

Table 15.2. Approximation errors of FACG with a constant covariance matrix for the Mackey-Glass data

R	σ	Training data NRMSE	Test data NRMSE	Rules	Time (s)
0.05	0.03	0.006	0.006	100	12
	0.05	0.009	0.008	100	12
	0.07	0.013	0.013	100	12
0.07	0.03	0.006	0.006	100	12
	0.05	0.008	0.008	100	12
	0.07	0.013	0.012	100	12

Table 15.3 shows the NRMSE's for $R = 0.05$ and $\sigma = 0.03$ for different numbers of fuzzy rules. For the same number of rules, the first row shows the results without tuning and the second row shows the results with tuning. The NRMSE for the test data decreased monotonically as the rules were generated and reached 0.004 when 150 rules were generated. When 60 rules were generated, the set $S - (S_1 \cup \cdots \cup S_{60})$ became empty and the centers of the subsequent rules were determined according the maximum approximation error. The effect of tuning was clear for small numbers of fuzzy rules, but as the number of fuzzy rules increased, the effect of tuning diminished and the tuning time became long.

Instead of setting the diagonal elements of the covariance matrix, we calculated the covariance matrix using the training data. Tables 15.4 and 15.5 show the NRMSE's when the diagonal and non-diagonal covariance matrices were used. In Table 15.4, the results for membership tuning are also included. Since the overfitting occurred during training, we terminated rule generation when the NRMSE for the test data began to increase. In Table 15.4 by tuning the NRMSE's for the training and test data were decreased but tuning required much time. Comparing Tables 15.4 and 15.5, FACG with a non-diagonal covariance matrix showed inferior performance while consuming more computation time. Therefore, in the following we do not consider non-diagonal covariance matrices for both FACG and FALC. Comparing Tables 15.2, 15.4, and 15.5, FACG with a constant covariance matrix showed the best performance. From Tables 15.1 and 15.4, FACG with a diagonal covariance matrix showed comparable performance with that of the cluster estimation-based method [115].

In all the cases, singularity of the matrix in determining parameters in (14.2) did not occur.

Performance of FALC. Setting the maximum number of fuzzy rules to be 100, and changing σ for a constant covariance matrix and changing R for a diagonal matrix, fuzzy rules were generated. Tables 15.6 and 15.7 show

Table 15.3. Approximation errors of FACG with a constant covariance matrix for the Mackey-Glass data for different numbers of fuzzy rules ($R = 0.05, \sigma = 0.03$)

Rules	Training data NRMSE	Test data NRMSE	Epochs	Time (s)
20	0.238	0.263	0	2
	0.051	0.053	20	37
30	0.020	0.021	0	3
	0.017	0.016	20	507
40	0.014	0.013	0	4
	0.012	0.012	20	649
60	0.010	0.010	0	7
	0.009	0.008	20	997
80	0.007	0.007	0	11
	0.007	0.006	20	1350
100	0.006	0.006	0	17
	0.005	0.005	20	1728
150	<u>0.004</u>	<u>0.004</u>	0	37
	<u>0.004</u>	<u>0.004</u>	20	2667

Table 15.4. Approximation errors of FACG with a diagonal covariance matrix for the Mackey-Glass data

R	Training data NRMSE	Test data NRMSE	Rules	Epochs	Time (s)
0.04	0.019	0.021	64	0	6
	<u>0.008</u>	0.013	64	16	806
0.05	0.013	0.012	44	0	3
	0.012	<u>0.011</u>	44	11	370
0.06	0.018	0.017	34	0	2
	0.014	0.014	34	19	545
0.07	0.022	0.022	27	0	2
	0.020	0.020	27	11	246
0.08	0.028	0.028	24	0	1
	0.022	0.021	24	13	290

Table 15.5. Approximation errors of FACG with a non-diagonal covariance matrix for the Mackey-Glass data

R	Training data NRMSE	Test data NRMSE	Rules	Time (s)
0.04	0.033	0.039	48	7
0.05	0.015	0.015	38	5
0.06	0.020	0.024	38	4
0.07	0.027	0.027	27	3
0.08	0.031	0.033	25	2

performance of FALC with constant and diagonal covariance matrices, respectively when the membership functions were not tuned. From the tables FALC with a diagonal covariance matrix showed slightly better performance. In Table 15.6, the number of rules was 81 when $\sigma = 0.20$. This was because the matrix associated with determining parameters $q_0, \ldots, q_{82}$ became singular.

From Tables 15.3, 15.4, 15.6, and 15.7, the smallest approximation errors by FACG and FALC were almost the same and the computation time was comparable. Table 15.8 shows the approximation errors of FALC with $R = 0.4$ for different numbers of fuzzy rules. The table also includes the results when membership functions were tuned. The calculation was terminated when the 125th rule was generated because of the singularity of the matrix associated with determining parameters $q_0, \ldots, q_{126}$ and the tuning of membership functions was not possible. The effect of tuning was very small and for FALC tuning was useless for this particular problem. From Tables 15.3 and 15.8, for the numbers of rules from 40 to 80, the approximation errors of FACG were smaller than those of FALC with and without tuning; this meant that FACG could realize the same approximation error that FALC did with a smaller number of fuzzy rules.

Table 15.6. Approximation errors of FALC with a constant covariance matrix for the Mackey-Glass data

σ	Training data NRMSE	Test data NRMSE	Rules	Time (s)
0.05	0.038	0.038	100	14
0.10	0.013	0.012	100	14
0.15	0.007	0.007	100	14
0.20	0.009	0.008	81	9

Table 15.7. Approximation errors of FALC with a diagonal covariance matrix for the Mackey-Glass data

R	Training data NRMSE	Test data NRMSE	Rules	Time (s)
0.1	0.028	0.028	100	14
0.2	0.009	0.009	100	14
0.3	0.008	0.007	100	14
0.4	0.006	0.006	100	14
0.5	0.007	0.006	100	14

Table 15.8. Approximation errors of FALC with a diagonal covariance matrix for the Mackey-Glass data for different numbers of fuzzy rules ($R = 0.4$)

Rules	Training data NRMSE	Test data NRMSE	Epochs	Time (s)
20	0.039	0.038	0	1
	0.039	0.038	3	169
40	0.024	0.024	0	2
	0.026	0.024	13	2339
60	0.014	0.014	0	4
	0.014	0.014	3	579
80	0.010	0.009	0	8
	0.010	0.009	11	4113
100	0.006	0.006	0	15
	0.006	0.006	7	3055
125	0.005	0.005	0	25

15.4.2 Water Purification Plant

In a water purification plant, to eliminate small particles floating in the water taken from a river, coagulant is added and the water is stirred while these small particles begin sticking to each other. As more particles stick together they form flocs which fall to the bottom of a holding tank. Potable water is obtained by removing the precipitated flocs and adding chlorine. Careful implementation of the coagulant injection is very important in obtaining high quality water. Usually an operator determines the amount of coagulant needed according to an analysis of the water qualities, observation of floc formation, and prior experience.

To automate this operation, as inputs for water quality, 1) turbidity, 2) temperature, 3) alkalinity, 4) pH, and 5) flow rate were used and to replace the operator's observation of floc properties by image processing, 1) floc diameter, 2) number of flocs, 3) floc volume, 4) floc density, and 5) illumination intensity were used [124].

The 563 input-output data, which were gathered over a one-year period, were divided into 478 stationary data and 95 nonstationary data according to whether turbidity values were smaller or larger than a specified value. Then each type of data were further divided into two groups to form a training data set and a test data set; division was done in such a way that both sets had similar distributions in the output space. The data sets used in this study were: 1) 241 training data and 237 test data for stationary data, and 2) 45 training data and 40 test data for nonstationary data.

To compare performance of FAMI, FACG, and FALC with that of the multilayer neural network, we calculated the average approximation error and the maximum approximation error. Since both the stationary and nonstationary data were noisy, overfitting occurred; thus we set the approximation error limit ε to 4.0 in the following study.

Performance Evaluation for the Stationary Data. Table 15.9 shows the best performance for the stationary data using the multilayer neural network (NN), FAMI, FACG, and FALC. The four methods were comparable; the maximum errors of FAMI, FACG, and FALC for the test data were smaller than that of NN but the average approximation error of FAMI for the test data was the largest but their differences were small.

Table 15.9. Approximation errors for the stationary data

Approximator	Training data		Test data	
	Ave.err (mg/l)	Max.err (mg/l)	Ave.err (mg/l)	Max.err (mg/l)
NN	0.84	4.75	0.99	6.95
FAMI	1.07	4.75	1.18	5.57
FACG[1]	0.91	5.06	1.05	5.33
FALC[2]	1.09	4.34	1.16	5.22

1: $R = 0.3$ and $\sigma = 0.2$ with tuning.
2: $R = 0.3$ and $\sigma = 0.1$.

In the following we discuss performance of FAMI, FACG, and FALC more in detail.

Performance of FAMI. The average and maximum approximation errors of FAMI were investigated for different numbers of divisions of the output range, which varied from 3 to 7. For one or two data in this test data set, the fuzzy system could not determine an output value for the sensitivity parameter $\gamma = 4$ because the inputs of these data were outside the ranges defined by the training data set. To overcome this problem, for these data the sensitivity parameter γ was changed from 4 to a smaller value by which the fuzzy region in the input space was enlarged. To check the effect of tuning the parameters o_i and w_i on approximation errors, the learning rate α in (14.25) and (14.26) was set to 0.01 and o_i and w_i were tuned for 100 epochs using the training data. Here the sensitivity parameter was set to 4. The approximation errors are listed in Table 15.10.

Table 15.10. Approximation errors of FAMI with parameter tuning for the stationary data ($\gamma = 4$, 100 epochs, $\alpha = 0$)

Divisions	Training data		Test data	
	Ave.err	Max.err	Ave.err	Max.err
	(mg/l)	(mg/l)	(mg/l)	(mg/l)
3	1.75	6.47	1.90	8.33
5	1.12	4.42	1.25	5.31
7	1.07	4.75	1.18	5.57

When the output range was divided into 7 regions, the best performance was obtained. Tuning the parameters for only 10 epochs was sometimes sufficient because good initial values of o_i and σ_i were used; and generally, tuning the parameters for more than 100 epochs did not further improve the performance.

Performance of FACG. Table 15.11 shows the approximation errors of FACG with a diagonal covariance matrix for different R fixing $\sigma = 0.1$. The table includes the results when membership functions were tuned. The numbers in the parentheses are the numbers of fuzzy rules with fixed outputs because of singularity of the matrix associated with determining parameters p_{ij}. The maximum error for the training data exceeded the approximation error limit ε of 4.0. This happened as follows. When a fuzzy rule was generated, the approximation error for the training datum that had been selected as a center of a fuzzy rule exceeded 4.0. But since this datum had been selected as the center, a reduction in the approximation error for this datum was not realized.

Except for $R = 0.1$, the average approximation errors for the test data were reduced but tuning was much slower than extracting fuzzy rules. The best performance was obtained for $R = 0.3$ and $\sigma = 0.1$. Then we evaluated performance changing σ and fixing $R = 0.3$. Table 15.12 shows the results.

282 15. Fuzzy Rule Generation

The reason why performance changed for different values of σ is that some of the diagonal elements of the covariance matrix became zero sometimes during iterations and σ^2 needed to be set to the diagonal elements. The minimum average error was achieved for $\sigma = 0.2$ with tuning.

Table 15.11. Approximation errors of FACG with a diagonal covariance matrix ($\varepsilon = 4.0$, $\sigma = 0.1$) for the stationary data

R	Training data		Test data		Rules	Epochs	Time
	Ave.err (mg/l)	Max.err (mg/l)	Ave.err (mg/l)	Max.err (mg/l)			(s)
0.1	0.55	4.06	2.15	123	43 (17)	0	2
	0.42	3.63	3.11	157	43 (17)	14	304
0.2	0.87	9.48	1.21	11.8	14 (7)	0	0.35
	0.72	3.08	1.14	11.8	14 (7)	14	175
0.3	1.02	7.37	1.14	5.40	7	0	0.2
	0.93	5.20	1.10	5.28	7	7	65
0.4	1.08	12.4	1.17	6.88	4 (1)	0	0.1
	1.04	11.8	1.11	7.77	4 (1)	4	19

(): The number of fuzzy rules with fixed outputs.

Table 15.12. Approximation errors of FACG with a diagonal covariance matrix ($\varepsilon = 4.0$, $R = 0.3$) for the stationary data

σ	Training data		Test data		Rules	Epochs	Time
	Ave.err (mg/l)	Max.err (mg/l)	Ave.err (mg/l)	Max.err (mg/l)			(s)
0.1	1.02	7.37	1.14	5.40	7	0	0.2
	0.93	5.20	1.10	5.28	7	7	65
0.2	1.01	7.84	1.10	5.46	7	0	0.2
	0.91	5.06	1.05	5.33	7	7	51
0.3	1.01	11.6	1.12	6.32	4	0	0.2
	0.97	10.6	1.08	6.29	4	6	37

(): The number of fuzzy rules with fixed outputs.

Performance of FALC. Table 15.13 shows approximation errors of FALC for $\sigma = 0.1$ and for different values of R. The results of tuning membership functions are also included. The average approximation errors for the test data were improved by tuning. The minimum approximation error for the test data was obtained for $R = 0.4$ with tuning membership functions.

Table 15.13. Approximation errors of FALC with a diagonal covariance matrix for the stationary data ($\varepsilon = 4.0$, $\sigma = 0.1$)

R	Training data		Test data		Rules	Epochs	Time
	Ave.err (mg/l)	Max.err (mg/l)	Ave.err (mg/l)	Max.err (mg/l)			(s)
0.1	1.32	4.26	1.48	7.86	11	0	0.2
	1.02	3.75	1.21	5.50	11	6	78
0.2	1.14	3.91	1.32	5.21	10	0	0.2
	1.08	3.72	1.26	5.12	10	4	102
0.3	1.16	5.20	1.24	5.94	12	0	0.2
	1.09	4.34	1.16	5.22	12	7	303
0.4	1.15	15.4	1.12	5.45	6	0	0.2
	1.07	14.8	1.05	5.52	6	10	469

Performance Evaluation for the Nonstationary Data. Table 15.14 shows performance of the multilayer neural network, FAMI, FACG, and FALC for the nonstationary data. FAMI showed the best approximation performance and the performance of the remaining three were comparable.

Table 15.14. Approximation errors for the nonstationary data

Approximator	Training data		Test data	
	Ave.err (mg/l)	Max.err (mg/l)	Ave.err (mg/l)	Max.err (mg/l)
NN	1.59	6.83	1.74	6.78
FAMI	1.56	7.20	1.46	4.97
FACG[1]	1.91	6.30	1.95	7.18
FALC[2]	1.63	5.79	1.92	6.30

1: $R = 0.22$ and $\sigma = 0.1$.
2: $R = 0.3$ and $\sigma = 0.1$.

Performance of FAMI. The range of output variable was divided into 5 divisions when the effect of the sensitivity parameter γ on the approximation error was evaluated. The approximation errors (in mg/l) after tuning the parameters for 100 epochs with the learning rate $\alpha = 0.01$ are listed in Table 15.15. The best results were obtained when $\gamma = 20$.

Table 15.15. Approximation errors of FAMI with parameter tuning for the non-stationary data (5 divisions of the output range, 100 epochs, $\alpha = 0.01$)

γ	Training data		Test data	
	Ave.err (mg/l)	Max.err (mg/l)	Ave.err (mg/l)	Max.err (mg/l)
12	1.60	7.75	1.66	6.28
16	1.57	7.64	1.64	5.90
20	1.56	7.20	1.46	4.97

Performance of FACG. Table 15.16 shows performance of FACG fixing $\sigma = 0.1$ and changing R. The tuning results are also included. Although the average approximation errors and the maximum approximation errors for the training data were decreased by tuning, those for the test data increased. Namely, overfitting occurred. Performance was sensitive to the value of R. For $R = 0.24$, the average approximation errors and maximum approximation errors were exceedingly large while those for the training data were not so large. This was caused by the small number of training data for fitting.

Table 15.16. Approximation errors of FACG with a diagonal covariance matrix for the nonstationary data ($\varepsilon = 4.0$, $\sigma = 0.1$)

R	Training data		Test data		Rules	Epochs	Time
	Ave.err (mg/l)	Max.err (mg/l)	Ave.err (mg/l)	Max.err (mg/l)			(s)
0.2	1.72	7.07	2.42	6.63	8 (6)	0	0.1
	1.13	3.88	2.71	15.4	8 (6)	8	2
0.22	1.91	6.30	1.95	7.18	9 (6)	0	0.1
	1.30	3.50	4.46	85.6	9 (6)	7	2
0.24	1.69	6.51	153	5760	8 (5)	0	0.1
	1.17	3.66	713	22000	8 (5)	20	4

(): The number of fuzzy rules with fixed outputs.

Performance of FALC. Table 15.17 shows performance of FALC fixing $\sigma = 0.1$ and changing R. The tuning results are also included. Except for $R = 0.1$, overfitting occurred and for $R = 0.1$, the improvement of the approximation error for the training data was small. The approximation performance was sensitive to the value of R and except for $R = 0.3$ the maximum approximation errors for the test data exceeded 10 mg/l.

Table 15.17. Approximation errors of FALC with a diagonal covariance matrix for the nonstationary data ($\varepsilon = 4.0$, $\sigma = 0.1$)

R	Training data		Test data		Rules	Epochs	Time
	Ave.err	Max.err	Ave.err	Max.err			
	(mg/l)	(mg/l)	(mg/l)	(mg/l)			(s)
0.1	0.94	3.41	2.73	14.0	15	0	0.1
	0.76	2.92	2.71	14.0	15	4	2
0.2	1.40	6.64	2.37	17.3	11	0	0.1
	1.22	6.64	2.44	17.3	11	3	2
0.3	1.62	6.77	2.12	8.66	11	0	0.1
	1.46	5.26	2.38	8.67	11	5	5
0.4	1.56	6.33	2.58	19.0	11	0	0.1
	1.47	6.33	2.55	19.0	11	7	19

Since the minimum average approximation error for the test data was achieved for $R = 0.3$, we fixed $R = 0.3$ and changed the value of σ. Table 15.18 shows the results including the tuning results. Overfitting occurred by tuning. The minimum average approximation error was achieved when $\sigma = 0.5$ without tuning membership functions.

15.4.3 Discussions

We evaluated performance of FAMI, FACG, and FALC using the data generated by the Mackey-Glass differential equation and the data gathered from the water purification plant. The former data were noise-free and the training data and the test data were quite similar. Thus performance of approximators was determined by how well the approximators fitted the training data. (In practice, 100 or 200 fuzzy rules for FACG an FALC may be impractical in expressing such a simple system. But here, the function approximators were tested for those numbers of fuzzy rules only to show the approximation power of the approximators.) On the other hand, the data obtained from the water purification plant were very noisy and the training data and the test data were not similar. Thus, we needed to avoid overfitting to obtain good performance for the test data. Both FACG and FALC performed well for both data

Table 15.18. Approximation errors of FALC with a diagonal covariance matrix for the nonstationary data ($\varepsilon = 4.0$, $R = 0.3$)

σ	Training data		Test data		Rules	Epochs	Time
	Ave.err	Max.err	Ave.err	Max.err			
	(mg/l)	(mg/l)	(mg/l)	(mg/l)			(s)
0.1	1.62	6.77	2.12	8.66	11	0	0.1
	1.46	5.26	2.38	8.67	11	5	5
0.3	1.63	6.53	2.06	6.70	10	0	0.1
	1.41	5.11	2.31	6.70	10	7	6
0.5	1.63	5.79	<u>1.92</u>	6.30	8	0	0.1
	1.54	4.96	1.98	6.30	8	5	4
0.7	1.31	4.49	2.17	<u>5.57</u>	10	0	0.1
	<u>1.25</u>	<u>3.91</u>	2.15	6.47	10	5	7

except for the nonstationary data obtained by the water purification plant. FAMI performed poorly for the Mackey-Glass data, but performed well for the water purification data.

For the Mackey-Glass data, the number of fuzzy rules generated by FACG was larger than that by ANFIS [111] but was compatible with that of the cluster estimation based method [115]. The methods discussed in this section are based on simpler algorithms and thus need more fuzzy rules to realize the same approximation error. The cluster estimation-based method and FACG are similar methods; the major differences are that the former method uses sophisticated preclustering while the latter adopts the simpler postclustering and that the latter method tunes membership functions. For the Mackey-Glass data the number of fuzzy rules of the cluster estimation-based method to realize the NRMSE of 0.014 was 25 [115]. From Table 15.3, the number of FACG to realize the NRMSE of 0.016 was 30.

Unlike pattern classification (cf. Section 5.2.3), for FACG and FALC we need to set the value of the average, one-dimensional distance from the center, R. In addition, we need to set the value of σ, to avoid the inverse of the covariance matrix from being singular. But since the computation time for generating fuzzy rules is very short, trial and error for determining the optimal values for parameters R and σ is not a problem.

The main reason that FACG and FALC did not perform well for the nonstationary data was that the number of data was very small: only 45 data samples. Thus a fraction of 45 data were used to generate fuzzy rules and overfitting occurred quite easily.

16. Robust Function Approximation

In developing a function approximator, we assume that the training data do not include outliers. But outliers may occur in many occasions and the detection of outliers is difficult especially for multi-dimensional data (cf. Section 8.1). In addition, if outliers are included, they affect the approximator's performance. Outliers may be excluded by preprocessing, but in this chapter we consider excluding outliers while generating fuzzy rules. We focus our discussions on a robust training method for FACG, namely the function approximator based on Takagi-Sugeno type model with the center-of-gravity defuzzification. In FACG, the parameters of the liner equation that defines the output value of the fuzzy rule are determined by the least-squares method. Therefore, if the training data include outliers, the method fails to determine the parameter values correctly. To overcome this problem we use the least-median-of-squares method. Among the original training data set, we randomly select training data more than the number of parameters, and determine the parameter values using the least-squares method. We repeat this for a specified number of times and determine the parameters with the smallest median of squared errors. We compare the proposed method with the least-squares method and the conventional least-median-of-squares method using the data generated by the Mackey-Glass differential equation.

16.1 Introduction

Fuzzy systems are suited for nonlinear function approximation and many fuzzy systems with training capability have been developed. In general, fuzzy systems have faster training capability than multilayer neural networks and have comparable generalization ability. But most fuzzy systems do not consider the situation when the training data include outliers which deteriorate approximation performance.

To overcome the problem of outliers, usually the error function is modified [76, 77, 125, 126]. Let the input-output pairs be $(\mathbf{x}_i, y_i), i = 1, \ldots, M$. Then in approximating a one-dimensional output y, the error function is given by

$$E = \frac{1}{M} \sum_{i=1}^{M} f(r_i), \qquad (16.1)$$

where $f(\cdot)$ is an error function and $r_i = y_i - \hat{y}_i$ is the residual. Usually the squared error function is used. Namely,

$$f(r) = \frac{1}{2}r^2. \tag{16.2}$$

By this error function, when the residuals are large, the adjustments of parameter values become large. Thus if the outliers are included, the estimate will be erroneously corrected to offset the residuals. To solve this problem, error functions other than the squared error function are used. In [76], the following log square function is used:

$$f(r) = \log\left(1 + \frac{1}{2}r^2\right). \tag{16.3}$$

In [77, 125, 126], using the function similar to the Hampel's error function (see Fig. 16.1) the adjustments of parameters are made only when the absolute residuals are smaller than a specified value.

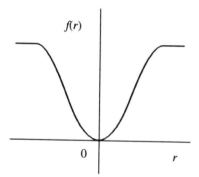

Fig. 16.1. Hampel's error function

In this chapter, we discuss robust function approximation based on FACG [127]. Since in the model the output is expressed by a linear combination of input variables, we determine the parameters by the least-median-of-squares method (LMS) [74, 128]. Since inclusion of outliers is usually not known, approximation performance should not be deteriorated when there are no outliers. Thus, unlike the conventional LMS, we repeat determining the parameters with the least-squares method (LS) using randomly selected training data and then select the parameters with the smallest median of squared errors. We call this method the least-squares method with partial data (LSPD).

In the following, we discuss robust fuzzy function approximation based on FACG. First we summarize the approximator and then we discuss the robust approximation method for the approximator. Finally, we compare the performance of the method with original FACG and FACG with the conventional LMS.

16.2 Fuzzy Function Approximator with Ellipsoidal Regions

16.2.1 Fuzzy Rule Representation

We discuss function approximation using fuzzy rules with ellipsoidal regions as follows:

$$R_i : \quad \text{If } \mathbf{x} \text{ is } \mathbf{c}_i \quad \text{then } y = o_i \quad \text{for} \quad i = 1, \dots, N, \tag{16.4}$$

where $\mathbf{x}$ is the m-dimensional input vector, $\mathbf{c}_i$ is the center of the ellipsoidal region for the ith fuzzy rule, y is the output for $\mathbf{x}$, o_i is the corresponding output, and N is the number of fuzzy rules. The degree of membership of the fuzzy rule R_i, $m_i(\mathbf{x})$, is given by

$$m_i(\mathbf{x}) = \exp(-d_i^2(\mathbf{x})), \tag{16.5}$$
$$d_i^2(\mathbf{x}) = (\mathbf{x} - \mathbf{c}_i)^t \, Q_i^{-1} \, (\mathbf{x} - \mathbf{c}_i), \tag{16.6}$$

where $d_i(\mathbf{x})$ is the weighted distance between $\mathbf{x}$ and $\mathbf{c}_i = (c_{i1}, \cdots, c_{im})^t$, Q_i is the $m \times m$ covariance matrix and is given by either of the following ways:

1. a constant diagonal matrix with the same diagonal element σ^2; and
2. the diagonal matrix calculated using the set of data around the center $\mathbf{c}_i$, S_i:

$$Q_{i,jj} = \frac{1}{|S_i|} \sum_{\mathbf{x} \in S_i} (x_j - c_{ij})^2, \tag{16.7}$$

where $|S_i|$ is the number of data in S_i.

In the Takagi-Sugeno type model, instead of using constant o_i in (16.4), the linear combination of input variables is used:

$$o_i = p_{i0} + p_{i1} x_1 + \cdots + p_{im} x_m \quad \text{for} \quad i = 1, \dots, N, \tag{16.8}$$

where $p_{i0}, \dots, p_{im}$ are constants and are determined by the least-squares method when robust approximation is not considered.

The output of fuzzy rules R_i $(i = 1, \dots, N)$, $\widehat{y}(\mathbf{x})$, for the input $\mathbf{x}$ can be synthesized by the center-of-gravity method:

$$\widehat{y}(\mathbf{x}) = \frac{\displaystyle\sum_{i=1}^{N} o_i \, m_i(\mathbf{x})}{\displaystyle\sum_{i=1}^{N} m_i(\mathbf{x})}, \tag{16.9}$$

where o_i are calculated by (16.8).

16.2.2 Fuzzy Rule Generation

We generate fuzzy rules until the approximation error meets the required error limit. For each rule we determine the subset of the training data that should be used to determine the parameters p_{ij}.

The center of the fuzzy rule needs to be at the center of a cluster where data gather. In [115], for each training datum, the potential is calculated and the training datum with the maximum potential is selected as the cluster center. To avoid consuming the calculation time in clustering, here we use the simplified version of the method. Namely, to generate the first fuzzy rule, for each training input, we count the number of the training inputs that are within the distance $R\,(> 0)$ from the training input in consideration. Then we select the training input that has the maximum number of the training inputs within the specified distance as the center of the first fuzzy rule and determine the parameters $p_{10}, p_{11}, \ldots, p_{1m}$ using the data within the distance R from the center.

The ith $(i > 1)$ fuzzy rule is generated as follows: We delete the training inputs that are within the specified distance from the center of the kth $(k = 1, \ldots, i - 1)$ fuzzy rule. For each of the remaining training inputs, we count the number of the training inputs that are within the distance R from the training input in consideration. Then we select the training input that has the maximum number of training inputs within the specified distance as the center of the ith fuzzy rule. We denote as S_i the set with the training data that are within the distance R from the center. Next, we determine the parameters $p_{i0}, p_{i1}, \ldots, p_{im}$ using the data in $S_1 \cup \cdots \cup S_i$. By deleting the training inputs that are within the distance R from the already selected centers, we can avoid selecting the training inputs that are near some of the centers.

16.3 Robust Parameter Estimation

In Takagi-Sugeno type model, the parameters $p_{i0}, \ldots, p_{im}$ are determined by the least-squares method. Thus if the training data include outliers, approximation performance is worsened. Therefore, to realize robust function approximation, we apply the least-median-of-squares method, instead of the lease-squares method, to the fuzzy function approximator with ellipsoidal regions.

16.3.1 Robust Estimation by the Least-median-of-squares Method

When we generate the ith fuzzy rule, the parameters $p_{k0}, \ldots, p_{km}$ $(k = 1, \ldots, i - 1)$ have already been determined. Thus from (16.9) the parameters $p_{i0}, \ldots, p_{im}$ satisfy

$$\widehat{y}(\mathbf{x}) = a_i(\mathbf{x})\, o_i + b_i(\mathbf{x})$$
$$= a_i(\mathbf{x})\, (1, \mathbf{x}^t)\, \mathbf{p}_i + b_i(\mathbf{x}) \qquad \text{for} \quad \mathbf{x} \in S_1 \cup \cdots \cup S_i, \tag{16.10}$$

where $\mathbf{p}_i = (p_{i0}, p_{i1}, \ldots, p_{im})^t$ and $a_i(\mathbf{x})$ and $b_i(\mathbf{x})$ are given by (15.7) and (15.8), respectively, and $\widehat{y}(\mathbf{x})$ is the estimate of the training output $y(\mathbf{x})$. We define the residual as $r(\mathbf{x}, \mathbf{p}_i) = y(\mathbf{x}) - \widehat{y}(\mathbf{x})$ and the set of the data for determining $\mathbf{p}_i$ as $Z_i = \{(\mathbf{x}, y) \mid \mathbf{x} \in S_1 \cup \cdots \cup S_i\}$. Then the estimate $\widehat{\mathbf{p}}_i(Z_i)$ of $\mathbf{p}_i$ by LMS is given by

$$\widehat{\mathbf{p}}_i(Z_i) = \arg\min_{\mathbf{p}_i}(r^2(\mathbf{x}, \mathbf{p}_i))_{h:|Z_i|}$$
$$= \arg\min_{\mathbf{p}_i} |r(\mathbf{x}, \mathbf{p}_i)|_{h:|Z_i|}, \tag{16.11}$$

where $|Z_i|$ is the number of the input-output pairs in Z_i and $r(\cdot)_{h:|Z_i|}$ stands for the h-th element of the $|Z_i|$ ordered outputs of the real valued function $r(\cdot)$. In the following we set $h = \lceil n/2 \rceil + 1$ where $\lceil a \rceil$ denotes the maximum integer that does not exceeds a.

In general, it is difficult to minimize the objective function given by (16.11). Therefore, in [74, 128], the following resampling algorithm is used.

1. Randomly select $m + 1$ input-output pairs from Z_i and solve (16.10) for $\mathbf{p}_i$.
2. Using $\mathbf{p}_i$, for $|Z_i|$ training data, calculate the squared errors between the estimated outputs and the training outputs, and calculate and memorize the median.
3. Repeat 1 and 2 for a specified number of times while keeping $\mathbf{p}_i$ with the smallest median of squared errors.

16.3.2 Robust Estimation by the Least-squares Method with Partial Data

Since in the above algorithm, $\mathbf{p}_i$ is determined by solving a linear equation, $\mathbf{p}_i$ may be specialized to the selected input-output pairs and approximation performance may not be good when there are no outliers. Thus to avoid overfitting, we select l data given by

$$l = r\,(m + 1), \tag{16.12}$$

where m is the number of input variables and r is the multiplier and takes on a natural number. Then from the set of data Z_i, we select a subset of data Z consisting of l data, and determine the parameters $\mathbf{p}_i$ by the least-squares method as follows:

1. Randomly select an l-element subset Z from Z_i and determine $\mathbf{p}_i$ by minimizing the sum of squared errors:

$$\frac{1}{2} \sum_{(\mathbf{x}, y) \in Z} (y - a_i(\mathbf{x})\, (1, \mathbf{x}^t)\, \mathbf{p}_i - b_i(\mathbf{x}))^2. \tag{16.13}$$

2. Using $\mathbf{p}_i$, for $|Z_i|$ training data, calculate the squared errors between the estimated outputs and the training outputs, and calculate and memorize the median.

3. Repeat 1 and 2 for a specified number of times while keeping $\mathbf{p}_i$ with the smallest median of squared errors.

We call this method the least-squares method with partial data (LSPD).

16.4 Performance Evaluation

Using the Mackey-Glass differential equation given by (15.37) on page 274, we evaluated LSPD, the conventional LMS, and the LS. The first 500 data from the time series data $x(118),\ldots,x(1117)$ were used to generate fuzzy rules and the remaining 500 data were used to test approximation performance.

If we calculate the covariance matrix using the training data, it will be affected by outliers. Thus in this study, in generating fuzzy rules, we used the constant covariance matrix with σ^2 as the diagonal elements.

To evaluate the performance of the function approximators, we used NRMSE and the median of the absolute approximation errors.

We measured the execution time using a SUN Ultra SPARC-II (360MHz) workstation.

In the following, first we determine the optimal r that determines the number of training data for LSPD. Then we compare the approximation performance of LSPD with conventional LS and LMS.

16.4.1 Determination of the Multiplier

Without Outliers. We set $R = 0.05$, $\sigma = 0.03$, $r = 5$, and evaluated the approximation performance of LSPD changing the number of samplings. Table 16.1 shows the results. As seen from the table, the approximation performance did not change when the number of samplings was larger than or equal to 3000. Thus, we sampled the training data 3000 times in our study.

Next, we examined the effect of the number of selected data l on the approximation performance when outliers were not included. Table 16.2 shows the results when the value of the multiplier r was changed from 2 to 6. From the table, when r was larger, namely, the number of selected data was larger, NRMSE and the median of absolute errors became smaller.

Setting $r = 5$, which gave relatively small values of NRMSE for both the test and training data (see Table 16.2), we evaluated the approximation performance changing σ. Table 16.3 shows the results. As seen from the table, when the value of σ was small, the median of the absolute errors was small.

Table 16.1. Approximation performance for the different numbers of samplings ($R = 0.05$, $\sigma = 0.03$, $r = 5$)

| No. | Training data | | Test data | | Time |
samplings	NRMSE	Median	NRMSE	Median	(s)
1000	0.0348	0.00098	0.0347	0.00143	375
3000	0.0276	0.00101	0.0261	0.00149	1152
5000	0.0276	0.00101	0.0261	0.00149	1922

Table 16.2. Approximation performance for the different numbers of training data when outliers were not included ($R = 0.05$, $\sigma = 0.03$)

| r | Training data | | Test data | |
	NRMSE	Median	NRMSE	Median
2	0.0518	0.00154	0.0528	0.00237
3	0.0415	0.00112	0.0393	0.00185
4	0.0364	0.00101	0.0352	0.00165
5	0.0276	0.00101	0.0261	0.00149
6	0.0301	0.00099	0.0295	0.00146
All data	0.0135	0.00144	0.0134	0.00127

Table 16.3. Approximation performance for the different values of σ when outliers were not included ($R = 0.05$, $r = 5$)

| σ | Training data | | Test data | |
	NRMSE	Median	NRMSE	Median
0.03	0.0276	0.00101	0.0261	0.00149
0.05	0.0364	0.00133	0.0351	0.00164
0.07	0.0266	0.00164	0.0267	0.00183
0.09	0.0311	0.00220	0.0316	0.00218

With Outliers. If we replace a datum in the time series data with an outlier, the five consecutive input-output pairs that are generated from the time series data include the outlier. Thus, since we replaced five data in the time series data with outliers, 25 data in the input-output pairs were replaced with outliers. We generated five training data sets randomly changing the locations of outliers.

To evaluate the approximation performance for the change of the values of the multiplier r, we trained FACG changing r from 2 to 6. We generated the outliers multiplying 1.5 to the original magnitudes of the data. The ap-

proximation performance shown in Table 16.4 is the average values for the
five data sets. From the table, we can see that as the number of the selected
data became larger, the median of the absolute errors became smaller. The
value of NRMSE was minimum when $r = 5$.

Then setting $r = 5$ and changing the value of σ, we evaluated the approximation performance. Table 16.5 shows the results. From the table, for
$\sigma = 0.03$, NRMSE and the median of the absolute errors were relatively
small.

Whether outliers were included or not, for $r = 5$, LSPD performed well.
Thus, in comparing the performance of LSPD with that of the other methods,
we set $r = 5$.

Table 16.4. Approximation performance of the test data for the different numbers
of training data when outliers were included ($R = 0.05$, $\sigma = 0.03$)

r	NRMSE	Median
2	0.0499	0.00327
3	0.0388	0.00201
4	0.0451	0.00169
5	<u>0.0348</u>	0.00154
6	0.0415	<u>0.00150</u>

Table 16.5. Approximation performance of the test data for the different values
of σ when outliers were included ($R = 0.05$, $r = 5$)

σ	NRMSE	Median
0.03	<u>0.0353</u>	0.00168
0.05	0.0456	<u>0.00167</u>
0.07	0.0404	0.00231

16.4.2 Performance Comparison

With Outliers. We evaluated the approximation performance changing the
magnitude of outliers included in the training data from 1.1 to 2 times the
original magnitude. Fig. 16.2 shows average values of NRMSE for five training data sets for LSPD, LS, and LMS. The outlier magnitude of 1 means that
no outliers were included. From the figure, as the magnitude of outliers became larger, the value of NRMSE for LS became larger, but those for LSPD

and LMS were nearly constant. Performance of LSPD was better than that of LMS, and LSPD showed the best performance when the magnitudes of outliers were larger than 1.2. Table 16.6 shows the values of NRMSE and the medians of the absolute errors of the test data for the three methods when the magnitudes of outliers were twice as large as those of the original data. The table also lists the execution time for the three methods. Compared with LS, LMS and LSPD required much computation time because of sampling.

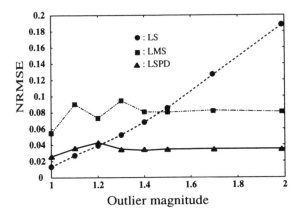

Fig. 16.2. NRMSE versus magnitude of outliers

Table 16.6. Comparison of approximation performance when outliers were included

Method	NRMSE	Median	Time (s)
LS	0.1900	0.00427	7
LMS	0.0806	0.00333	1095
LSPD	0.0353	0.00168	1125

Fig. 16.3 shows the prediction results for the test data. FACG's were trained using one of the five training data sets contaminated with outliers in which magnitudes of the original data were multiplied by two. While the prediction by LS was worsened by the effect of outliers, LSPD predicted correctly without the effect of outliers.

Without Outliers. Table 16.7 shows the best approximation performance of the three methods when the outliers were not included. For LS, we set $R = 0.05$, $\sigma = 0.03$, and for LMS $R = 0.05$, $\sigma = 0.07$, and for LSPD

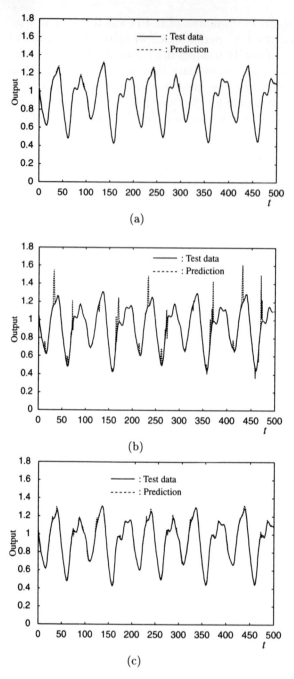

Fig. 16.3. Comparison of the actual outputs of test data with approximated outputs by (a) LSPD, (b) LS, and (c) conventional LMS

$R = 0.05$, $\sigma = 0.03$, and $r = 5$. In Tables 16.6 and 16.7, 40 fuzzy rules were generated. The table also includes the execution time. For LMS and LSPD, the execution took time because of 3000 samplings.

For the training data, NRMSE of LSPD was larger than that of LS, but the median of LSPD was the smallest. For the test data, NRMSE of LSPD was larger than that of LS but smaller than that of LS and the median of LSPD was comparable with that of LS.

Table 16.7. Comparison of approximation performance between different methods when outliers were not included

No. samplings	Training data NRMSE	Median	Test data NRMSE	Median	Time (s)
LS	0.0135	0.00144	0.0134	0.00127	7
LMS	0.0400	0.00280	0.0400	0.00330	1093
LSPD	0.0276	0.00101	0.0261	0.00149	1152

16.5 Discussions

In this chapter we discussed robust function approximation using the least-median-of-squares method. To avoid overfitting, we used the least-squares method for the data selected from the training data and repeated this process many times to obtain the solution with the least median of squared errors. The simulation results for the data generated from the Mackey-Glass differential equation showed approximation performance robuster than that of the conventional least-median-of-squares method. But when outliers were not included, the approximation performance was a little worse than that of the least-squares method.

Part III

Appendices

A. Conventional Classifiers

A.1 Bayesian Classifiers

Bayesian classifiers are based on probability theory and give the theoretical basis for pattern classification.

Let ω be a random variable and take one of n states: $\omega_1, \ldots, \omega_n$, where ω_i indicates class i, and an m-dimensional feature vector $\mathbf{x}$ be a random variable vector. We assume that we know the *a priori* probabilities $P(\omega_i)$ and conditional densities $p(\mathbf{x} \mid \omega_i)$. Then when $\mathbf{x}$ is observed, the *a posteriori* probability of ω_i, $P(\omega_i \mid \mathbf{x})$ is calculated by the Bayesian rule:

$$P(\omega_i \mid \mathbf{x}) = \frac{p(\mathbf{x} \mid \omega_i)\, P(\omega_i)}{p(\mathbf{x})}, \tag{A.1}$$

where

$$p(\mathbf{x}) = \sum_{i=1}^{n} p(\mathbf{x} \mid \omega_i)\, P(\omega_i). \tag{A.2}$$

Assume that the cost c_{ij} is given when $\mathbf{x}$ is classified into class i although it is class j. Then the expected conditional cost in classifying $\mathbf{x}$ into class i, $C(\omega_i \mid \mathbf{x})$, is given by

$$C(\omega_i \mid \mathbf{x}) = \sum_{j=1}^{n} c_{ij}\, P(\omega_j \mid \mathbf{x}). \tag{A.3}$$

The conditional cost is minimized when $\mathbf{x}$ is classified into the class

$$\arg \min_{i=1,\ldots,n} C(\omega_i \mid \mathbf{x}). \tag{A.4}$$

This rule is called the Bayesian decision rule.

In diagnosis problems, usually there are normal and abnormal classes. Misclassification of normal data into the abnormal class is less fatal than misclassification of abnormal data into the normal class. In such a situation, we set a smaller cost to the former than the latter.

If we want to minimize the average probability of misclassification, we set the cost as follows:

$$c_{ij} = \begin{cases} 0 & \text{for } i = j, \\ 1 & \text{for } i \neq j, \end{cases} \quad i, j = 1, \ldots, n. \tag{A.5}$$

Then, from (A.1) and (A.2) the conditional cost given by (A.3) becomes

$$C(\omega_i \,|\, \mathbf{x}) = \sum_{\substack{j=1 \\ j \neq i}}^{n} P(\omega_j \,|\, \mathbf{x})$$

$$= 1 - P(\omega_i \,|\, \mathbf{x}). \tag{A.6}$$

Therefore, the Bayesian decision rule given by (A.4) becomes

$$\arg \max_{i=1,\ldots,n} P(\omega_i \,|\, \mathbf{x})$$

$$= \arg \max_{i=1,\ldots,n} p(\mathbf{x} \,|\, \omega_i)\, P(\omega_i). \tag{A.7}$$

Now, we assume that the conditional densities $p(\mathbf{x} \,|\, \omega_i)$ are normal:

$$p(\mathbf{x} \,|\, \omega_i) = \frac{1}{\sqrt{(2\pi)^n \det(Q_i)}} \exp\left(-\frac{(\mathbf{x} - \mathbf{c}_i)^t\, Q_i^{-1}\, (\mathbf{x} - \mathbf{c_i})}{2} \right), \tag{A.8}$$

where $\mathbf{c}_i$ is the mean vector and Q_i is the covariance matrix of the normal distribution for class i. If the *a priori* probabilities $P(\omega_i)$ are the same for $i = 1,\ldots,n$, $\mathbf{x}$ is classified into class i if $p(\mathbf{x} \,|\, \omega_i)$ given by (A.8) is the maximum.

A.2 Nearest Neighbor Classifiers

A.2.1 Classifier Architecture

Nearest neighbor classifiers use all the training data as templates for classification. In the simplest form, for a given input vector, the nearest neighbor classifier searches the nearest template and classifies the input vector into the class to which the template belongs. In the complex form the classifier treats k nearest neighbors. For a given input vector, the k nearest templates are searched and the input vector is classified into the class with the maximum number of templates. The classifier architecture is simple but as the number of training data becomes larger, the classification time becomes longer. Therefore many methods for speeding up classification are studied [65, pp. 181–191], [129, pp. 191–201]. One uses the branch-and-bound method [108, pp. 360–362] and another edits the training data, i.e., select or replace the data with the suitable templates. It is proved theoretically that as the number of templates becomes larger, the expected error rate of the nearest neighbor classifier is bounded by twice that of the Bayesian classifier [93, pp. 159–175].

Usually the Euclidean distance is used for measuring the distance between two data $\mathbf{x}$ and $\mathbf{y}$:

$$d(\mathbf{x} - \mathbf{y}) = \sqrt{\sum_{i=1}^{m} (x_i - y_i)^2} \tag{A.9}$$

but other distances, such as the Manhattan distance:

$$d(\mathbf{x}, \mathbf{y}) = \sum_{i=1}^{m} |x_i - y_i| \tag{A.10}$$

are used. It is clear from the architecture that the recognition rate of the training data for the 1-nearest neighbor classifier is 100%. But for the k-nearest neighbor classifier with $k > 1$, the recognition rate of the training data is not always 100%.

Since the distances such as the Euclidean and Manhattan distances are not invariant in scaling, classification performance varies according to the scaling of input ranges.

The fuzzy min-max classifier discussed in Section 9.1 is equivalent to the 1-nearest neighbor classifier with the Manhattan distance when $\theta = 0$, i.e., a fuzzy rule is defined for each datum, and the sensitivity parameter γ is set so that the degrees of membership of each datum are non-zero.

A.2.2 Performance Evaluation

We evaluated the performance of the k-nearest neighbor classifier using the data sets listed in Table 1.1 on page 19. Since the maximum recognition rates of the test data using the Euclidean distance and the Manhattan distance did not differ significantly for different k's, we used the Euclidean distance. We coded k-nearest neighbor classifier without using any speedup method and ran the optimized Fortran code on a Sun UltraSPARC IIi 333MHz workstation. The time listed in the following tables is the time for evaluating the recognition rate of the test data.

Iris Data. Table A.1 lists the recognition rates of the test (training) data and the execution time. For the training data, the numbers of misclassified data varied from 0 to 3 and for the test data the numbers varied from 4 to 6. The 1-nearest neighbor classifier showed the maximum recognition rates both for the test and the training data. Thus overfitting was not occurred.

Table A.1. Performance for the iris data

k	Rates (%)	Time (s)
1	<u>94.67</u> (100)	0.02
3	92.00 (96.00)	0.01
5	92.00 (97.33)	0.01
7	93.33 (98.67)	0.03

Numeral Data. Table A.2 shows the results for the numeral data. The 1-nearest neighbor classifier showed the maximum recognition rates both for the test and training data; 4 data among the 820 test data were misclassified. In this case also overfitting was not occurred.

Table A.2. Performance for the numeral data

k	Rates (%)	Time (s)
1	<u>99.51</u> (100)	0.39
3	99.02 (99.63)	0.43
5	99.15 (99.26)	0.46
7	98.90 (99.26)	0.46

Thyroid Data. Table A.3 lists the performance for the thyroid data. The recognition rate of the test data for the 1-nearest neighbor classifier was 91.98%. Since 92% of the data belong to one class, the recognition rate was very bad. This might be due to the use of the Euclidean distance for the mostly digital features. The 5-nearest neighbor classifier showed the maximum recognition rate of 93.67% for the test data, but it was still very bad. This was the worst recognition rate among the classifiers evaluated in this book for the thyroid data as listed in Table 10.3 on page 200. For the thyroid data overfitting occurred.

Table A.3. Performance for the thyroid data

k	Rates (%)	Time (s)
1	91.98 (100)	8
3	93.55 (95.71)	9
5	<u>93.67</u> (94.96)	10
7	93.58 (94.19)	10

Blood Cell Data. Table A.4 lists the performance for the blood cell data. The recognition rates of the training data decreased monotonically as k increased. The drop of the recognition rate was the largest among the benchmark data sets. This might indicate the heavy overlap between classes. But for the recognition rates of the test data, there was not so much difference among the classifiers and the 5-nearest neighbor classifier showed the maximum recognition rate of 90.13%.

Table A.4. Performance for the blood cell data

k	Rates (%)	Time (s)
1	89.90 (100)	5
3	89.84 (95.12)	6
5	<u>90.13</u> (93.51)	6
7	89.84 (93.32)	6

Hiragana Data. Table A.5 lists the results for the hiragana-50 data. The 1-nearest neighbor classifier showed the maximum recognition rates of the test and training data. Thus no overfitting occurred.

Table A.5. Performance for the hiragana-50 data

k	Rates (%)	Time (s)
1	<u>97.16</u> (100)	39
3	96.57 (98.94)	40
5	95.77 (97.79)	41
7	94.49 (96.70)	42

Table A.6 lists the performance for the hiragana-105 data. For the recognition rates of the test data, although there was no much difference among the classifiers, the 3-nearest neighbor classifier showed the maximum recognition rate of 99.99%.

Table A.6. Performance for the hiragana-105 data

k	Rates (%)	Time (s)
1	99.96 (100)	358
3	<u>99.99</u> (99.84)	363
5	99.90 (99.73)	365
7	99.68 (99.63)	359

Table A.7 lists the performance for the hiragana-13 data. The 1-nearest neighbor classifier showed the maximum recognition rates both for the test and training data. The recognition rates for the test and training data decreased monotonically as k increased. The hiragana-13 data were obtained from the hiragana-105 and the maximum recognition rate of the test data was comparable with that of the hiragana-105 data.

Table A.7. Performance for the hiragana-13 data

k	Rates (%)	Time (s)
1	<u>99.55</u> (100)	42
3	99.21 (99.50)	47
5	98.68 (98.88)	50
7	98.24 (98.33)	48

Discussions. The nearest neighbor classifier uses all the training data as templates and thus training is not necessary. Thus the classification time even for the hiragana-105 was not so long (about 6 minutes). But the problem is that a large number of templates must be stored for classification.

For four data sets among seven data sets, the 1-nearest neighbor classifier performed best without overfitting. But for the thyroid and blood cell data the 5-nearest neighbor classifier performed best and for the hiragana-105 data the 3-nearest neighbor classifier did. Thus the best classifier depends on the classification problem.

As compared in Chapter 10, the classification performance of the k-nearest neighbor classifier was not stable. Namely, the classification performance was good for the hiragana data sets but was the worst for the iris and thyroid data sets. To improve classification performance, proper scaling might be necessary [129, pp.197-198], [130, p. 71].

B. Matrices

B.1 Matrix Properties

In this section, we summarize the matrix properties used in this book. For more detailed explanation, see, e.g. [47].

Vectors $\mathbf{x}_1, \ldots, \mathbf{x}_n$ are *linearly independent* if

$$a_1 \mathbf{x}_1 + \cdots + a_n \mathbf{x}_n = 0 \tag{B.1}$$

holds only when $a_1 = \cdots = a_m = 0$. Otherwise, namely, at least one a_i is nonzero, $\mathbf{x}_1, \ldots, \mathbf{x}_n$ are *linearly dependent*.

Let A be an $m \times m$ matrix:

$$A = \begin{pmatrix} a_{11} & \cdots & a_{1m} \\ \cdots\cdots\cdots\cdots \\ a_{m1} & \cdots & a_{mm} \end{pmatrix}. \tag{B.2}$$

Then the *transpose* of A denoted by A^t is

$$A^t = \begin{pmatrix} a_{11} & \cdots & a_{m1} \\ \cdots\cdots\cdots\cdots \\ a_{1m} & \cdots & a_{mm} \end{pmatrix}. \tag{B.3}$$

If A satisfies $A = A^t$, A is a *symmetric matrix*. If A satisfies $A^t A = A A^t = I$, A is an *orthogonal matrix*.

The $m \times m$ unit matrix I is

$$I = \begin{pmatrix} 1 & 0 & \cdots & 0 \\ 0 & 1 & \cdots & 0 \\ \cdots\cdots\cdots \\ 0 & 0 & \cdots & 1 \end{pmatrix}. \tag{B.4}$$

If $m \times m$ matrices A and B satisfies $A B = I$, B is called the *inverse* of A and is denoted by A^{-1}. If A has the inverse, A is *nonsingular*. Otherwise, A is singular.

The *determinant* of an $m \times m$ matrix $A = \{a_{ij}\}$, $\det(A)$, is defined recursively by

$$\det(A) = \sum_{i=1}^{m} (-1)^{i+1} a_{1i} \det(A_{1i}), \tag{B.5}$$

where A_{1i} is the $(m-1) \times (m-1)$ matrix obtained by deleting the first row and the ith column from A. When $m = 1$, $\det(A) = a_{11}$.

If the $m \times m$ matrix A satisfies

$$A\mathbf{x} = \lambda\mathbf{x}, \tag{B.6}$$

where $\mathbf{x}$ is a non-zero, m-dimensional vector, λ is a constant and is called an *eigenvalue*, and $\mathbf{x}$ is called an *eigenvector*. Rearranging (B.6) gives

$$(A - \lambda I)\mathbf{x} = 0. \tag{B.7}$$

Thus, (B.7) has nonzero $\mathbf{x}$, when

$$\det(A - \lambda I) = 0, \tag{B.8}$$

which is called a *characteristic equation*.

Theorem B.1.1. *All the eigenvalues of a real symmetric matrix are real.*

Theorem B.1.2. *Eigenvectors associated with different eigenvalues for a real symmetric matrix are orthogonal.*

For an m-dimensional vector $\mathbf{x}$ and an $m \times m$ symmetric matrix A, $Q = \mathbf{x}^t A \mathbf{x}$ is called a *quadratic form*. If for any nonzero $\mathbf{x}$, $Q = \mathbf{x}^t A \mathbf{x} \geq 0$, Q is *positive semi-definite*. Matrix Q is *positive definite*, if the strict inequality holds. Let L be an $m \times m$ orthogonal matrix. By $\mathbf{y} = L\mathbf{x}$, $\mathbf{x}$ is transformed into $\mathbf{y}$. This is the transformation from one orthonormal base into another orthonormal basis. The quadratic form Q is

$$\begin{aligned} Q &= \mathbf{x}^t A \mathbf{x} \\ &= \mathbf{y}^t L A L^t \mathbf{y}. \end{aligned} \tag{B.9}$$

Theorem B.1.3. *The characteristic equations for A and LAL^t are the same.*

Theorem B.1.4. *If an $m \times m$ real symmetric matrix A is diagonalized by L:*

$$LAL^t = \begin{pmatrix} \lambda_1 & 0 & \cdots & 0 \\ 0 & \lambda_2 & \cdots & 0 \\ \multicolumn{4}{c}{\dotfill} \\ 0 & 0 & \cdots & \lambda_m \end{pmatrix}, \tag{B.10}$$

$\lambda_1, \ldots, \lambda_m$ *are the eigenvalues of A and the ith row of L is the eigenvector associated with λ_i.*

If all the eigenvalues of A are positive, A is *positive definite*. If all the eigenvalues are non-negative, A is *positive semi-definite*.

B.2 Least-squares Method and Singular Value Decomposition

Assume that we have M input-output pairs $\{(\mathbf{a}'_1, b_1), \ldots, (\mathbf{a}'_M, b_M)\}$ in the $(n-1)$-dimensional input space $\mathbf{x}'$ and one-dimensional output space y. Now using the least-squares method, we determine the linear relation of the input-output pairs:

$$y = \mathbf{p}^t \mathbf{x}' + q, \tag{B.11}$$

where $\mathbf{p}$ is the $(n-1)$-dimensional vector, q is a scalar constant, and $M \geq n$.

Rewriting (B.11), we get

$$(\mathbf{x}'^t, 1) \begin{pmatrix} \mathbf{p} \\ q \end{pmatrix} = y. \tag{B.12}$$

Substituting $\mathbf{a}'_i$ and b_i into $\mathbf{x}'$ and y of (B.12), respectively, and replacing $(\mathbf{p}^t, q)^t$ with the n-dimensional parameter vector $\mathbf{x}$, we obtain

$$\mathbf{a}_i^t \mathbf{x} = b_i \qquad \text{for} \quad i = 1, \ldots, M, \tag{B.13}$$

where $\mathbf{a}_i = (\mathbf{a}'^t_i, 1)^t$.

We determine the parameter vector $\mathbf{x}$ so that the sum of squared errors:

$$E = (A\mathbf{x} - \mathbf{b})^t (A\mathbf{x} - \mathbf{b}) \tag{B.14}$$

is minimized, where A is an $M \times n$ matrix and $\mathbf{b}$ is an M-dimensional vector:

$$A = \begin{pmatrix} \mathbf{a}_1^t \\ \mathbf{a}_2^t \\ \vdots \\ \mathbf{a}_M^t \end{pmatrix}, \quad \mathbf{b} = \begin{pmatrix} b_1 \\ b_2 \\ \vdots \\ b_M \end{pmatrix}. \tag{B.15}$$

Here, if the rank of A is smaller than n, there is no unique solution. In that situation, we determine $\mathbf{x}$ so that the Euclidean norm of $\mathbf{x}$ is minimized.

Matrix A is decomposed into singular values [47]:

$$A = USV^t, \tag{B.16}$$

where U and V are $M \times M$ and $n \times n$ orthogonal matrices, respectively, and S is an $M \times n$ diagonal matrix given by

$$S = \begin{pmatrix} \sigma_1 & & 0 \\ & \ddots & \\ 0 & & \sigma_n \\ \hline & 0_{M-n,n} & \end{pmatrix}. \tag{B.17}$$

Here, σ_i are singular values and $\sigma_1 \geq \sigma_2 \geq \cdots \geq \sigma_n \geq 0$, and $0_{M-n,n}$ is the $(M-n) \times n$ zero matrix.

It is known that the columns of U and V are the eigenvectors of AA^t and $A^t A$, respectively, and the singular values correspond to the square roots of the eigenvalues of AA^t which are the same with those of A^tA [17, pp. 434–435]. Thus when A is a symmetric square matrix, $U = V$ and $A = USU^t$. This is similar to the diagonalization of the square matrix given by Theorem B.1.4 on page 308. The difference is that the singular values A are the absolute values of the eigenvalues of A. Thus, if A is a positive (semi-)definite matrix, the both decompositions are the same.

Rewriting (B.14), we get [47, p. 256]

$$
\begin{aligned}
E &= (A\mathbf{x} - \mathbf{b})^t(A\mathbf{x} - \mathbf{b}) \\
&= (USV^t\mathbf{x} - UU^t\mathbf{b})^t(A\mathbf{x} - \mathbf{b}) \\
&= (SV^t\mathbf{x} - U^t\mathbf{b})^t(SV^t\mathbf{x} - U^t\mathbf{b}) \\
&= \sum_{i=1}^{n}(\sigma_i \mathbf{v}_i^t \mathbf{x} - \mathbf{u}_i^t \mathbf{b})^2 + \sum_{i=n+1}^{M}(\mathbf{u}_i^t \mathbf{b})^2,
\end{aligned}
\tag{B.18}
$$

where $U = (\mathbf{u}_1,\ldots,\mathbf{u}_M)$ and $V = (\mathbf{v}_1,\ldots,\mathbf{v}_M)$. Assuming the rank of A is $r\,(\leq n)$, (B.18) is minimized when

$$
\begin{aligned}
\sigma_i \mathbf{v}_i^t \mathbf{x} &= \mathbf{u}_i^t \mathbf{b} &\quad \text{for} \quad i = 1,\ldots,r, \tag{B.19}\\
\mathbf{v}_i^t \mathbf{x} &= 0 &\quad \text{for} \quad i = r+1,\ldots,n. \tag{B.20}
\end{aligned}
$$

Equation (B.20) is imposed to obtain the minimum Euclidean norm solution. From (B.19) and (B.20), we obtain

$$
\mathbf{x} = VS^+U^t\mathbf{b} = A^+\mathbf{b},
\tag{B.21}
$$

where S^+ is the $n \times M$ diagonal matrix given by

$$
S^+ = \begin{pmatrix} \begin{array}{ccc} \frac{1}{\sigma_1} & & 0 \\ & \ddots & \\ & & \frac{1}{\sigma_r} \\ 0 & & 0 \end{array} & \Big| & 0 \end{pmatrix}.
\tag{B.22}
$$

We call A^+ the pseudo-inverse of A. We must bear in mind that in calculating the pseudo-inverse, we replace the reciprocal of 0 with 0, not infinity. This ensures the minimum norm solution.

From (B.16) and (B.21),

$$
\begin{aligned}
A^+A &= VS^+U^tUSV^t \\
&= VS^+SV^t \\
&= V\begin{pmatrix} I_r & 0_{r,n-r} \\ 0_{n-r,r} & 0_{n-r} \end{pmatrix}V^t \\
&= \begin{pmatrix} I_r & 0_{r,n-r} \\ 0_{n-r,r} & 0_{n-r} \end{pmatrix}, \tag{B.23}\\
AA^+ &= USS^+U^t \\
&= \begin{pmatrix} I_r & 0_{r,M-r} \\ 0_{M-r,r} & 0_{M-r} \end{pmatrix}, \tag{B.24}
\end{aligned}
$$

where I_r is the $r \times r$ unit matrix, 0_i is the $i \times i$ zero matrix, $0_{i,j}$ is the $i \times j$ zero matrix. Therefore, if A is a square matrix with rank n, $A^+A = AA^+ = I$. Namely, the pseudo-inverse of A coincides with the inverse of A, A^{-1}. If $M > n$ and the rank of A is n, $A^+A = I$ but $AA^+ \neq I$. In this case A^+ is given by

$$A^+ = (A^tA)^{-1}A^t. \tag{B.25}$$

This is obtained by taking the derivative of (B.14) with respect to $\mathbf{x}$ and equating the result to zero.

When $M > n$ and the rank of A is smaller than n, $A^+A \neq I$ and $AA^+ \neq I$.

Even when A^tA is nonsingular, it is recommended to calculate the pseudo-inverse by singular value decomposition, not using (B.25). Because if A^tA is near singular, $(A^tA)^{-1}A^t$ is vulnerable to the small singular values [131, pp. 59–70].

B.3 Covariance Matrix

Let $\mathbf{x}_1, \ldots, \mathbf{x}_M$ be M samples of the m-dimensional random variable X. Then the sample covariance matrix of X is given by

$$Q = \frac{1}{M} \sum_{i=1}^{M} (\mathbf{x}_i - \mathbf{c})(\mathbf{x}_i - \mathbf{c})^t, \tag{B.26}$$

where $\mathbf{c}$ is the mean vector:

$$\mathbf{c} = \frac{1}{M} \sum_{i=1}^{M} \mathbf{x}_i. \tag{B.27}$$

To get the unbiased covariance matrix, we replace M with $M - 1$ in (B.26), but in this book we use (B.26) as the sample covariance matrix.

Let

$$\mathbf{y}_i = \mathbf{x}_i - \mathbf{c}. \tag{B.28}$$

Then, (B.26) becomes

$$Q = \frac{1}{M} \sum_{i=1}^{M} \mathbf{y}_i \mathbf{y}_i^t. \tag{B.29}$$

From (B.27) and (B.28), $\mathbf{y}_1, \ldots, \mathbf{y}_M$ are linearly dependent.

According to the definition, the covariance matrix Q is symmetric. Matrix Q is positive (semi-)definite as the following theorem shows.

Theorem B.3.1. *The covariance matrix Q given by (B.29) is positive definite if $\mathbf{y}_1, \ldots, \mathbf{y}_M$ have at least m linearly independent data. Matrix Q is positive semi-definite, if any m data from $\mathbf{y}_1, \ldots, \mathbf{y}_M$ are linearly dependent.*

Proof. Let $\mathbf{z}$ be an m-dimensional nonzero vector. From (B.29),

$$
\mathbf{z}^t Q\,\mathbf{z} = \mathbf{z}^t \left(\frac{1}{M} \sum_{i=1}^{M} \mathbf{y}_i\,\mathbf{y}_i^t \right) \mathbf{z}
$$

$$
= \frac{1}{M} \sum_{i=1}^{M} \left(\mathbf{z}^t\,\mathbf{y}_i \right) \left(\mathbf{z}^t\,\mathbf{y}_i \right)^t
$$

$$
= \frac{1}{M} \sum_{i=1}^{M} \left(\mathbf{z}^t\,\mathbf{y}_i \right)^2 \geq 0. \tag{B.30}
$$

Thus Q is positive semi-definite. If there are m linearly independent data in $\{\mathbf{y}_1,\ldots,\mathbf{y}_M\}$, they span the m-dimensional space. Since any $\mathbf{z}$ is expressed by a linear combination of these data, the strict inequality holds for (B.30).

Since $\mathbf{y}_1,\ldots,\mathbf{y}_M$ are linearly dependent, at least $m+1$ samples are necessary so that Q becomes positive definite. (Q.E.D.)

Assuming that Q is positive definite, the following theorem holds.

Theorem B.3.2. *If Q is positive definite, the mean square weighted distance for $\{\mathbf{y}_1,\ldots,\mathbf{y}_M\}$ is m:*

$$
\frac{1}{M} \sum_{i=1}^{M} \mathbf{y}_i^t\, Q^{-1}\,\mathbf{y}_i = m. \tag{B.31}
$$

Proof. Let P be the orthogonal matrix that diagonalizes Q. Namely,

$$
P\,Q\,P^t = \mathrm{diag}(\lambda_1,\ldots,\lambda_m), \tag{B.32}
$$

where diag denotes the diagonal matrix, and $\lambda_1,\ldots,\lambda_m$ are the eigenvalues of Q. From (B.32),

$$
Q = P^t\mathrm{diag}(\lambda_1,\ldots,\lambda_m)P, \tag{B.33}
$$

$$
Q^{-1} = P^t\mathrm{diag}(\lambda_1^{-1},\ldots,\lambda_m^{-1})P. \tag{B.34}
$$

Let

$$
\tilde{\mathbf{y}}_i = P\,\mathbf{y}_i. \tag{B.35}
$$

Then from (B.29) and (B.35), (B.32) becomes

$$
\frac{1}{M} \sum_{i=1}^{M} \tilde{\mathbf{y}}_i\,\tilde{\mathbf{y}}_i^t = \mathrm{diag}(\lambda_1,\ldots,\lambda_m). \tag{B.36}
$$

Thus for the diagonal elements of (B.36),

$$
\frac{1}{M} \sum_{i=1}^{M} \tilde{y}_{ik}^2 = \lambda_k \quad \text{for} \quad k = 1,\ldots,m, \tag{B.37}
$$

where $\tilde{y}_{ik}$ is the kth element of $\tilde{\mathbf{y}}_i$. From (B.34) and (B.35), the left hand side of (B.31) becomes

$$\frac{1}{M} \sum_{i=1}^{M} \mathbf{y}_i^t \, Q^{-1} \, \mathbf{y}_i = \frac{1}{M} \sum_{i=1}^{M} \tilde{\mathbf{y}}_i^t \, \text{diag}(\lambda_1^{-1}, \ldots, \lambda_m^{-1}) \, \tilde{\mathbf{y}}_i$$

$$= \frac{1}{M} \sum_{i=1}^{M} \sum_{k=1}^{m} \lambda_k^{-1} \, \tilde{y}_{ik}^2. \tag{B.38}$$

Thus from (B.37) and (B.38), the theorem holds. (Q.E.D.)

References

1. J. J. Buckley, Y. Hayashi, and E. Czogala. On the equivalence of neural networks and fuzzy expert systems. In *Proceedings of International Joint Conference on Neural Networks*, volume 2, pages 691–695, Baltimore, MD, June 1992.
2. S. Abe. *Neural networks and fuzzy systems: Theory and applications.* Kluwer Academic Publishers, Boston, MA, 1996.
3. N. K. Kasabov. *Foundations of neural networks, fuzzy systems, and knowledge engineering.* MIT Press, Cambridge, MA, 1996.
4. S. K. Pal and S. Mitra. *Neuro-fuzzy pattern recognition: Methods in soft computing.* John Wiley & Sons, New York, NY, 1999.
5. N. K. Kasabov and R. Kozma, editors. *Neuro-fuzzy techniques for intelligent information systems.* Physica-Verlag, Hidelberg, Germany, 1999.
6. R. Fullér. *Introduction to neuro-fuzzy systems.* Physica-Verlag, Hidelberg, Germany, 1999.
7. C.-T. Lin and C. S. George Lee. Neural-network-based fuzzy logic control and decision system. *IEEE Transactions on Computers*, 40(12):1320–1336, 1991.
8. S. Mitra and S. K. Pal. Fuzzy multi-layer perceptron, inferencing and rule generation. *IEEE Transactions on Neural Networks*, 6(1):51–63, 1995.
9. S. Mitra, R. K. De, and S. K. Pal. Knowledge-based fuzzy MLP for classification and rule generation. *IEEE Transactions on Neural Networks*, 8(6):1338–1350, 1997.
10. S. Mitra and S. K. Pal. Logical operation based fuzzy MLP for classification and rule generation. *Neural Networks*, 7(2):353–373, 1994.
11. C.-T. Lin and Y.-C. Lu. A neural fuzzy system with fuzzy supervised learning. *IEEE Transactions on Systems, Man, and Cybernetics—Part B*, 26(5):744–763, 1996.
12. V. N. Vapnik. *Statistical learning theory.* John Wiley & Sons, New York, NY, 1998.
13. T. H. Reiss. *Recognizing planar objects using invariant image features.* Springer-Verlag, Berlin, 1993.
14. H. Takenaga, S. Abe, M. Takatoo, M. Kayama, T. Kitamura, and Y. Okuyama. Input layer optimization of neural networks by sensitivity analysis and its application to recognition of numerals. *Electrical Engineering in Japan*, 111(4):130–138, 1991.
15. G. L. Cash and M. Hatamian. Optical character recognition by the method of moments. *Computer Vision, Graphics, and Image Processing*, 39:291–310, 1989.
16. M.-S. Lan, H. Takenaga, and S. Abe. Character recognition using fuzzy rules extracted from data. In *Proceedings of Third IEEE International Conference on Fuzzy Systems*, volume 1, pages 415–420, Orlando, FL, June 1994.
17. V. Cherkassky and F. Mulier. *Learning from data: Concepts, theory, and methods.* John Wiley & Sons, New York, 1998.

18. S. Young and T. Downs. CARVE—A constructive algorithm for real-valued examples. *IEEE Transactions on Neural Networks*, 9(6):1180–1190, 1998.

19. C.-T. Lin and C. S. George Lee. *Neural fuzzy systems: A neuro-fuzzy synergism to intelligent systems*. Prentice Hall, Upper Saddle River, NJ, 1996.

20. M. Russo. FuGeNeSys—A fuzzy genetic neural system for fuzzy modeling. *IEEE Transactions on Fuzzy Systems*, 6(3):373–388, 1998.

21. Y. Shi, R. Eberhart, and Y. Chen. Implementation of evolutionary fuzzy systems. *IEEE Transactions on Fuzzy Systems*, 7(2):109–119, 1999.

22. H. Ishibuchi, T. Nakashima, and T. Murata. Performance evaluation of fuzzy classifier systems for multidimensional pattern classification problems. *IEEE Transactions on Systems, Man, and Cybernetics—Part B*, 29(5):601–618, 1999.

23. S. Abe, M.-S. Lan, and R. Thawonmas. Tuning of a fuzzy classifier derived from data. *International Journal of Approximate Reasoning*, 14(1):1–24, 1996.

24. R. A. Fisher. The use of multiple measurements in taxonomic problems. *Annals of Eugenics*, 7:179–188, 1936.

25. J. C. Bezdek, J. M. Keller, R. Krishnapuram, L. I. Kuncheva, and N. H. Pal. Will the real iris data please stand up? *IEEE Transactions on Fuzzy Systems*, 7(3):368–369, 1999.

26. S. M. Weiss and I. Kapouleas. An empirical comparison of pattern recognition, neural nets, and machine learning classification methods. In *Proceedings of the Eleventh International Joint Conference on Artificial Intelligence*, pages 781–787, Detroit MI, August 1989.

27. A. Hashizume, J. Motoike, and R. Yabe. Fully automated blood cell differential system and its application. In *Proceedings of the IUPAC Third International Congress on Automation and New Technology in the Clinical Laboratory*, pages 297–302, Kobe, Japan, September 1988.

28. K. Funahashi. On the approximate realization of continuous mappings by neural networks. *Neural Networks*, 2(3):183–192, 1989.

29. K. Hornik, M. Stinchcombe, and H. White. Multilayer feedforward networks are universal approximators. *Neural Networks*, 2(5):359–366, 1989.

30. C. M. Bishop. *Neural networks for pattern recognition*. Clarendon Press, Oxford, 1995.

31. G. J. Gibson and C. F. N. Cowan. On the decision regions of multilayer perceptron. *Proceedings of the IEEE*, 78(10):1590–1594, 1990.

32. J. Makhoul, A. El-Jaroudi, and R. Schwartz. Formation of disconnected decision regions with a single hidden layer. In *Proceedings of International Joint Conference on Neural Networks*, volume 1, pages 455–460, Washington, D. C, June 1989.

33. P. Ruján. A geometric approach to learning in neural networks. In *Proceedings of International Joint Conference on Neural Networks*, volume 2, pages 105–109, Washington, D. C., June 1989.

34. S. Abe, M. Kayama, and H. Takenaga. How neural networks for pattern recognition can be synthesized. *Journal of Information Processing*, 14(3):344–350, 1991.

35. D. E. Rumelhart, J. L. McClelland, and the PDP Research Group. *Parallel distributed processing*. volumes 1 and 2, MIT Press, Cambridge, MA, 1986.

36. A. J. Shepherd. *Second-order methods for neural networks: Fast and reliable training methods for multi-layer perceptrons*. Springer-Verlag, London, 1997.

37. S. Abe, M. Kayama, and H. Takenaga. Acceleration of learning and improvement of generalization ability for pattern classification networks. *Transactions of the Institute of Electronics, Information and Communication Engineers D-II*, J76-DII(3):647–652, 1993 (in Japanese).

38. M. H. Hassoun and J. Song. Adaptive Ho-Kashyap rules for perceptron training. *IEEE Transactions on Neural Networks*, 3(1):51–61, 1992.

39. N. Tsuchiya, S. Ozawa, and S. Abe. Training three-layered neural network classifiers by solving inequalities. In *Proceedings of the International Joint Conference on Neural Networks*, volume 3, pages 555–560, Como, Italy, July 2000.

40. N. Tsuchiya, S. Ozawa, and S. Abe. Fast training of three-layered neural network classifiers by solving inequalities. *Transactions of the Institute of Systems, Control and Information Engineers*, 13(6):276–283, 2000 (in Japanese).

41. S. Ergezinger and E. Thomsen. An accelerated learning algorithm for multilayer perceptrons: Optimization layer by layer. *IEEE Transactions on Neural Networks*, 6(1):31–42, 1995.

42. R. Lengellé and T. Denœux. Training MLPs layer by layer using an objective function for internal representations. *Neural Networks*, 9(1):83–87, 1996.

43. G.-J. Wang and C.-C. Chen. A fast multilayer neural network training algorithm based on the layer-by-layer optimizing procedures. *IEEE Transactions on Neural Networks*, 7(3):768–775, 1996.

44. B. Ph. van Milligen, V. Tribaldos, J. A. Jiménez, and C. Santa Cruz. Comments on "An accelerated learning algorithm for multilayer perceptrons: Optimizing layer by layer". *IEEE Transactions on Neural Networks*, 9(2):339–341, 1998.

45. N. S. Rubanov. The layer-wise method and the backpropagation hybrid approach to learning a feedforward neural network. *IEEE Transactions on Neural Networks*, 11(2):295–305, 2000.

46. R. O. Duda and P. E. Hart. *Pattern classification and scene analysis*. John Wiley & Sons, New York, NY, 1973.

47. G. H. Golub and C. F. Van Loan. *Matrix computation*. The Johns Hopkins University Press, Baltimore, MD, third edition, 1996.

48. B. Schölkopf, C. J. C. Burges, and A. J. Smola, editors. *Advances in kernel methods: Support vector learning*. The MIT Press, Cambridge, MA, 1999.

49. U. H.-G. Kreßel. Pairwise classification and support vector machines. In B. Schölkopf, C. J. C. Burges, and A. J. Smola, editors, *Advances in kernel methods: Support vector learning*, pages 255–268. The MIT Press, Cambridge, MA, 1999.

50. J. A. K. Suykens and J. Vandewalle. Training multilayer preceptron classifiers based on a modified support vector method. *IEEE Transactions on Neural Networks*, 10(4):907–911, 1999.

51. C. Saunders, M. O. Stitson, J. Weston, L. Bottou, B. Schölkopf, and A. Smola. Support vector machine reference manual. Technical Report CSD-TR-98-03, Royal Holloway, University of London, London, 1998.

52. T. Joachims. Making large-scale support vector machine learning practical. In B. Schölkopf, C. J. C. Burges, and A. J. Smola, editors, *Advances in kernel methods: Support vector learning*, pages 169–184. The MIT Press, Cambridge, MA, 1999.

53. O. Barzilay and V. L. Brailovsky. On domain knowledge and feature selection using a support vector machine. *Pattern Recognition Letters*, 20:475–484, 1999.

54. L. Hermes and J. M. Buhmann. Feature selection for support vector machines. In *Proceedings of 15th International Conference on Pattern Recognition*, volume 2, pages 716–719, Barcelona, Spain, September 2000.

55. F. Uebele, S. Abe, and M.-S. Lan. A neural network–based fuzzy classifier. *IEEE Transactions on Systems, Man, and Cybernetics*, 25(2):353–361, 1995.

56. S. Abe and R. Thawonmas. A fuzzy classifier with ellipsoidal regions. *IEEE Transactions on Fuzzy Systems*, 5(3):358–368, 1998.

57. S. Abe. Dynamic cluster generation for a fuzzy classifier with ellipsoidal regions. *IEEE Transactions on Systems, Man, and Cybernetics—Part B*, 28(6):869–876, 1998.

58. I. H. Suh, J. H. Kim, and F. C.-H. Rhee. Convex-set-based fuzzy clustering. *IEEE Transactions on Fuzzy Systems*, 7(3):271–285, 1999.

59. Y. Ohta, Y. Nagai, and L. Gong. Beneath-beyond method and construction of Lyapunov functions. In *Proceedings of the International Symposium on Nonlinear Theory and its Applications (NOLTA'97)*, pages 353–356, 1997.

60. R. Katayama, M. Watanabe, K. Kuwata, Y. Kajitani, and Y. Nishida. Performance evaluation of self generating radial basis function for function approximation. In *Proceedings of 1993 International Joint Conference on Neural Networks*, volume 1, pages 471–474, Nagoya, Japan, October 1993.

61. M. T. Musavi, W. Ahmed, K. H. Chan, K. B. Faris, and D. M. Hummels. On the training of radial basis function classifiers. *Neural Networks*, 5(4):595–603, 1992.

62. T. Takigawa. Effect of membership functions on generalization ability. Bachelor's thesis, Electrical and Electronics Engineering, Kobe University, Kobe, Japan, March 2000 (in Japanese).

63. T. Uhicda. Clustering suited for pattern classification. Bachelor's thesis, Electrical and Electronics Engineering, Kobe University, Kobe, Japan, March 2000 (in Japanese).

64. M. P. Windham. Cluster validity for the fuzzy c-means clustering algorithm. *IEEE Transactions on Pattern Analysis and Machine Intelligence*, PAMI-4(4):357–363, 1982.

65. J. C. Bezdek, J. Keller R. Krishnapuram, and N. R. Pal. *Fuzzy Models and algorithms for pattern recognition and image processing*. Kluwer Academic Publishers, Boston, MA, 1999.

66. T. Kohonen. *Self-organizing Maps*. Springer-Verlag, Berlin, second edition, 1997.

67. P. K. Simpson. Fuzzy min-max neural networks—Part 2: Clustering. *IEEE Transactions on Fuzzy Systems*, 1(1):32–45, 1993.

68. R. Krishnapuram and J. M. Keller. A possibilistic approach to clustering. *IEEE Transactions on Fuzzy Systems*, 1(2):98–110, 1993.

69. J. C. Bezdek, R. J. Hathaway, M. J. Sabin, and W. T. Tucker. Convergence theory for fuzzy c-means: Counterexamples and repairs. *IEEE Transactions on Systems, Man, and Cybernetics*, 17(5):873–877, 1987.

70. H. Ritter, T. Martinetz, and K. Schulten. *Neural computation and self-organizing maps: An introduction*. Addison-Wesley, Reading, MA, 1992.

71. H.-S. Dai. System identification using neural networks. Technical Report CSD-930002, University of California, Los Angeles, CA, 1992.

72. S. Abe. Generalization improvement of a fuzzy classifier with pyramidal membership functions. In *Proceedings of 15th International Conference on Pattern Recognition*, volume 2, pages 211–214, Barcelona, Spain, September 2000.

73. K. Sakaguchi. Tuning locations of a fuzzy classifier with ellipsoidal regions. Bachelor's thesis, Electrical and Electronics Engineering, Kobe University, Kobe, Japan, March 2000 (in Japanese).

74. P. J. Rousseeuw and A. M. Leroy. *Robust regression and outlier detection*. John Wiley & Sons, New York, NY, 1987.

75. R. R. Wilcox. *Introduction to robust estimation and hypothesis testing*. Academic Press, San Diego, CA, 1997.

76. K. Liano. Robust error measure for supervised neural network learning with outliers. *IEEE Transactions on Neural Networks*, 7(1):246–250, 1996.

77. D. S. Chen and R. C. Jain. A robust back propagation learning algorithm for function approximation. *IEEE Transactions on Neural Networks*, 5(3):467–479, 1994.

78. A. G. Borş and I. Pitas. Median radial basis function neural network. *IEEE Transactions on Neural Networks*, 7(6):1351–1364, 1996.

79. T. Eguchi, T. Tamaki, and S. Abe. A robust fuzzy classifier with ellipsoidal regions. *Transactions of the Institute of Systems, Control and Information Engineers*, 13(9):433–440, 2000 (in Japanese).

80. P. K. Simpson. Fuzzy min-max neural networks—Part 1: Classification. *IEEE Transactions on Neural Networks*, 3(5):776–786, 1992.

81. S. Abe and M.-S. Lan. A method for fuzzy rules extraction directly from numerical data and its application to pattern classification. *IEEE Transactions on Fuzzy Systems*, 3(1):18–28, 1995.

82. R. Thawonmas and S. Abe. Extraction of fuzzy rules for classification based on partitioned hyperboxes. *Journal of Intelligent Fuzzy Systems*, 4(3):215–226, 1996.

83. M. Shimizu and S. Abe. On input invariance of fuzzy classifiers with learning capability. *Transactions of the Institute of Systems, Control and Information Engineers*, 12(12):739–746, 1999 (in Japanese).

84. H. A. Malki and A. Moghaddamjoo. Using the Karhunen-Loe've transformation in the back-propagation training algorithm. *IEEE Transactions on Neural Networks*, 2(1):162–165, 1991.

85. C. Chatterjee and V. P. Roychowdhury. On self-organizing algorithms and networks for class-separability features. *IEEE Transactions on Neural Networks*, 8(3):663–678, 1997.

86. C. Lee and D. A. Landgrebe. Feature extraction based on decision boundaries. *IEEE Transactions on Pattern Analysis and Machine Intelligence*, 15(4):388–400, 1993.

87. C. Lee and D. A. Landgrebe. Decision boundary feature extraction for nonparametric classification. *IEEE Transactions on Systems, Man, and Cybernetics*, 23(2):433–444, 1993.

88. C. Lee and D. A. Landgrebe. Decision boundary feature extraction for neural networks. *IEEE Transactions on Neural Networks*, 8(1):75–83, 1997.

89. J. Mao and A. K. Jain. Artificial neural networks for feature extraction and multivariate data projection. *IEEE Transactions on Neural Networks*, 6(2):296–317, 1995.

90. J. Lampinen and E. Oja. Distortion tolerant pattern recognition based on self-organizing feature extraction. *IEEE Transactions on Neural Networks*, 6(3):539–547, 1995.

91. S. De Backer, A. Naud, and P. Scheunders. Non-linear dimensionality reduction techniques for unsupervised feature extraction. *Pattern Recognition Letters*, 19:711–720, 1998.

92. C. Santa Cruz and J. R. Dorronsoro. A nonlinear discriminant algorithm for feature extraction and data classification. *IEEE Transactions on Neural Networks*, 9(6):1370–1376, 1998.

93. E. Gose, R. Johnsonbaugh, and S. Jost. *Pattern recognition and image analysis*. Prentice Hall, Upper Saddle River, NJ, 1996.

94. J. Kittler. Feature selection and extraction. In T. Y. Young and K. S. Fu, editors, *Handbook of Pattern Recognition and Image Processing*, pages 59–83. Academic Press, San Diego, CA, 1986.

95. S. K. Pal and B. Chakraborty. Fuzzy set theoretic measure for automatic feature evaluation. *IEEE Transactions on Systems, Man, and Cybernetics*, 16(5):754–760, 1986.

96. R. Battiti. Using mutual information for selecting features in supervised neural net learning. *IEEE Transactions on Neural Networks*, 5(4):537–550, 1994.

97. R. Thawonmas and S. Abe. A novel approach to feature selection based on analysis of fuzzy regions. *IEEE Transactions on Systems, Man, and Cybernetics—Part B*, 27(2):196–207, 1997.

98. S. Abe, R. Thawonmas, and Y. Kobayashi. Feature selection by analyzing class regions approximated by ellipsoids. *IEEE Transactions on Systems, Man, and Cybernetics—Part C*, 28(2):282–287, 1998.

99. E. D. Karnin. A simple procedure for pruning back-propagation trained neural networks. *IEEE Transactions on Neural Networks*, 1(2):239–242, 1990.

100. R. Setiono and H. Liu. Neural-network feature selector. *Transactions on Neural Networks*, 8(3):654–662, 1997.

101. M. Egmont-Petersen, J. L. Talmon, A. Hasman, and A. W. Ambergen. Assessing the importance of features for multi-layer perceptrons. *Neural Networks*, 11(4):623–635, 1998.

102. A. M. Fraser and H. L. Swinney. Independent coordinates for strange attractors from mutual information. *Physical Review A*, 33(2):1134–1140, 1986.

103. S. Salzberg. A nearest hyperrectangle learning method. *Machine Learning*, 6:251–276, 1991.

104. W. Siedlecki and J. Sklansky. Constrained genetic optimization via dynamic reward-penalty balancing and its use in pattern recognition. In *Proceedings of the Third International Conference on Genetic Algorithms*, pages 141–150, Arlington, VA, 1989.

105. F. Z. Brill, D. E. Brown, and W. N. Martin. Fast genetic selection of features for neural network classifiers. *IEEE Transactions on Neural Networks*, 3(2):324–328, 1992.

106. B. Efron and R. J. Tibshirani. *An introduction to the bootstrap (Monographs on statistics and applied probability 57)*. Chapman & Hall/CRC Press, New York, NY, 1993.

107. A. J. Owens. Empirical modeling of very large data sets using neural networks. In *Proceedings of the International Joint Conference on Neural Networks*, volume 6, pages 302–307, Como, Italy, July 2000.

108. K. Fukunaga. *Introduction to statistical pattern recognition*. Academic Press, San Diego, CA, second edition, 1990.

109. S. M. Weiss and N. Indurkhya. *Predictive data mining*. Morgan Kaufmann, San Francisco. CA, 1998.

110. M. Kobayashi, M. Kimata, and S. Abe. Improvement of generalization ability of a fuzzy classifier with ellipsoidal regions by the optimum division of data. *Transactions of the Institute of Systems, Control and Information Engineers*, 13(2):87–94, 2000 (in Japanese).

111. J.-S. R. Jang. ANFIS: Adaptive-network-based fuzzy inference system. *IEEE Transactions on Systems, Man, and Cybernetics*, 23(3):665–685, 1993.

112. L.-X. Wang and J. M. Mendel. Generating fuzzy rules by learning from examples. *IEEE Transactions on Systems, Man, and Cybernetics*, 22(6):1414–1427, 1992.

113. R. Thawonmas and S. Abe. Function approximation based on fuzzy rules extracted from partitioned numerical data. *IEEE Transactions on Systems, Man, and Cybernetics—Part B*, 29(4), 1999.

114. J. Mitchell and S. Abe. Fuzzy clustering networks: Design criteria for approximation and prediction. *Transactions of Institute of Electronics, Information and Communication Engineers of Japan*, E79-D(1):63–71, 1996.

115. S. L. Chiu. Fuzzy model identification based on cluster estimation. *Journal of Intelligent and Fuzzy Systems*, 2:267–278, 1994.

116. S. Chen, C. F. N. Cowan, and P. M. Grant. Orthogonal least squares learning algorithm for radial basis function networks. *IEEE Transactions on Neural Networks*, 2(2):302–309, 1991.

117. J. A. Dickerson and B. Kosko. Fuzzy function approximation with ellipsoidal rules. *IEEE Transactions on Systems, Man, and Cybernetics—Part B*, 26(4):542–560, 1996.

118. S. Abe and M.-S. Lan. Fuzzy rules extraction directly from numerical data for function approximation. *IEEE Transactions on Systems, Man, and Cybernetics*, 25(1):119–129, 1995.

119. S. Abe. Fuzzy function approximators with ellipsoidal regions. *IEEE Transactions on Systems, Man, and Cybernetics—Part B*, 29(5):654–661, 1999.

120. C.-F. Juang and C.-T. Lin. An on-line self-constructing neural fuzzy inference network and its applications. *IEEE Transactions on Fuzzy Systems*, 6(1):12–32, 1998.

121. T. Takagi and M. Sugeno. Fuzzy identification of systems and its applications to modeling and control. *IEEE Transactions on Systems, Man, and Cybernetics*, 15(1):116–132, 1985.

122. A. E. Gaweda and J. M. Zurada. Fuzzy neural network with relational fuzzy rules. In *Proceedings of the International Joint Conference on Neural Networks*, volume 5, pages 3–7, Como, Italy, July 2000.

123. R. S. Crowder. Predicting the Mackey-Glass time series with cascade-correlation learning. In *Proceedings of 1990 Connectionist Models Summer School*, pages 117–123, Carnegie Mellon University, 1990.

124. K. Baba, I. Enbutsu, and M. Yoda. Explicit representation of knowledge acquired from plant historical data using neural network. In *Proceedings of International Joint Conference on Neural Networks*, volume 3, pages 155–160, San Diego, CA, June 1990.

125. W.-Y. Wang, T.-T. Lee, C.-L. Liu, and C.-H. Wang. Function approximation using fuzzy neural networks with robust learning algorithm. *IEEE Transactions on Systems, Man, and Cybernetics—Part B*, 27(4):740–747, 1997.

126. C.-C. Lee, P.-C. Chung, J.-R. Tsai, and C.-I Chang. Robust radial basis function neural networks. *IEEE Transactions on Systems, Man, and Cybernetics—Part B*, 29(6):674–685, 1999.

127. H. Kubota, H. Tamaki, and S. Abe. Robust function approximation using fuzzy rules with ellipsoidal regions. In *Proceedings of the International Joint Conference on Neural Networks*, volume 6, pages 529–534, Como, Italy, July 2000.

128. P. J. Rousseeuw and M. Hubert. Recent developments in PROGRESS. In Y. Dodge, editor, L_1-*statistical procedures and related topics*, pages 201–214. Institute of Mathematical Statistics Lecture Notes-Monograph Series 31, Hayward, CA, 1997.

129. B. D. Ripley. *Pattern recognition and neural networks*. Cambridge University Press, Cambridge, UK, 1996.

130. S. M. Weiss and C. A. Kulikowski. *Computer systems that learn: Classification and prediction methods from statistics, neural nets, machine learning, and expert systems*. Morgan Kaufmann, San Francisco. CA, 1991.

131. W. H. Press, S. A. Teukolsky, W. T. Vetterling, and B. P. Flannery. *Numerical recipes in c: The art of scientific computing*. Cambridge University Press, Cambridge, UK, second edition, 1996.

Index